Nanomaterials and Nanotechnologies for Enhanced Oil Recovery and Gas Separation

Nanomaterials and Nanotechnologies for Enhanced Oil Recovery and Gas Separation

Dr. Vikas Mittal
Editor and Lead Author

CWP
Central West Publishing

This edition has been published by Central West Publishing, Australia

For more information about the books published by Central West Publishing, please visit https://centralwestpublishing.com

Disclaimer

NATIONAL LIBRARY OF AUSTRALIA

A catalogue record for this book is available from the National Library of Australia

ISBN (print): 978-1-922617-16-3

Contents

Preface

Nanomaterials and nanotechnologies have impacted all engineering, science and technology sectors owing to their superior properties and performance profiles as compared to the conventional systems. As a result, a large number of advanced materials and technologies have been developed for specific applications. The book aims to present a variety of such advanced systems developed specifically for enhanced oil recovery and gas separation by providing both basic information as well as recent developments. All aspects related to synthesis, properties and applications have been reviewed.

Chapter 1 investigates a variety of applications of nanotechnology in the upstream petroleum industry, particularly in the area of enhanced oil recovery (EOR). Chapter 2 serves as a precursor to the research in the field of synthesis and characterization of associative polymers for achieving enhanced oil recovery. In Chapter 3, the different types of ionic liquids (ILs) employed for the efficient removal of acid gases during the natural gas sweetening process have been discussed. In Chapter 4, the application of the biopolymer xanthan gum in the oil and gas industries is focused. The structure, properties and production method of xanthan gum are also briefly outlined. In Chapter 5, gas barrier and packaging applications of different polymer nanocomposites (PNCs) based on the nanoscale fillers, such as graphene, carbon nanotubes (CNT), montmorillonite (MMT) clay, silicon dioxide (SiO_2) and zinc oxide (ZnO), are analyzed. Chapter 6 outlines the state of the art on the deep eutectic solvent (DES) types, properties as well as application in the oil and gas industries ranging from gas absorption and enhanced oil recovery to the preparation of materials as well as metal extraction and metal oxides processing. In Chapter 7, various literature studies specifically reporting the recent developments in the synthesis and structure-property correlations of the polyurethane membranes for specific applications as gas separation materials have been reviewed in detail, along with presenting the future prospects. In Chapter 8, recent studies on the development of biopolymer schizophyllan, especially related to its application (and application potential) in EOR, have been focused. An overview of the structure, synthesis as well as the different physico-chemical properties of schizophyllan has also been presented. To further explore the usefulness of metal organic frameworks (MOFs) for developing high performance membranes, mixed metal membranes (MMMs) involving PEBAX polymer and HKUST-1 (in varying weight fractions) were developed and analyzed for their mechanical and structural characteristics in Chapter 9. The membranes developed in Chapter 9 were also analyzed for their environmental and water stability as well as CO_2 permeation characteristics in Chapter 10.

The support of all the chapter contributors is greatly acknowledged towards the successful accomplishment of the book. The book is dedicated to my family for their constant support.

Vikas MITTAL

1

Application of Nanotechnology for Enhanced Oil Recovery

1.1 Introduction

Nanotechnology refers to the technology where at least one of the constituents is sized at the nanometer scale, generally from a few nanometres to a few hundred nanometers. One of the fundamental concepts encompassed by nanotechnology is that of the nanoparticle. A nanoparticle is generally defined as an aggregation of between 10 and 100,000 atoms, with a radius of 1-100 nm [1]. The unique benefit of nanotechnology, in contrast to micro- and macro-scale technologies, is the change in properties of matter that takes place at the nanometre scale. Scaling the dimension of a macroscopic material down in size initially does not affect the material's properties, but eventually causes minor property changes, and finally causes major property changes once the size reduces below 100 nm [1]. Due to this unique change in properties, the potential offered by the materials designed at the nanometer scale is significant. In this respect, the application and potential offered by nanotechnology specifically for enhanced oil recovery has developed manifold in the last few decades.

Enhanced oil recovery (EOR) is defined by Morrow and Heller [2] as oil recovery brought about through means beyond that of the natural forces present in the reservoir. This definition includes tertiary oil recovery, which is oil recovery that takes place after the completion of primary (natural reservoir forces) and secondary (water or gas flooding) oil recovery methods. There are a variety of physical, chemical, and biological processes that have been theorised and tested for tertiary recovery methods. Researchers have explored, among other methods, microbial bio-surfactants, aqueous chemical flooding, polymer flooding, and CO_2 injection for tertiary oil recovery [3-6]. The goal of tertiary recovery methods is to increase the amount of oil that can be recovered economically. As mentioned earlier, this is the area where nanotechnology has a sig-

Arjun Dhillon, Ali Elkamel and Vikas Mittal, Khalifa University of Science and Technology, Abu Dhabi, UAE*
**Current address: University of Waterloo, Canada*

nificant benefit. It has been mentioned that primary and secondary oil recovery generally leave between 50% and 65% of the original oil in place (OOIP) [7]. Nanoparticle-based technologies can be used to replace or augment existing EOR methods for greater economic recovery efficiencies. This chapter explores some promising prospects for such applications, as well as some other methods for improving oil recovery.

1.2 Improved CO_2 Foams

A popular and established form of tertiary oil recovery is carbon dioxide flooding. Carbon dioxide is miscible in many hydrocarbons present in crude oil, particularly those with between five and twenty carbon atoms, at sufficiently high pressures, as noted by Holm [8]. Also, upon dissolution of carbon dioxide in crude oil, the oil reduces its viscosity by a factor of 5 to 10 as it swells by 10% to 60% [8]. By reducing the viscosity of the oil, the carbon dioxide improves the oil's mobility and enables higher oil recovery, as a result. Martin and Taber also reported that the vaporizing-gas-drive mechanism enables miscibility, although carbon dioxide solubility at reservoir pressures also helps to swell oil volume at an earlier point [9].

A main weakness of gas injection EOR techniques such as CO_2 flooding is said to be the relatively high mobility of gases [10]. According to Aronofsky and Ramey, the efficiency and success of oil extraction depends on the mobility ratio, which is the ratio of the mobility of the displacing phase (CO_2) to that of the displaced phase (oil/water) [11]. A lower mobility ratio implies a greater sweep efficiency and, thus, is desirable when comparing injection processes. Since the mobility ratio is directly proportional to the mobility of the displacing phase (CO_2), the high mobility of carbon dioxide results in a higher mobility ratio, which has an adverse effect on the sweep efficiency of this oil recovery method. Heller has identified the cause of this to be the formation of viscous fingers, where the carbon dioxide is able to move faster than fluids such as oil and water, with the given pressure gradient, which reduces the sweep efficiency [12]. A method known as water-alternating-gas (WAG) has been employed frequently to try to address the issue of mobility-related problems arising from the carbon dioxide. This method involves alternating between injection of water and carbon dioxide. Although this method is preferable to carbon dioxide injection alone, it is also reported to be inadequate and justified only in the absence of an acceptable alternative [13]. It is for this reason that CO_2 foams are being explored as a potential solution for tertiary oil recovery.

It has been frequently reported that injecting the carbon dioxide gas as foam reduces its mobility [10,14-16]. Making the gas into foam causes it to lose much of its mobility, while its efficiency remains unaffected. The viscosity of the carbon dioxide is increased, which results in a mobility ratio that is more favourable for oil recovery. Experimentation has shown that carbon dioxide foams created using a surfactant known as alpha olefin sulfonate (AOS) were successful in improving oil recovery and efficient exploitation of the carbon dioxide [10]. As shown in Figure 1.1, the comparison of miscible gas and miscible foam indicated a significant increase in the oil production when foam was used.

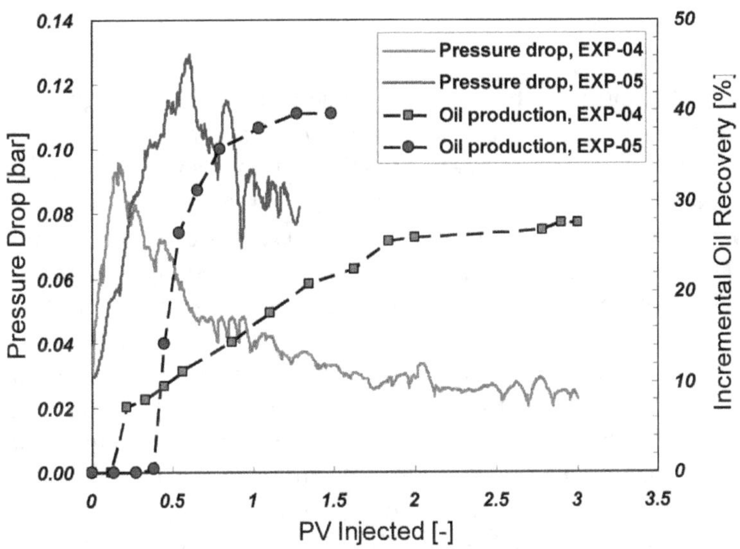

Figure 1.1 Pressure drop and oil production profiles for the miscible gas (EXP-04) and miscible foam (EXP-05) experiments. Reproduced from Reference 10 with permission from American Chemical Society.

The application of nanotechnology in this area exists in the form of improving carbon dioxide foams with nanoparticles. Worthen *et al.* [17] identified the potential for nanoparticle reinforcement of carbon dioxide foams for improved foam stability and longevity. The authors successfully created carbon dioxide foam using 50% SiOH silica (SiO_2) nanoparticles in the place of surfactants or polymers. When compared to carbon dioxide foams formed using nanoparticles coated with polyethylene glycol (PEG), as well as those formed by the surfactant Tergitol™ 15-S-20, the carbon dioxide foams created using the silica nanoparticles were observed to be much more stable. Espinosa *et al.* [18] used 5 nm silica nanoparticles to form a carbon dioxide foam in water. The supercritical foam underwent PEG surface treatment and was found to be very stable, even at high reservoir temperatures. The foam was produced at temperatures between 50 °C and 95 °C successfully, demonstrating its tolerance of high temperatures. There are currently a large number of ongoing research efforts in this area with a goal of augmenting carbon dioxide foam EOR methods with nanoparticles [19,20]. In summary, carbon dioxide foam flooding is a tertiary oil recovery method with a lot of potential offered as a result of nanotechnology.

1.3 EOR Surfactants

Surfactant flooding, a tertiary oil recovery process that involves the addition of surfactants to waterflooding formulations to improve its overall oil recovery, has been investigated under various conditions as an EOR technique for many years [21-24]. A surfactant is a chemical whose molecular structure is such that it is capable of reducing interfacial tension (IFT). In the context of oil recovery, surfactants are typically amphiphilic molecules that reduce the oil/water IFT. Uren and Fahmy [25] stated that reducing the interfacial tension between oil and injected water should generally improve the oil recovery brought about through waterflooding. Surfactants possess both hydrophilic and hydrophobic components, indicating that the molecules are able to interact with both oil and water, thereby, reducing the IFT that is present between the two liquid phases. As stated by Levitt *et al.* [26], lowering IFT allows more oil to be released from the formation's capillary forces that would otherwise prevent the oil from being extracted.

The utility of surfactants for EOR comes from their molecular properties, which can be modified at the nanometer scale. Kong and Ohadi [27] mentioned that nanotechnology has the techniques and methods necessary to specifically tailor surfactants for necessary conditions or functions. The authors further reported that there is potential for a more controlled addition of these surfactants to the oil reservoir, implying a better overall recovery. Amanullah and Al-Tahini [28] described the additional benefit of nano-sized surfactants, referencing the fact that nanoscale particles have a much greater accessible surface area for a given mass or volume, compared to their microscale counterparts. The fact that nanoparticles have a much higher specific surface area enables them to interact to a much greater extent than larger particles. The result of this increased reactivity is a reduced required concentration for nanosized surfactants to achieve the same result as a higher concentration of larger surfactants. Kong and Ohadi [27] also mentioned that surfactant flooding, like carbon dioxide flooding, can suffer from a poor mobility ratio. The authors suggested this issue can be resolved using nanoparticles to modify the viscosity of the injected solution. Dexter and Middelberg [29] investigated biological surfactants derived from peptides (proteins) that are capable of lowering IFT at the oil/water interface. These nanoscale surfactants are an attractive alternative to synthetic surfactants for oil recovery.

The benefit of nanotechnology to improvement of surfactant flooding exists primarily in the form of enhanced surfactants. Using nanoscale chemical and biological synthesis methods, it becomes possible to create surfactants that can drastically reduce the oil/water interfacial tension in low concentrations. The ongoing research and experimentation in this field has the potential to make surfactant flooding more efficient and economical.

1.4 Nano-fluids for Improving Oil Recovery

Nano-fluids are defined by Hendraningrat *et al.* [30] as suspension mixtures of solid nanoparticles contained within water, oil, ethylene glycol, or other fluids. It has been reported that nano-fluids exhibit unique properties such as high thermal conductivity and rapid heat transfer [31]. These enhanced thermal properties of nano-fluids can be useful in various applications throughout the upstream oil and gas industry, including enhanced oil recovery. Another significant advantage of nano-fluids is the fact that these can act as a mechanism for delivering nanoparticles into oil reservoirs to perform specific functions. These important features of nano-fluids enable their usage for a variety of different applications in enhanced oil recovery.

An important physicochemical characteristic of reservoir rock is its wettability. Abdallah *et al.* [32] defined this property as the affinity of a particular fluid to a given solid surface. This property is the result of the intermolecular forces acting at the surfaces and interface of the fluid and solid. A fluid that is considered wetting with respect to a surface will increase its contact with the surface by spreading out and will displace a non-wetting fluid that will reduce its own contact with the surface. In oil recovery, the focus is on the relative wettabilities of water and oil, with respect to the formation rock. Water-wet rock will be hydrophilic, while oil-wet rock will be hydrophobic, implying that water-wet and oil-wet conditions are opposite extremes on the wettability continuum. Laboratory experiments have indicated that oil recovery is most successful when wettability is neutral, meaning that the rock does not preferentially attract water over oil, or vice versa [33]. It was also stated that reservoir rock wettability influences the recovery factor, which has a role in determining how commercially feasible an oil field is [34]. These findings have led to numerous studies related to wettability alteration procedures under various conditions with the goal of improving oil recovery [35-38]. Hendraningrat *et al.* [30] created a nano-fluid by suspending silica nanoparticles (7 nm average size) in synthetic brine and tested its oil recovery potential with Berea sandstone. In this experiment, it was found that the nanoparticles reduced the interfacial tension and increased oil recovery in both the secondary and tertiary phases of EOR. The additional recovery was nearly 2% of the original oil in place (OOIP) for tertiary nano-fluid flooding and was nearly 8% OOIP for secondary nano-fluid flooding (compared to brine flooding). Li *et al.* [39] also reported that nano-fluids containing silica nanoparticles reduced interfacial tension at the oil/water interface, and further noted that the nano-fluid was able to alter the rock wettability, making it more water-wet. Ju *et al.* [40] reported the mechanism for this phenomenon to be the adsorption of nanoparticles onto the pore walls, resulting from Brownian motion, which made their surface more hydrophilic. It was also discovered by Li *et al.* [39] that increasing the concentration of silica nanoparticles in the fluid caused an increase in oil recovery until a maximum point of 0.05 wt%, after which an increase in nanoparticle concentration resulted in a decrease in oil recovery.

Nanoparticles can be added to fluids that are already in use in oil recovery to form nano-fluids with enhanced properties. Amanullah and Al-Tahini [28] emphasized the utility that specifically functionalized and tailored nanoparticles could potentially have in oil recovery fluids such as drilling fluids, completion fluids, and stimulation fluids. The authors suggested that the addition of nanoparticles to these various fluids will be of importance when facing production and environmental obstacles. According to Amanullah and Al-Tahini, the quantity of solid particles in drilling fluids impacts the level of formation damage, the level of productivity, and the rate of penetration (ROP). Due to the high specific surface area, and high resulting reactivity of nanoparticles, low concentrations of nanoparticles in fluids can achieve similar results to those brought about by higher concentrations of their larger-scale counterparts. The use of nanoparticles would, therefore, reduce the problems associated with the solids content in drilling fluids. The authors further identified drilling fluid filtrate loss as an important additional issue faced by existing drilling fluids that do not incorporate nanotechnology. Amanullah [41] described the process of filtrate loss, also known as fluid invasion, as the result of a pressure differential within the oil reservoir. When drilling fluids apply a significantly higher amount of pressure on the reservoir than the internal reservoir pressure, such a differential exists and results in fluid filtrate and particles forcibly entering the formation. As stated by Donaldson and Chernoglazov [42], filtrate migration occurs radially through the walls of the wellbore. It occurs to varying degrees depending on formation porosity and displaces water and hydrocarbons along its path. Donaldson and Chernoglazov [42] also specified that fluid invasion can span between 5 and 10 cm in high porosity formations, and up to 3 m in low porosity formations. According to Amanullah [41], the consequences of fluid invasion can be severe, including alteration of the permeability of the formation and reduction or termination of the flow of reservoir fluids toward the production well. These consequences are somewhat prevented by the drilling fluid through the formation of a filter cake. Elkatatny *et al.* [43] explained that a filter cake forms on the face of the formation as particles of different sizes adhere to it, reducing its permeability. The filter cake formation can help to reduce fluid invasion into the reservoir formation, whereas the filter cake formed by standard drilling fluids does not completely prevent this problem. It is for this reason that nanotechnology has the potential to improve upon existing drilling fluid technologies. Amanullah and Al-Tahini [28] also suggested that the inclusion of nanoparticles in drilling fluids has the potential to solve the problem of fluid invasion as a result of unique nanoparticle properties. A drilling fluid incorporating nanotechnology would be thin, yet essentially impermeable and resistant to erosion. This lack of permeability would help to prevent the adverse effects of fluid invasion into the formation such as formation damage and loss of productivity. An additional benefit to such drilling fluids, as noted by Amanullah and Al-Tahini [28], is the ease with which their filter cake can be removed. The high reactivity resulting from the high specific surface area of nanoparticles enables efficient chemical removal of the filter cake from the wellbore.

Research has recently been conducted on a nano-fluid made up of hydrophobic and lipophilic polysilicon (HLP) nanoparticles in ethanol. It has been reported that the nanoparticles caused an alteration in rock wettability from a water-wet condition to a more neutral wettability [44]. In addition to the wettability alteration, a decrease in oil/water interfacial tension was noted as a result of flooding with this HLP nano-fluid. In the core-flood experiments conducted on water-wet sandstone, the authors observed a reduction in oil/water IFT by a factor of ten, a decrease in water-wetness denoted by a 24° decrease in contact angle, and a resulting 19.31% increase in oil recovery after injection of the HLP nano-fluid. Core-flood experiments conducted by Onyekonwu and Ogolo [45] also demonstrated an increase in the extent of oil recovery from the ethanol-based HLP nanofluid, as well as an increase in oil recovery from the neutrally wet polysilicon nanoparticles dispersed in ethanol.

Some nano-fluids are also capable of improving oil recovery directly through a mechanism known as disjoining pressure. Wasan and Nikolov [46] described this as the formation of a wedge-like transition area that exists at the interface between an aqueous nano-fluid, a phase such as oil, and a solid. This is the result of nanoparticles assembling at the interface between the solid and oil phases, as shown in Figure 1.2 [47]. As stated by McElfresh *et al.* [48], significant quantities of nanoparticles in nano-fluids have a tendency to move toward the phase boundary between an oil droplet and the surrounding water and exert a force on the oil droplet. This ordering of nanoparticles is favorable due to increased overall entropy, according to Kondiparty *et al.* [47], as it increases the freedom of nanoparticles in the bulk solution. As nanoparticles fill the area between the droplet and the solid it is adhered to, this force acts at the vertex (point of contact) of the droplet and solid to push the droplet away from this surface. In this way, oil droplets can be forced away from the solid rock of the formation by nanoparticles, leading to a greater amount of recovered oil. Hendraningrat *et al.* [30] also observed that when using hydrophilic silica nanoparticles in nano-fluid flooding, the disjoining pressure mechanism in conjunction with the reduction of interfacial tension resulted in improved oil recovery.

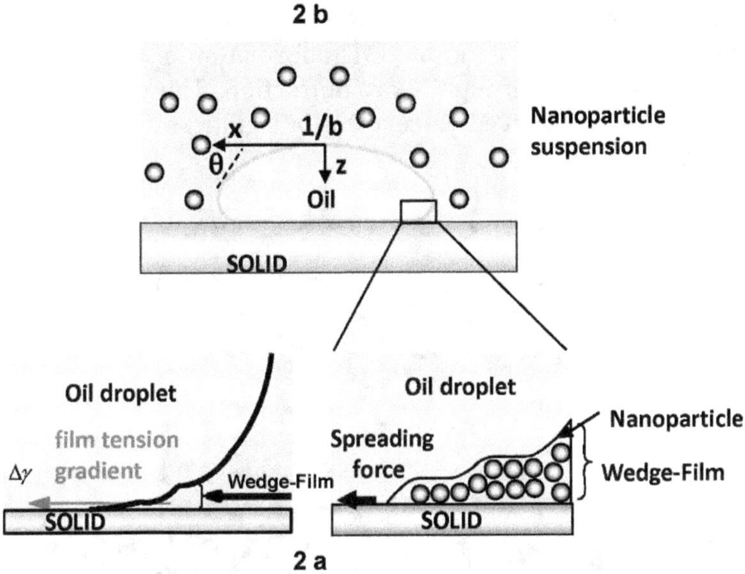

Figure 1.2 Disjoining pressure mechanism caused by nano-fluid mobilizing adsorbed oil droplet from solid. Reproduced from Reference 47 with permission from American Chemical Society.

Zirconium oxide nano-fluids have also been investigated for EOR purposes. Karimi *et al.* [49] added zirconium oxide (ZrO_2) nanoparticles to non-ionic surfactant fluid mixtures to form a nano-fluid and used this nano-fluid to conduct EOR tests on carbonate rock. This rock was initially found to be strongly oil-wet, having a great wetting preference for oil over water, but was changed to water-wet after exposure to the zirconium oxide nano-fluid. The authors stated that the wettability alteration occurred as a result of adsorption of the zirconium oxide nanoparticles on the surface of the rock. The authors further mentioned that the adsorption process was slow, requiring a minimum of two days to complete. Figure 1.3 also shows the scanning electron microscopy (SEM) images of surface of the rock aged in fluids 1-3. Fluid 1 had no nanoparticles, whereas fluid 2 and 3 differed in nanoparticle content. For fluid 1, the morphology of the rock surface resembled that of clean surface. In the case of fluids 2 and 3, formation of nanostructures on the rock surface was observed. The authors identified these nanostructures as nano-sized ribbons (fluid 2) and nano-flower (fluid 3). The nanostructures formed on the rock surface were suggested to have altered the wettability of an oil-wet rock to significantly water-wet nature. Thus, nanoparticle concentration was observed to have significant effect on the performance of nano-fluid. The formation of nanostructures was also confirmed through X-ray diffraction. The obtained nano-fluids were also observed to result in a significantly enhanced oil recovery.

Figure 1.3 SEM images of oil-wet carbonate rock aged in fluid 1 (A), fluid 2 (B), and fluid 3 (C). Reproduced from Reference 49 with permission from American Chemical Society.

Similar to zirconium oxide nanoparticles, alumina (Al_2O_3) nanoparticles have also been shown to change rock wettability from strongly oil-wet to strongly water-wet in nano-fluid form. Giraldo *et al.* [50] created nano-fluids using anionic surfactants and alumina nanoparticles and tested their wettability alteration potential on sandstone that had been previously made strongly oil-wet. The nano-fluid was able to change the wettability of the sandstone materials to strongly water-wet, which is also demonstrated in Figure 1.4.

In summary, nano-fluids offer a variety of EOR applications, especially as research continues on functionalizing nanoparticles in more complex and useful ways. Nanoparticles can be used both to create new fluids and to enhance conventional fluids, making nano-fluids research quite expansive.

Figure 1.4 Contact angles for the treatment with a nanoparticle concentration of 100 ppm: (a) oil/air/rock system before treatment, (b) oil/air/system after treatment, (c) water/air/system before treatment, (d) water/air/system after treatment. Reproduced from Reference 50 with permission from American Chemical Society.

1.5 Emulsions and Nano-emulsions for EOR

Emulsions and nano-emulsions have been the topics of numerous studies for the purpose of enhancing oil recovery. Mandal *et al.* [51] defined emulsions as suspension mixtures involving droplets of one liquid in another, where the two liquids are immiscible and the droplet size is greater than 0.1 μm. For EOR purposes, the two main immiscible liquids that are of concern are oil and water. Emulsions of oil and water, both oil-in-water and water-in-oil, have been investigated for EOR. For instance, Figure 1.5 demonstrates the performance of emulsion with 20% oil content.

Oil-in-water emulsions are used in the process of emulsion flooding, a process that involves the usage of these emulsions as displacing fluid for oil recovery. The mechanism for this EOR process, as described by Thomas and Farouq Ali [52], is the blocking of permeable paths for the fluids injected into the reservoir. The oil droplets dispersed in the emulsion can block these preferred fluid paths, thereby, forcing the injection fluids to areas that have not been contacted. This result can be made possible under the right circumstances, making emulsion flooding a potential EOR technique. Another benefit to oil-in-water emulsions is their altered viscosity. As previously mentioned, the success of oil recovery is partially dependent on mobility ratio, which in turn is partially dependent on the viscosity of the displacing fluid, making this viscosity an important factor for EOR. Pal [53] concluded that there was a relationship between the viscosity of an emulsion and the size of its constituent droplets. It was found that decreasing the droplet size resulted in an increase in the emulsion viscosity. This further adds to the potential of emulsion flooding as an effective EOR method.

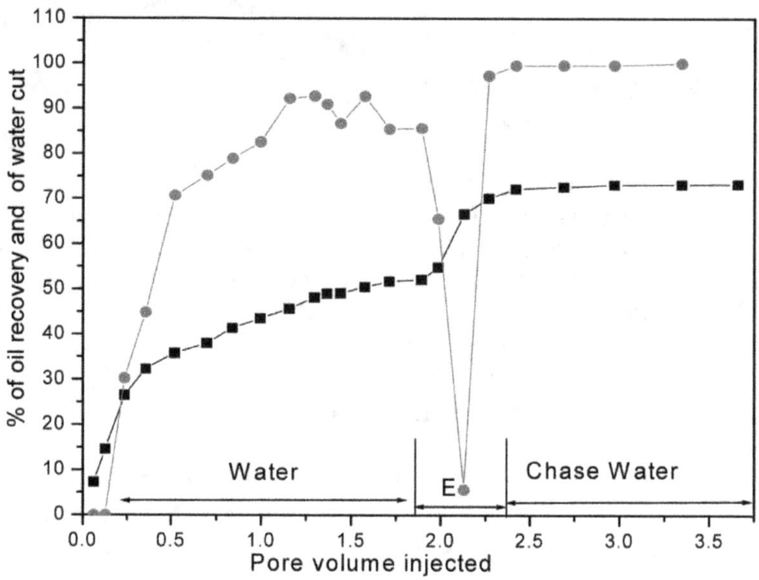

Figure 1.5 Production performance of emulsion (20% oil) flooding: ■, % oil recovery; •, % water cut. Reproduced from Reference 51 with permission from American Chemical Society.

A major issue faced by emulsion flooding is the inherent instability of emulsions. Since oil and water are immiscible, droplets of oil within water tend to coalesce in order to form a separated phase, thereby minimizing the contact surface between the oil and water phases. It is for this reason that certain conditions must be met for emulsions to be stable. This condition, according to Chen and Tao [54], is the presence of an emulsifying agent or emulsifier. The authors defined such an emulsifier as a surfactant that is capable of film formation at the oil/water interfaces in order to prevent coalescence. Nanoparticles have been investigated as potential alternatives to conventional surfactants for stabilizing emulsions. Zhang *et al.* [55] observed that it was possible to stabilize oil/water emulsions with silica nanoparticles which had undergone surface functionalization. The emulsions generated with these nanoparticles were 2-4 μm in size on average and were reported to reduce permeability of the rock with respect to the injected water. The uniformity of the nanoparticles is reflected in the tight, organized interfacial barrier they form at the oil/water interface. According to Zhang *et al.* [55], this results in significantly greater emulsion stability, especially at high temperatures, which is an excellent quality for an EOR agent. In the experiments performed by Zhang *et al.* [55], the emulsions formed using nanoparticles remained stable for weeks under conditions involving both ambient and elevated temperatures. Further research by Zhang *et al.* [56] used functionalized silica 5 nm nanoparticles to stabilize emulsions for several months after their formation.

Peptide surfactants mentioned previously in this paper also offer a unique advantage to the creation and manipulation of oil/water emulsions for oil recovery. Dexter and Middelberg [29] stated that proteins can be modified with hydrophilic and hydrophobic groups along their polyamide chain. This enables the creation of protein-based surfactants, which could be used to lower IFT at oil/water interfaces. According to Dexter and Middelberg [29], the peptides adsorb at the oil/water interface and cross-link with one another to stabilize an emulsion. Unlike typical surfactants, these surfactants arrange themselves parallel to the interface rather than perpendicular, as their hydrophilic and hydrophobic groups are along their chain rather than at the ends of it. The authors further stated that the cross-linking can be deactivated and reactivated chemically with a change in pH or a change in the concentration of certain metal ions present, as seen in Figure 1.6. Since it is possible to chemically manipulate these peptide surfactants in order to promote or hinder the formation of interfacial films around emulsion droplets, it is possible to create or break down emulsions. Another attractive feature of peptide surfactants is the potential for inexpensive production, as it has been shown by Morreale *et al.* [57] that such peptides could be produced biologically through a process known as bioprocess-centred molecular design. Furthermore, according to Malcolm *et al.* [58], these surfactants, as biomolecules, are biodegradable, making them more environmentally friendly option for stabilizing emulsions.

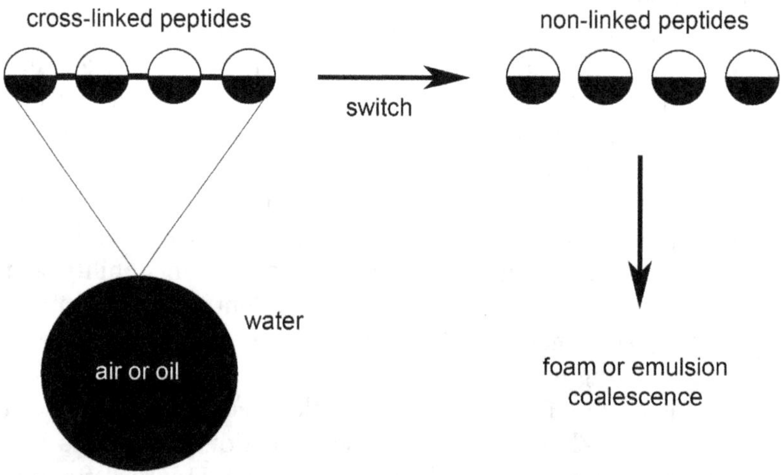

Figure 1.6 Switchable cross-linking of peptide surfactants to control formation and collapse of emulsions. Reproduced from Reference 29 with permission from American Chemical Society.

The potential applications for nano-emulsions for EOR are similar to those of emulsions. Nano-emulsions are smaller droplets than typical emulsions, with droplet size varying within the range of 50-500 nm, according to Mandal *et al.* [59]. These nano-emulsions, similarly to other emulsions, can be used as displacing fluid for emulsion flooding. The comparative benefit of nano-emulsion flooding over traditional emulsion flooding comes from the unique properties held by nano-emulsions. Nano-emulsions can be transported and stored without loss of utility due to processes such as sedimentation or creaming. Robins [60] defined creaming as the buoyancy-driven movement and separation of droplets from an emulsion to the top of a container. This process is mentioned to be similar to sedimentation, but in the opposite direction. According to Mandal *et al.* [59], it is due to the small droplet size of nano-emulsions that separation driven by density differences does not occur significantly, resulting in the prevention of both sedimentation and creaming. Additionally, the small size of the droplets allows them access to the pores within reservoir rock, which should be expected to improve oil recovery when nano-emulsion flooding is used. Research studies have also reported the preparation of oil-in-water nano-emulsions from mineral oil using non-ionic surfactants [59]. Flooding tests performed using these nano-emulsions exhibited an increase of more than 30% OOIP recovered in comparison to waterflooding alone.

Thus, nanotechnology offers numerous ways to develop emulsion flooding as a tertiary oil recovery technique. Nanoparticles and other nano-engineered technology can be used to tailor better surfactants and interfacial films to stabilize emulsions, whereas nano-emulsions offer unique benefits for oil recovery.

1.6 Ferro-fluids

Ferro-fluids are nano-fluids that are composed of nanoparticles possessing magnetic properties, according to Odenbach [61]. Odenbach [61] further stated that the particles used in ferro-fluids are normally magnetite (Fe_3O_4) nanoparticles with an average size of 10 nm which have also been modified with a surfactant. The role of the surfactant is said to preven coagulative aggregation, which would otherwise be the result of the van der Waals forces that cause interactions between particles. The author further mentioned that it is, therefore, possible to produce ferro-fluids with adequate stability, in liquids such as water, oil, and kerosene. The author attributed the unique superparamagnetic characteristic of ferro-fluids on the macroscopic scale to the nanoscale magnetism of their constituent nanoparticles. This property is what makes the ferro-fluids unique in their behavior, especially in the presence of magnetic fields.

Kothari *et al.* [62] identified the magneto-rheological effect as the applicable capability of ferro-fluids that is the result of their magnetic properties. This effect gives rise to the ability of ferro-fluids to alter their viscosity as a consequence of exposure to an external magnetic field. Since viscosity is a significant factor that influences fluid mobility in enhanced oil recovery, the potential for viscosity alteration using ferro-fluids is promising. The authors described an application that manipulated this capability of ferro-fluids in surfactant flooding. Ferro-fluids used in addition to surfactant flooding reduced the viscosity of reservoir fluids after coming into contact with them, once a magnetic field was applied. The authors opined that this decrease in reservoir fluid viscosity can help to improve fluid flow [62]. Improved flow of the fluids within the reservoir can be expected to have a benefit on the oil recovery process of surfactant flooding, making ferro-fluid-enhanced surfactant flooding EOR more effective than conventional surfactant flooding.

Pramana *et al.* [63] explored the use of ferro-fluids for heating heavy oil. To successfully recover heavy oil, the viscosity must be reduced, and this is normally achieved through thermal methods. The authors investigated the potential of electromagnetic induction heating as a method of thermally reducing the viscosity of heavy oil in the oil reservoir. Ferro-fluids were used as part of this process because they have been found to heat up more rapidly under these circumstances than fluids that are not magnetic. According to the authors, this is due to a phenomenon known as hysteresis, which results in hysteresis heating. The ferro-fluid used in this experiment was found to heat at a faster rate than both graphite fluid and brine, making it preferential as a fluid for this type of thermal EOR method.

In another study, Melle *et al.* [64] also investigated the application of ferro-fluids for the stabilization of oil/water emulsions, also as demonstrated in Figure 1.7. The mechanism for emulsion stabilization involving ferro-fluids is essentially the same as the mechanism used by silica nanoparticles for the same purpose. The magnetic nanoparticles in ferro-fluids are able to adsorb at the interface between the oil and water phases, allowing a sufficient reduction in interfacial tension and the resulting stabilization of the oil/water emulsion. The benefit to using ferro-fluids as opposed to standard nano-fluids for stabilizing emulsions is the ability to manipulate the stability and motion of droplets within the emulsion using an external magnetic field. The authors observed that specific manipulation of the magnetic field caused the separation of the emulsion into its constituent phases. As with peptide surfactants that were mentioned previously, these ferro-fluids are able to destabilize and stabilize oil/water emulsions at will. Additionally, it was found that manipulating the magnetic field in a different way enabled transportation of the emulsion droplets through the continuous phase, while the emulsions remained stable. The movement of these droplets was controllable through further manipulation of the magnetic field. It was also observed that small changes in the magnetic field caused the emulsion droplets to increase their length to a small degree. These features of ferro-fluid-stabilized emulsions make them very attractive as EOR agents for emulsion flooding. These emulsions can be controlled in various ways using magnetic fields, making them much more versatile than typical oil/water emulsions used in conventional emulsion flooding.

1.7 Metallic Nanoparticles for Heavy Oil Recovery

Metallic nanoparticles have been the subject of interest for some investigations of EOR involving heavy oil. The United States Geological Survey defines heavy oil as a class of crude oil that is heavier than light crude oil, possessing an API gravity of no more than 22° and a viscosity of no less than 100 cP [65].

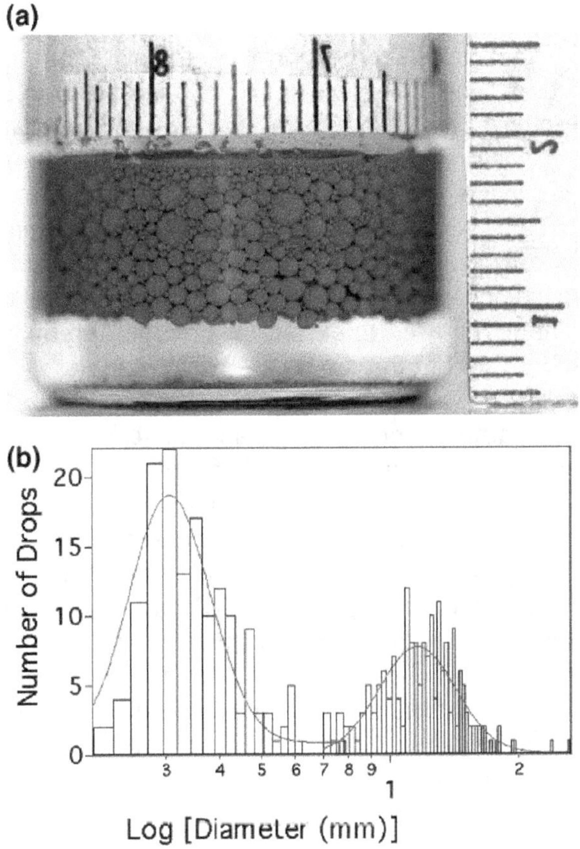

Figure 1.7 (a) Oil-in-water emulsion stabilized with carbonyl iron paramagnetic particles (4 wt%). and (b) histogram of the distribution of drops sizes corresponding to the emulsion in Figure 1.7a. Reproduced from Reference 64 with permission from American Chemical Society.

The high density and viscosity of heavy oil make its extraction and transportation much less economical, in comparison to light oil. Due to the challenging economics of heavy oil extraction, there is a lot of heavy oil that is not being exploited in oil reservoirs [65]. It is for this reason that the oil industry has been looking into novel nanotechnology solutions to this issue.

Presently, heavy oil extraction is performed mainly through thermal methods that involve heating the oil to reduce its viscosity. Shokrlu and Babadagli [66] identified the injection of air or steam as conventional thermal methods, and electrical or electromagnetic heating as unconventional thermal methods. The authors tested several different types of metal nanoparticles and observed that they helped to enhance the property of heat transfer within the heavy oil. By improving the heat transfer, thermal methods of viscosity reduction can take place more efficiently, thereby making heavy oil recovery more economical. Furthermore, the authors observed that viscosity reduction of the heavy oil at various temperatures, which occurred through a mechanism known as aquathermolysis, was catalyzed by the same metallic nanoparticles that improved heat transfer. According to the authors, this catalytic effect of the nanoparticles can be expected to reduce the amount of thermal energy input required to extract heavy oil, thereby making the process even more economical.

Hashemi *et al.* [67] have also performed experiments to explore the catalytic capabilities of metallic nanoparticles. The authors observed that trimetallic nanoparticles, composed of tungsten, nickel, and molybdenum, were able to catalyze chemical reactions that upgraded heavy oil, reducing its overall viscosity. This concept was called an "underground refinery" approach. Using the nanoparticles in the form of a hot vacuum gas oil (VGO) nano-fluid, the heavy oil was broken into lighter components through catalytic hydrocracking. The resulting components had lower viscosities and densities than heavy oil, making them easier to extract. The benefits to oil recovery are summarized in Figures 1.8. Additionally, by hydrocracking heavy oil within the reservoir, less processing was required during oil refining, making the economics of heavy oil recovery even more favorable.

In another study, Ehtesabi *et al.* [68] studied the recovery of heavy oil in sandstone cores using titanium dioxide nanoparticles. Oil recovery was increased to 80% with a low nanoparticle content, as compared to 49% achieved by using only water for coreflooding.

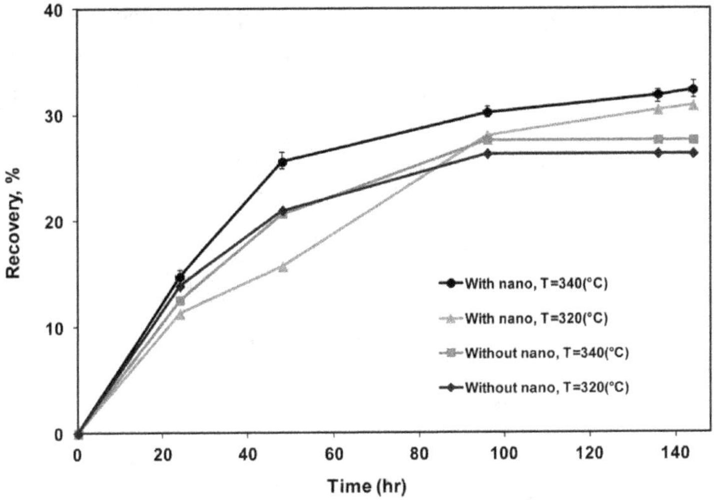

Figure 1.8 Increase in heavy oil recovery due to nano-fluid catalyzed hydrocracking. Reproduced from Reference 67 with permission from American Chemical Society.

Rock surface was observed to change in wettability for oil-wet to water-wet after treatment with nanoparticles. As also seen in Figure 1.9, nanoparticles formed homogenous deposition on the surface of core plug and nanorods with average diameter of 60 nm were observed.

1.8 Conclusions

Nanotechnology has been explored in recent years as a potential solution to a variety of problems involving oil recovery, particularly with respect to enhanced oil recovery. Nanotechnology involves manipulation of matter at the nanoscale level, generally dealing with particles 1-100 nm in size. This review has investigated a variety of applications of nanotechnology in the upstream petroleum industry, particularly in the area of enhanced oil recovery. The unique properties of nanoparticles as a result of their small size enable a wide range of potential applications within a number of different EOR techniques. For instance, carbon dioxide flooding has been made significantly more practical with the addition of nanoparticles, which can reduce mobility-related issues of CO_2 foams. Surfactant flooding can benefit from nanoscale manipulation of EOR surfactants as well. Nano-fluids containing various functional nanoparticles are versatile and can be used for a variety of EOR-related purposes. Nanotechnology also helps stabilize emulsions of oil and water for emulsion flooding. Superparamagnetic ferro-fluids have uses

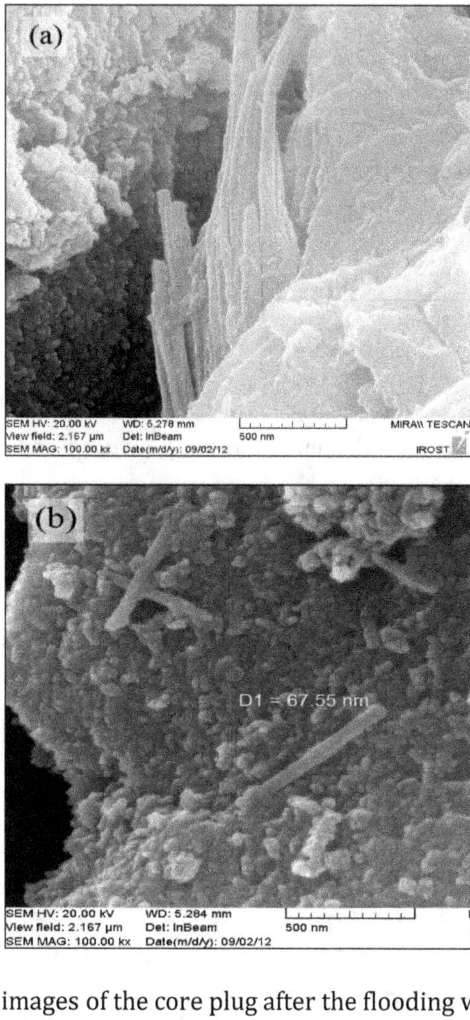

Figure 1.9 SEM images of the core plug after the flooding with a 1% concentration of TiO_2 nanoparticles. Reproduced from Reference 68 with permission from American Chemical Society.

in EOR due to their behavior in magnetic fields. Metallic nanoparticles have significant catalytic abilities, which is useful in heavy oil EOR. Thus, engineering at the nanoscale level can result in a variety of novel materials with enhanced properties that are optimized for various purposes. With nanotechnology, it is possible to tailor the properties of nanoparticles and nano-fluids to perform optimally under reservoir conditions, making them promising as a way of improving EOR methods. It is likely that nanotechnology will remain of interest to the oil and gas industry, as new applications of nanomaterials for both the upstream and downstream petroleum industry are continually revealed.

References

1. Bhushan, B. (2007) *Springer Handbook of Nanotechnology*, 2nd edition, Springer, New York.
2. Morrow, N. R., and Heller, J. P. (1985) Fundamentals of enhanced recovery. In: *Enhanced Oil Recovery, I, Fundamentals and Analyses*, Donaldson, E. C., Chilingarian, G. V., and Yen, T. F. (eds.), Elsevier Science Publishers B. V., The Netherlands, pp. 47-74.
3. Kamath, K., Zammerilli, A. M., Comberiati, J. R., Taylor, B. D., and Slagle, F. D. (1982) Mechanism of Tertiary Oil Recovery by Aqueous Chemical Flooding. *Society of Petroleum Engineers Oilfield and Geothermal Chemistry Symposium*, USA. Online: https://www.onepetro.org/conference-paper/SPE-10599-MS (assessed 15th March 2017).
4. Lang, W., and Rehage, G. (1980) Recent Results on the Use of Polymers in Tertiary Oil Recovery in Brines of High Salinity. *Society of Petroleum Engineers Oilfield and Geothermal Chemistry Symposium*, USA. Online: https://www.onepetro.org/conference-paper/SPE-8983-MS (assessed 16th March 2017).
5. Maudgalya, S., Knapp, R. M., McInerney, M. J., Folmsbee, M., and Nagle, D. P. (2005) Tertiary Oil Recovery with Microbial Biosurfactant Treatment of Low-Permeability Berea Sandstone Cores. *Society of Petroleum Engineers Production Operations Symposium*, USA. Online: https://www.onepetro.org/conference-paper/SPE-94213-MS (assessed 21st March 2017).
6. Wang, G., and Locke, D. (1980) A laboratory study of the effects of CO_2 injection sequence on tertiary oil recovery. *SPE Journal*, **20**(4), 278-280.
7. Rao, D., and Hughes, R. (2011) Current research and challenges pertaining to CO_2 flooding and sequestration. *The Way Ahead*, **7**(2), 17-19.
8. Holm, L. W. (1982) CO_2 flooding: Its time has come. *Journal of Petroleum Technology*, **34**(12), 2739-2745.
9. Martin, F. D., and Taber, J. J. (1992) Carbon dioxide flooding. *Journal of Petroleum Technology*, **44**(4), 396-400.
10. Farajzadeh, R., Andrianov, A., and Zitha, P. L. (2010) Investigation

of immiscible and miscible foam for enhancing oil recovery. *Industrial & Engineering Chemistry Research*, 49(4), 1910-1919.

11. Aronofsky, J. S., and Ramey, H. J., Jr. (1956) Mobility ratio - Its influence on injection or production histories in five-spot water flood. *Journal of Petroleum Technology*, **8**(9), 205-210.
12. Heller, J. P. (1984) Reservoir Application of Mobility Control Foams in CO_2 Floods. *Society of Petroleum Engineers Enhanced Oil Recovery Symposium*, USA. Online: https://www.onepetro.org/conference-paper/SPE-12644-MS (assessed 10th March 2017).
13. Kulkarni, M., and Rao, D. (2005) Experimental investigation of miscible and immiscible Water-Alternating-Gas (WAG) process performance. *Journal of Petroleum Science and Engineering*, **48**(1), 1-20.
14. Bernard, G. G., Holm, L. W., and Harvey, C. P. (1980) Use of surfactant to reduce CO_2 mobility in oil displacement. *Society of Petroleum Engineers Journal*, **20**(4), 281-292.
15. Khalil, F., and Asghari, K. (2006) Application of CO_2-foam as a means of reducing carbon dioxide mobility. *Journal of Canadian Petroleum Technology*, **45**(5), 37-42.
16. Lee, H. O., and Heller, J. P. (1990) Laboratory measurements of CO_2-foam mobility. *SPE Reservoir Engineering*, **5**(2), 193-197.
17. Worthen, A., Bagaria, H., Chen, Y., Bryant, S. L., Huh, C., and Johnston, K. P. (2012) Nanoparticle Stabilized Carbon Dioxide in Water Foams for Enhanced Oil Recovery. *Society of Petroleum Engineers Improved Oil Recovery Symposium*, USA. Online: https://www.onepetro.org/conference-paper/SPE-154285-MS (assessed 21st March 2017).
18. Espinosa, D. A., Johnston, K. P., Bryant, S. L., and Huh, C. (2010), 'Nanoparticle-Stabilized Supercritical CO_2 Foams for Potential Mobility Control Applications' in *Society of Petroleum Engineers Improved Oil Recovery Symposium*. Tulsa, Oklahoma. 24-28 April 2010. Texas: Society of Petroleum Engineers.
19. National Energy Technology Laboratory (2013) *Nanoparticle-stabilized CO_2 Foam for CO_2-EOR Application*. Online: http://www.netl.doe.gov/technologies/oil-gas/Petroleum/projects/EP/ImprovedRec/FE0005979-NMIMT.html (Accessed 1st August 2015).
20. National Energy Technology Laboratory (2013) *Use of Engineered Nanoparticle-stabilized CO_2 Foams to Improve Volumetric Sweep of CO_2 EOR Processes*. Online: http://netldev.netl.doe.gov/research/oil-and-gas/project-summaries/enhanced-oil-recovery/fe0005917-uta (Accessed 1st August 2015).
21. Adams, W. T., and Schievelbein, V. H. (1987) Surfactant flooding

carbonate reservoirs. *SPE Reservoir Engineering*, **2**(4), 619-626.

22. Hirasaki, G. J., Miller, C. A., and Puerto, M. (2008), Recent Advances in Surfactant EOR. *Society of Petroleum Engineers Annual Technical Conference and Exhibition*, USA. Online: https://www.onepetro.org/journal-paper/SPE-115386-PA (assessed 27th March 2017).
23. Mattax, C. C., Blackwell, R. J., and Tomich, J. F. (1983) Recent Advances in Surfactant Flooding. *11th World Petroleum Conference*, UK. Online: https://www.onepetro.org/conference-paper/WPC-20220 (assessed 13th March 2017).
24. Ziegler, V. M. (1988) Laboratory investigation of high temperature surfactant flooding. *SPE Reservoir Engineering*, **3**(2), 586-596.
25. Uren, L. C., and Fahmy, E. H. (1927) Factors influencing the recovery of petroleum from unconsolidated sands by waterflooding. *Transactions of the AIME*, **77**(1), 318-335.
26. Levitt, D., Jackson, A., Heinson, C., Britton, L. N., Malik, T., Dwarakanath, V., and Pope, G. A. (2006) Identification and Evaluation of High-Performance EOR Surfactants. *Society of Petroleum Engineers/Department of Energy Symposium on Improved Oil Recovery*, USA. Online: https://www.onepetro.org/journal-paper/SPE-100089-PA (assessed 10th March 2017).
27. Kong, X., and Ohadi, M. (2010) Applications of Micro and Nano Technologies in the Oil and Gas Industry – Overview of the Recent Progress. *Abu Dhabi International Petroleum Exhibition and Conference*, UAE. Online: https://www.onepetro.org/conference-paper/SPE-138241-MS (assessed 19th March 2017).
28. Amanullah, M., and Al-Tahini, A. (2009) Nano-Technology - Its Significance in Smart Fluid Development for Oil and Gas Field Application. *Society of Petroleum Engineers Saudi Arabia Section Technical Symposium*, Saudi Arabia. Online: https://www.onepetro.org/conference-paper/SPE-126102-MS (assessed 10th March 2017).
29. Dexter, A., and Middelberg, A. (2008) Peptides as functional surfactants. *Industrial & Engineering Chemistry Research*, **47**(17), 6391-6398.
30. Hendraningrat, L., Li, S., and Torsæter, O. (2013) A Coreflood Investigation of Nanofluid Enhanced Oil Recovery in Low-Medium Permeability Berea Sandstone. *Society of Petroleum Engineers International Symposium on Oilfield Chemistry*, USA. Online: https://www.onepetro.org/conference-paper/SPE-164106-MS (assessed 9th March 2017).
31. Murshed, S., Leong, K., and Yang, C. (2008) Thermophysical and electrokinetic properties of nanofluids – A critical review. *Applied Thermal Engineering*, **28**(17), 2109-2125.
32. Abdallah, W., Buckley, J. S., Carnegie, A., Edwards, J., Herold, B.,

Fordham, E., Graue, A., Habashy, T., Seleznev, N., Signer, C., Hussain, H., Montaron, B., and Ziauddin, M. (2007) Fundamentals of wettability. *Oilfield Review*, 19(2), 44-61.

33. Morrow, N. R. (1990) Wettability and its effect on oil recovery. *Journal of Petroleum Technology*, **42**(12), 1476-1484.
34. Aziz, A.K. (2011) Impact of Wettability Alteration on Recovery Factor. *SPE/DGS Saudi Arabia Section Technical Symposium and Exhibition*, Saudi Arabia. Online: https://www.onepetro.org/conference-paper/SPE-149044-MS (assessed 21st March 2017).
35. Chen, P., and Mohanty, K. K. (2014) Wettability Alteration in High Temperature Carbonate Reservoirs. *Society of Petroleum Engineers Improved Oil Recovery Symposium*, USA. Online: https://www.onepetro.org/conference-paper/SPE-169125-MS (assessed 29th March 2017).
36. Gupta, R., and Mohanty, K. K. (2008) Wettability Alteration of Fractured Carbonate Reservoirs. *Society of Petroleum Engineers/Department of Energy Improved Oil Recovery Symposium*, USA. Online: https://www.onepetro.org/conference-paper/SPE-113407-MS (assessed 21st March 2017).
37. Nasralla, R. A., Bataweel, M. A., and Nasr-El-Din, H.A. (2011) Investigation of Wettability Alteration by Low Salinity Water in Sandstone Rock. *Society of Petroleum Engineers Offshore Europe Oil and Gas Conference and Exhibition*, UK. Online: https://www.onepetro.org/conference-paper/SPE-146322-MS (assessed 11th March 2017).
38. Tang, G., and Kovscek, A. R. (2002) Wettability Alteration of Diatomite Induced by Hot-Fluid Injection. *Society of Petroleum Engineers Annual Technical Conference and Exhibition*, USA. Online: https://www.onepetro.org/conference-paper/SPE-77461-MS (assessed 19th March 2017).
39. Li, S., Hendraningrat, L., and Torsaeter, O. (2013) Improved Oil Recovery by Hydrophilic Silica Nanoparticles Suspension: 2-Phase Flow Experimental Studies. *International Petroleum Technology Conference*, China. Online: https://www.onepetro.org/conference-paper/IPTC-16707-MS (assessed 19th February 2017).
40. Ju, B., Fan, T., and Ma, M. (2006) Enhanced oil recovery by flooding with hydrophilic nanoparticles. *China Particuology*, **4**(1), 41-46.
41. Amanullah, M. (2003) A Novel Method of Assessment of Spurt and Filtrate Related Formation Damage Potential of Drilling and Drilling-in Fluids. *Society of Petroleum Engineers Asia Pacific Oil and Gas Conference and Exhibition*, Indonesia. Online: https://www.onepetro.org/conference-paper/SPE-80484-MS (assessed 20th March 2017).
42. Donaldson, E. C., and Chernoglazov, V. (1987) Characterization of

drilling mud fluid invasion. *Journal of Petroleum Science and Engineering*, **1**(1), 3-13.

43. Elkatatny, S., Mahmoud, M., and Nasr-El-Din, H. (2012) Characterization of filter cake generated by water-based drilling fluids Using CT scan. *SPE Drilling & Completion*, **27**(2), 282-293.
44. Shahrabadi, A., Bagherzadeh, H., Roustaei, A., and Golghanddashti, H. (2012) Experimental Investigation of HLP Nanofluid Potential to Enhance Oil Recovery: A Mechanistic Approach. *Society of Petroleum Engineers International Oilfield Nanotechnology Conference*, The Netherlands. Online: https://www.onepetro.org/conference-paper/SPE-156642-MS (assessed 19th March 2017).
45. Onyekonwu, M., and Ogolo, N. (2010) Investigating the Use of Nanoparticles in Enhancing Oil Recovery. *34th Annual Society of Petroleum Engineers International Conference and Exhibition*, Nigeria. Online: https://www.onepetro.org/conference-paper/SPE-140744-MS (assessed 29th March 2017).
46. Wasan, D., and Nikolov, A. (2003) Spreading of nanofluids on solids. *Nature*, 423(6936), 156-159.
47. Kondiparty, K., Nikolov, A., Wu, S., and Wasan D. (2011) Wetting and spreading of nanofluids on solid surfaces driven by the structural disjoining pressure: Statics analysis and experiments. *Langmuir*, 27(7), 3324-3335.
48. McElfresh, P., Holcomb, D., and Ector, D. (2012), Application of Nanofluid Technology to Improve Recovery in Oil and Gas Wells. *Society of Petroleum Engineers International Oilfield Nanotechnology Conference*, The Netherlands. Online: https://www.onepetro.org/conference-paper/SPE-154827-MS (assessed 19th February 2017).
49. Karimi, A., Fakhroueian, Z., Bahramian, A., Khiabani, N. P., Darabad, J. B., Azin, R., and Arya, S. (2012) Wettability alteration in carbonates using zirconium oxide nanofluids: EOR implications. *Energy & Fuels*, **26**(2), 1028-1036.
50. Giraldo, J., Benjumea, P., Lopera, S., Cortes, F. B., and Ruiz, M. A. (2013) Wettability alteration of sandstone cores by alumina-based nanofluids. *Energy & Fuels*, **27**(7), 3659-3665.
51. Mandal, A., Samanta, A., Bera, A., and Ojha, K. (2010) Characterization of oil-water emulsion and its use in enhanced oil recovery. *Industrial & Engineering Chemistry Research*, 49(24), 12756-12761.
52. Thomas, S., and Farouq Ali, S.M. (1989) Flow of emulsions in porous media, and potential for enhanced oil recovery. *Journal of Petroleum Science and Engineering*, **3**(1-2), 121-136.
53. Pal, R. (1998) A novel method to correlate emulsion viscosity data. *Colloids and Surfaces A: Physicochemical and Engineering Aspects*, **137**(1-3), 275-286.
54. Chen, G., and Tao, D. (2005) An experimental study of stability of

oil-water emulsion. *Fuel Processing Technology*, **86**(5), 499-508.

55. Zhang, T., Roberts, M., Bryant, S. L., and Huh, C. (2009) Foams and Emulsions Stabilized with Nanoparticles for Potential Conformance Control Applications. *Society of Petroleum Engineers International Symposium on Oilfield Chemistry*, USA. Online: https://www.onepetro.org/conference-paper/SPE-121744-MS (assessed 19th March 2017).
56. Zhang, T., Davidson, D., Bryant, S. L., and Huh, C. (2010) Nanoparticle-Stabilized Emulsions for Applications in Enhanced Oil Recovery. *Society of Petroleum Engineers Improved Oil Recovery Symposium*, USA. Online: https://www.onepetro.org/conference-paper/SPE-129885-MS (assessed 21st March 2017).
57. Morreale, G., Lee, E. G., Jones, D. B., and Middelberg, A. P. (2004) Bioprocess-centred molecular design (BMD) for the efficient production of an interfacially active peptide. *Biotechnology and Bioengineering*, **87**(7), 912-923.
58. Malcolm, A., Dexter, A., and Middelberg, A. (2007) Peptide surfactants (Pepfactants) for switchable foams and emulsions. *Asia-Pacific Journal of Chemical Engineering*, **2**(5), 362-367.
59. Mandal, A., Bera, A., Ojha, K., and Kumar, T. (2012) Characterization of Surfactant Stabilized Nanoemulsion and Its Use in Enhanced Oil Recovery. *Society of Petroleum Engineers International Oilfield Nanotechnology Conference*, The Netherlands. Online: https://www.onepetro.org/conference-paper/SPE-155406-MS (assessed 28th March 2017).
60. Robins, M. (2000) Emulsions - creaming phenomena. *Current Opinion in Colloids and Interface Science*, **5**(5-6), 265-272.
61. Odenbach, S. (2003) Ferrofluids - magnetically controlled suspensions. *Colloids and Surfaces A: Physicochemical and Engineering Aspects*, **217**(1-3), 171-178.
62. Kothari, N., Raina, B., Chandak, K. B., Iyer, V., and Mahajan, H. P. (2010) Application of Ferrofluid for Enhanced Surfactant Flooding in EOR. *Society of Petroleum Engineers EUROPEC/EAGE Annual Conference and Exhibition*, Spain. Online: https://www.onepetro.org/conference-paper/SPE-131272-MS (assessed 19th March 2017).
63. Pramana, A. A., Abdassah, D., Rachmat, S., and Mikrajuddin, A. (2010) Electromagnetic induction heat generation of nano-ferrofluid and other stimulants for heavy oil recovery. *AIP Conference Proceedings*, **1284**(1), 163-166.
64. Melle, S., Lask, M., and Fuller, G. (2005) Pickering emulsions with controllable stability. *Langmuir*, **21**(6), 2158-2162.
65. United States Geological Survey (2013). *Heavy Oil and Natural Bitumen - Strategic Petroleum Resources*. Online: http://pubs.usgs.gov/fs/fs070-03/fs070-03.html. (Accessed 21st

Aug 2016).

66. Shokrlu, Y., and Babadagli, T. (2010) Effects of Nano-Sized Metals on Viscosity Reduction of Heavy Oil/Bitumen During Thermal Applications. *Canadian Unconventional Resources & International Petroleum Conference,* Canada. Online: https://www.onepetro.org/conference-paper/SPE-137540-MS (assessed 19th March 2017).
67. Hashemi, R., Nassar, N., and Almao, P. (2013) Enhanced heavy oil recovery by in-situ prepared ultradispersed multimetallic nanoparticles: A study of hot fluid flooding for athabasca bitumen recovery. *Energy & Fuels,* **27**(4), 2194-2201.
68. Ehtesabi, H., Ahadian, M. M., Taghikhani, V., and Ghazanfari, M. H. (2014) Enhanced heavy oil recovery in sandstone cores using TiO_2 nanofluids. *Energy & Fuels,* **28**, 423-430.

2

Associative Polymer Applications in Chemical Injection for Enhanced Oil Recovery

2.1 Introduction

In the recent years, enhanced oil recovery (EOR) has become a prime field of research especially due to rapid increase in the energy demand. EOR is defined as implementation of various techniques to increase the amount of crude extracted from an oil reservoir. For this purpose, EOR is also called as improved oil recovery or tertiary oil recovery. This oil recovery technique may increase the amount of oil recovery from the reservoir by up to 40%. There are three primary techniques used for EOR applications, which are broadly categorized as gas injection, thermal injection and chemical injection techniques.

2.1.1 Gas Injection

This technique is the most commonly used technique for EOR operations. In the North America, gas injection amounts to more than 60% of EOR production. Gas injection usually involves injection of gases such as nitrogen, carbon dioxide (CO_2) or natural gas [1]. Since this technique uses miscible gases to flood the reservoir, the technique is also called as miscible flooding. The reservoir pressure is maintained due to the miscible displacement. Since the interfacial tension between oil and water is reduced in the reservoir, this technique in turn improves oil displacement, which results in improved oil recovery. This technique allows for total displacement efficiency since the interface between the two interacting fluids is removed [2]. CO_2 is the most commonly used gas for miscible displacement since it is cost effective and also has the ability to increase the viscosity of oil. Oil recovery by miscible displacement primarily depends on the phase behavior of crude oil and gas. The phase behavior characteristics of these fluids are significantly affected by the

Hariharan Krithivasan and Vikas Mittal, Khalifa University of Science and Technology, Abu Dhabi, UAE*
**Current address: University of Waterloo, Waterloo, Canada*

many factors such as crude oil composition, temperature and pressure of the reservoir, among others [2].

2.1.2 Thermal Injection

Thermal injection technique amounts for 40% of EOR operations in the North America. This technique involves introduction of heat into the reservoir, which in turn enhances the oil recovery from the reservoir. The most common techniques used are steam flooding and cyclic steam injection. The introduction of steam results in reduced oil viscosity and/or vaporization of part of the crude oil, thus, decreasing the mobility ratio. In addition to this, the increased heat also increases the permeability of the oil due to reduction in the surface tension. In the year 2011, solar power was used for thermal injection in Oman and California, USA. This method involves the same technique as mentioned above except solar power was used to produce steam.

2.1.3 Chemical Injection

Chemical injection is another EOR production method that uses injection of polymers to improve the effectiveness of water flooding. This technique is used to increase the viscosity of water during the water flooding process. The water/oil mobility ratio increases during the chemical injection process, which in turn improves the vertical and areal sweep efficiency of the reservoir. The surface tension between the oil and water can be further reduced by introducing surfactants with the polymers. This improves the microscopic sweep efficiency process by reduction of residual oil saturation. In order to improve the stability of the surfactant polymer formation, co-surfactants, activity boosters and co-solvents are also usually added.

2.2 Scope of Polymers in EOR

The polymer injection or polymer flooding technique was not commonly used compared to the gas and thermal injection techniques until the past decade. This technique involves injection of an alkaline solution, surfactant along with co-surfactants and solvents, and polymers. This method is usually called as alkaline surfactant polymer (ASP) technique [3]. The polymer flooding has proven to be very effective in improving the mobility ratio and, hence, resulting in effective recovery of unswept oil. Many case studies for polymer flooding in the reservoirs can be studied in the literature [4]. Polymer flooding technique has also proven to be cost effective compared to the gas and thermal injection. The efficiency of the polymer injection process falls in the range of 0.75 to 1.75 lb of polymer used for per barrel of oil produced [4].

2.2.1 Working Principle

Hydrogel polymers or associative polymer hydrogels play a very significant role in the EOR process. Firstly, this procedure involves injection of polymer (polymer flooding) for years until 33-50 percent of the pore volume of the reservoir is achieved. Secondly, the oil and the polymer slug is driven towards the production well by flooding the reservoir with water. As the water is injected, it naturally seeks the region of highest permeability. This, in turn, results in the reduced pressure inside the production well. The mobility ratio of the reservoir is one of the primary parameters to be analyzed. This parameter can be used for various screening processes. The *mobility ratio* (*M*) is a parameter that helps in the understanding of the effects of relative permeability (k) and viscosity (μ) of oil (o) and water (w) in the reservoir on a fractional flow (f). This is represented by Dake's law [5] as

$$f_0 = \frac{1}{1+M} = \frac{1}{1+\frac{\mu_0 k_w}{\mu_w k_0}}$$

From this, it can be seen that if the viscosity of the injected water is lower than that of the oil, then the sweep efficiency will be reduced. Polymer injection solves this issue by increasing the viscosity of water. The increased viscosity of water will reduce relative permeability of injected water in the reservoir. This will increase the fractional flow, hence, resulting in increased oil recovery. The displacement of oil in the reservoir by injected water is efficient when the mobility ratio is one or slightly less than one. On the other hand, if the mobility ratio is greater than one, the injected water is likely to leave behind unswept oil, thus, reducing the overall efficiency of the process [6]. When the mobility ratio in the reservoir crosses one, the polymer is injected to increase the viscosity of the water. This will result in reducing the permeability of water through rock pores resulting in increased efficiency [7]. This process is very efficient in reservoirs that have high mobile oil saturation.

2.2.2 Commonly Used Polymers for EOR

The most common characteristics that are expected from a polymer used for EOR are cost effectiveness, high injectivity, low mechanical degradation, effectiveness when mixed with the reservoir brine solutions, resistance to temperature over a long period of time, resistance to microbial degradation, low retention in porous rocks, sensitivity to oxygen (O_2), hydrogen sulfide (H_2S) and other reservoir and oilfield chemicals [8]. The two major polymers used for EOR production are partially hydrolyzed polyacrylamide (HPAM) and xanthan gum. The chain extension and physical entanglement of the solvated chains in these polymers are used for the viscosity enhancement mechanism.

HPAM results in higher viscosity because of its ability to cause chain expansion as a result of repulsion in the ionic groups. The molecular weight of HPAM is directly proportional to the viscosity of the solution. If the molecular weight of the HPAM increases, it results in an increase in the viscosity of the solution [9]. On the other hand, the higher molecular weight HPAM tends to degrade irreversibly during higher shear rates as the backbone of the polymer breaks down. Due to this issue, the higher molecular weight HPAM tends to degrade in pumps and near the bore well regions [10,11]. The EOR process, however, requires high molecular weight HPAM to achieve high viscosity at low concentration of polymers. In addition to degradation, the HPAM also results in rapid reduction of viscosity of the solution in saline environment [12]. In the saline environment, the HPAM's polymer chains contract due to the shielding of ionic groups in HPAM. There is also a possibility of polymer precipitation in the presence of divalent ions [13], thus, resulting in loss of performance.

Xanthan gum is a rigid polysaccharide used for EOR production. Unlike the aforementioned HPAM, xanthan gum is not degraded due to shear and is insensitive to saline environment and divalent ions [14]. On the other hand, the cost of xanthan gum is high as compared to HPAM. Xanthan gum is also susceptible to biodegradation inside the reservoirs. In addition, there is also a potential of microbial berries from the manufacturing process. Due to its properties, this polysaccharide is used exclusively in drilling fluids [15]. However, the solutions of both xanthan gum and HPAM show reduced viscosity at higher temperatures.

2.3 Associating Polymers

The very first associating polymers were developed to study the behavior of proteins [16]. Later, associating polymers were developed for polymer coatings applications [17]. The development of hydrophobically associating polymers was not undertaken until two decades later. The first hydrophobically associating polymers were prepared by partial esterification of maleic anhydride/styrene copolymers with non-ionic ethoxylated alcohol surfactants [18]. The use of these polymers was limited though, due to the factors like thermal degradation and alkaline hydrolysis. The other issues faced by long chain polyethers were the decomposition in the presence of O_2 and alkaline environment. To address this issue, further works led to the development of maleic anhydride/styrene/vinyl benzyl polyglycol ether copolymers [19,20]. Later to meet the demands of enhanced oil recovery, in the 1980's, acrylamide based hydrophobically associating polymers were developed extensively, leading to the generation of advanced category of EOR polymers. In a recent study, development of associating self-assembly polymer (SAP) systems were derived from HPAM and xanthan gum (XG) using an EOR surfactant and β-cyclodextrin [21]. The polymers, named as SAP-HPAM and SAP-XG, were analyzed for EOR operation. Figure 2.1 demonstrates differential pressure along the sandpack during heavy oil displacement by polymer flooding for these polymers in comparison with HPAM and commercial hydrophobically modified polyacrylamide (HMSPAM). It was observed that SAP-HPAM exhibited effective propagation of these polymers through the sandpack system, which was better in performance than SAP-XG. In another study, novel hydrophobically associating acrylamide polymer incorporating cucurbit[7]uril ester (ACE) and hexadecyl dimethyl allyl ammonium chloride (C_{16}DMAAC) was reported for EOR purposes, as shown in Figure 2.2.

2.3.1 Synthesis

The preparation of associating polymers can be broadly categorized into two techniques. In the first method, the associating polymer can

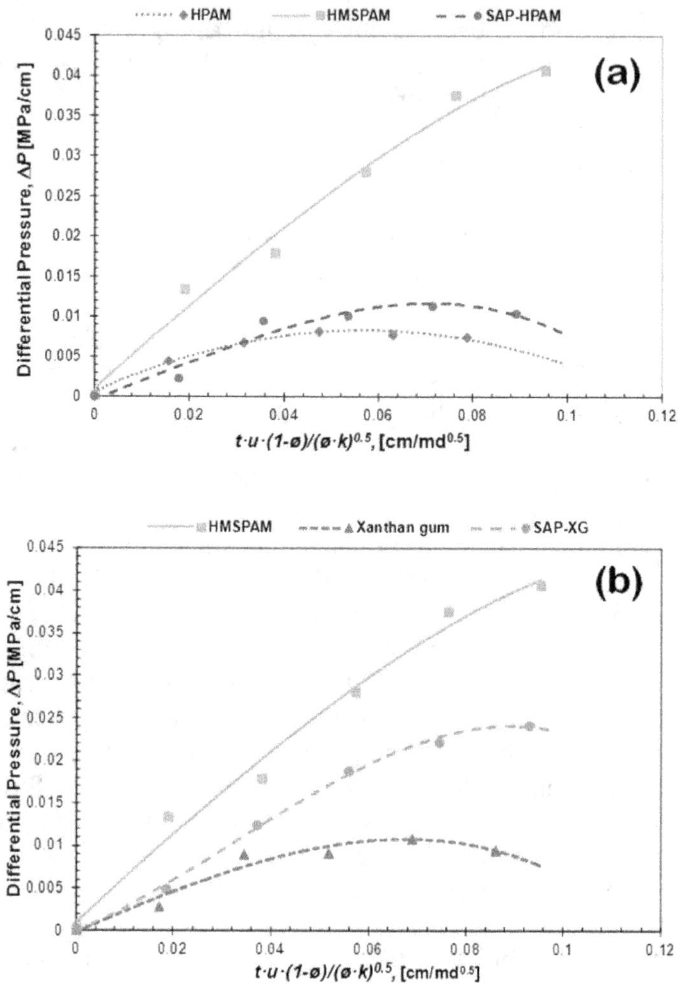

Figure 2.1 Differential pressure along the sandpack during heavy oil displacement by polymer flooding for (a) HPAM, SAP-HPAM, and HMSPAM and (b) xanthan gum, SAP-XG, and HMSPAM. Reproduced from Reference 21 with permission from American Chemical Society.

be obtained by co-polymerization of water-soluble and hydrophobic monomers. In the second method, the polymers are modified after polymerization process to introduce the hydrophobic or hydrophilic groups. The first method where the associating polymers are obtained by co-polymerization using micellar polymerization in free-radical mode is the most commonly used method. In this technique, sodium dodecyl sulphate (SDS) is used as a surfactant in the aqueous solution to solubilize hydrophobic monomer. This results in micelles of SDS containing hydrophobic monomer. Acrylamide is commonly used as the water-soluble monomer because of its ability to produce water-soluble copolymers that are effective at polymer concentrations below 1% mass [23-25]. Associating acrylamide

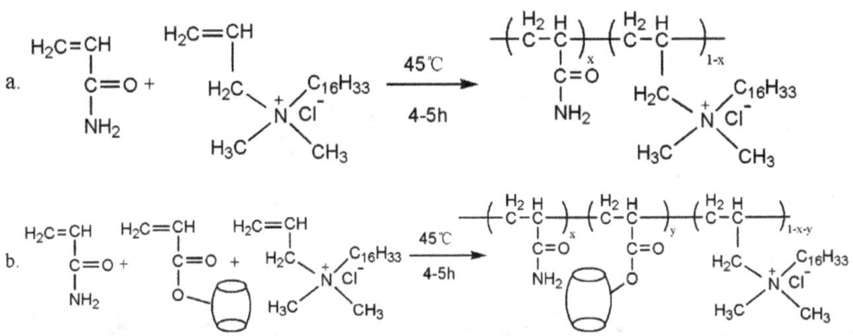

Figure 2.2 Synthesis of (a) AM/C_{16}DMAAC, and (b) AM/ACE/C_{16}DMAAC associating copolymers. Reproduced from Reference 22 with permission from American Chemical Society.

polymers are, thus, synthesized by the copolymerization of acrylamide with a hydrophobic monomer and other monomers like acrylic acid. Alternatively, in order to incorporate carboxylic acid groups in the associating polyacrylamide base, hydrolysis can also be achieved after the polymerization process. However, to achieve better compositional heterogeneity, copolymerization technique is predominantly used. The rheological properties of the resulting polymers depend largely on the compositional heterogeneity. Figure 2.3 shows an example of polymer nanoparticles generated by copolymerization of styrene with poly(n-isopropylacrylamide) [26]. Figure 2.4 also shows the kinetics of swelling and de-swelling of the various copolymer materials generated [26]. Similarly, Figure 2.5 also shows the schematics of growth of poly(N-isopropylacrylamide) from the surface of polystyrene particles using controlled polymerization [27]. The different monomers used for the free radical polymerization of acrylamide based associating polymers are also listed in the Table 2.1.

As mentioned above, associating polymers are also prepared by modification of the polymer after polymerization process. The major drawback in this process is the difficulty in mixing and achieving reaction homogeneity during reactions that involve viscous polymer solutions. For instance, the associating polymers prepared by modification of styrene/maleic anhydride copolymers have been reported [43]. The water solubility is achieved by fully sulfonating the copolymer. Then, the hydrophobic group can be incorporated by a reaction with alkyl amine.

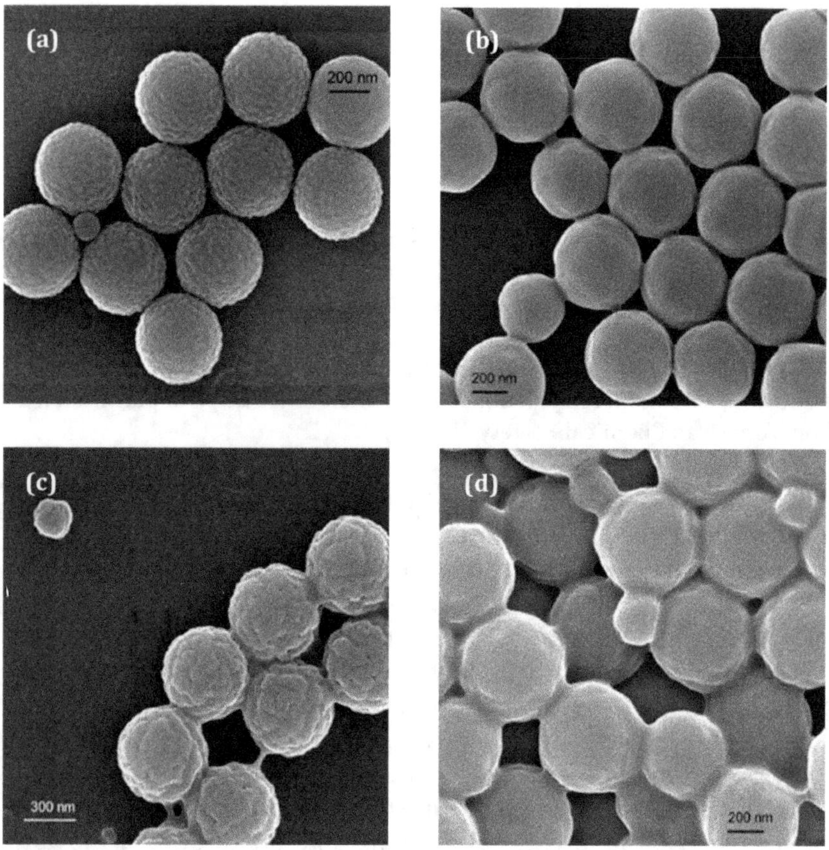

Figure 2.3 Microscope images of the particles formed by emulsion copolymerization of styrene and N-isopropylacrylamide. Reproduced from Reference 26 with permission from Wiley.

The efficiency of the process of incorporating hydrophobic monomer into the polymer has been reported in a large number of studies related to medical drug delivery research in the past two decades [44-48]. These research applications can also be extended for application in the EOR process. From these studies, it can be seen that the reaction rate for the acrylamide polymerization in the aqueous solution is very similar to that of the muscular solution. A positive increase in the rate of incorporation of the hydrophobic group in the polymer can be seen due to the solubilization of the hydrophobic group monomer within the micelles. The rate of incorporation is directly proportional to the number of hydrophobic monomers per micelle. This also means that the hydrophobic monomer can be depleted before complete polymerization, if the value of the ratio of hydrophobic monomer to micelle is greater than one. Figure 2.6 demonstrates an example of the synthesis of copolymer of hydrophobic l-lactide with hydrophilic ethylene glycol (PLA/PEG; (b) PLA/PEG/PLA; (c) a multiblock copolymer of L-lactide and ethylene oxide) [44].

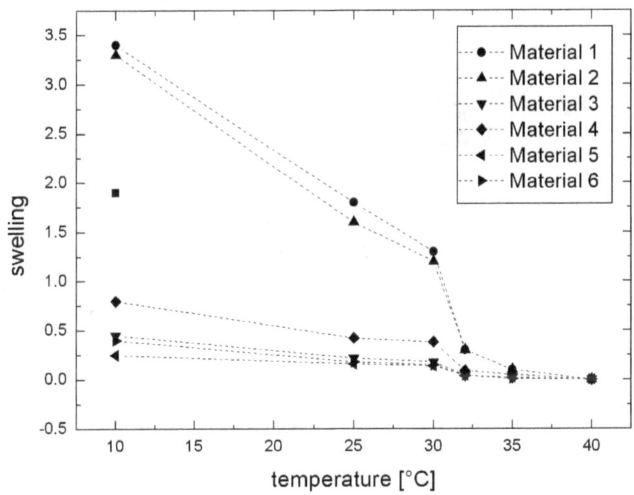

Figure 2.4 Swelling de-swelling kinetics of emulsion copolymerization of styrene and N-isopropylacrylamide. Reproduced from Reference 26 with permission from Wiley.

The final stage of synthesis is the polymer purification. The polymers have to be purified after the micellar polymerization process. The most basic method used is the dilution of the polymer solution and using it without further modification [49-51]. Another common technique used is the precipitation of polymer in a non-solvent like methanol. Subsequently, the polymer is filtered and dried under vacuum [52,53]. In some cases, the polymer is dialyzed with distilled water for up to three days. Afterwards, the low molecular weight impurities are removed by dilution.

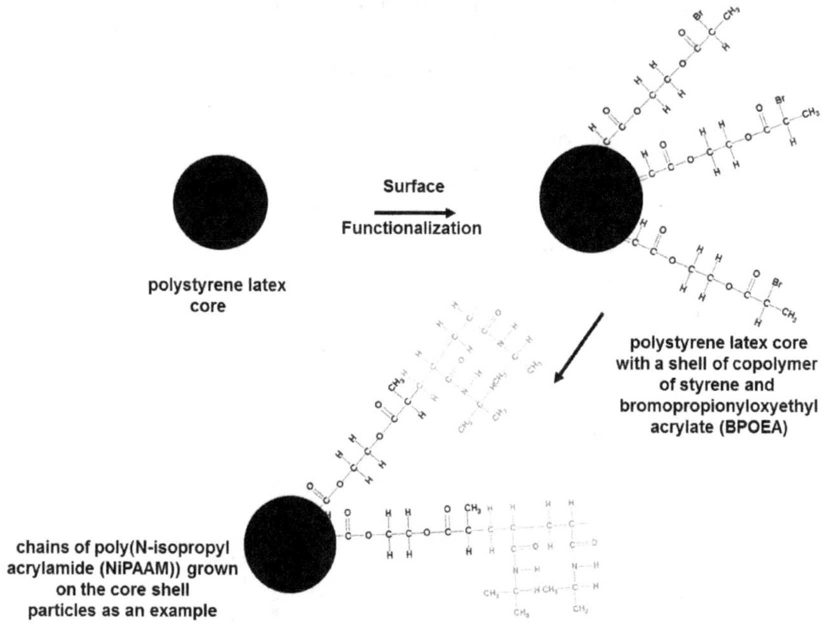

Figure 2.5 Representation of BPOEA functionalization of particles and subsequent growth of poly(N-isopropylacrylamide) brushes by controlled polymerization. Reproduced from Reference 27 with permission from Elsevier.

Table 2.1 Monomers for free radical polymerization

Monomer	Reference(s)
Vinyl sulfonate	[28]
2-Acrylamido-2-methyl-1-propanesulfonic acid (AMPS)	[29]
N-vinyl pyrolidinone	[30-32]
N-alkyl acrylates and methacrylates	[33,34]
Acrylamides	[23,26,27,35,36]
Styrene	[26,27,37-39]
Anionic monomers	[40,41]
Cationic monomers	[42]

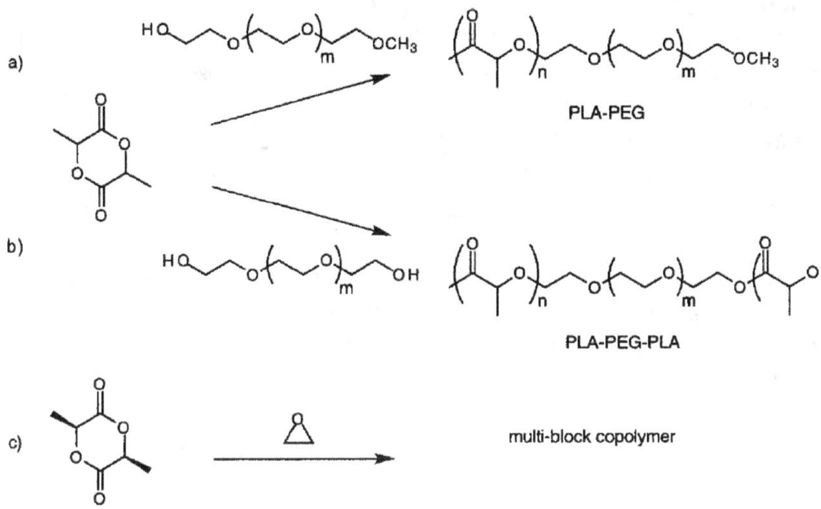

Figure 2.6 Synthesis of copolymers: (a) PLA/PEG; (b) PLA/PEG/PLA; (c) a multiblock copolymer of L-lactide and ethylene oxide. Reproduced from Reference 44 with permission from American Chemical Society.

2.4 Characterization of Polymer Interactions

2.4.1 Hydrophobic Interactions

The most common technique used for the characterization of hydrophobic interactions is by probing the hydrophobic groups with fluorescent probes. The most successful probe reported in a large number of research studies is pyrene [54-61]. The advantages of pyrene are low solubility in water and a characteristic fluorescence spectrum. The use of pyrene results in an increase in the fluorescence and a change in the emission spectrum, as it is solubilized by the micelles or the associating polymers [60,61]. An excimer is formed when there are two or more pyrene molecules present in a hydrophobic region. The excimer has a different fluorescence emission spectrum compared to the pyrene's emission spectrum. Based on this difference in the spectra, it has been calculated that a minimum of ten hydrophones are required to form a hydrophobic region in an acrylamide/dodecyl methacrylate copolymer. From these measurements, it is also seen that for the aggregation to occur, the concentration should be lower than the critical aggregation concentration. Above the critical aggregation concentration, no more aggregation can be seen. In addition to this, it is observed that the polymer behavior in a solution is dominated by the intermolecular hydrophobic associations.

The characterization of hydrophobic interactions leads to classification of the concentration regions for the associating polymers into three parts. Firstly, at low concentration, no more intermolecular association can be seen. Secondly, at intermediate concentration, sporadic association can be seen, but there is no evidence of networking. Finally, at the high concentration region, the viscosity is increased due to extensive intramolecular associations and networking.

2.4.2 Stability under Reservoir Conditions

The hydrolytic stability of the polymers is another important characteristic to be analyzed. This characterization can be used to analyze the stability of the polymers at reservoir conditions. The hydrolytic stability of some associating polymers were examined by Bock *et al.* [62] and Hickner *et al.* [63]. Polymer solutions at 2000 ppm in brine solution were heated at 93 °C for 100 days. On incorporating 2-acrylamido-2-methyl-1-propanesulfonic acid (AMPS) in the polymer, it was observed that the degree of hydrolysis was much lower. For example, a polymer containing polyacrylamide (PAM) group, 78% of amide groups were hydrolyzed as opposed to 36% of amide groups in polymers with AMPS. N-vinyl-2-pyrrolidinone (NVP) exhibited even lower degree of hydrolysis as only 5% of amide groups were hydrolyzed for polymers with NVP under the same conditions. The rate of hydrolysis could further be reduced by the associating groups in the polymer compared to an acrylamide/NVP copolymer.

2.5 Rheology of Associating Polymers

Several factors contribute to the rheological properties of the associating polymers. The major factors are the average molecular weight, hydrophobe distribution, degree of hydrolysis, type of hydrophobe used and degree of incorporation of the hydrophobe [64,65]. An increase in the hydrophobe content reduces the solubility of the associating polymers. In addition, an increase in the molecular weight of the associating polymer increases the chain length, due to which the chain length of the hydrophobe also increases. The increase in the chain length of the hydrophobe reduces the amount of hydrophobe required to make the polymer insoluble. In addition to this, the solubility of the associating polymer is further reduced when fluorocarbon containing hydrophobic groups are used. To overcome this issue, ionic character is introduced in the backbone of the polymer chain, which, in turn, increases the solubility of the polymer. Along with solubility, the rheological properties of the associating polymers can also be affected due to the introduction of ionic groups.

The flow behavior of the associating polymers is also affected due to the introduction of hydrophobic groups [66]. The change in the flow characteristics depends mainly on the intermolecular and intramolecular associations. In this work by English *et al.* [66], rheological properties of hydrophobically modified alkali-swellable emulsion (HASE) polymer were studied in alkaline media containing nonionic ethoxylated surfactants (NP6 and NP10). As shown in Figure 2.7, NP10 due to its miscibility formed mixed junction domains with the polymer hydrophobes, whereas in the case of NP6, the system segregated into polymer-rich and surfactant-rich phases. The work by Schulz *et al.* [67] established the relation between viscosity and polymer concentration. On the plot between reduced viscosity against associating and non-associating polymer concentration, it was observed that the associating polymer enhanced the viscosity after a critical concentration point. If the hydrophobic groups were introduced below this critical concentration point, there was a slight decrease in the viscosity due to intramolecular associations.

It is clear that the effect of hydrophobic group depends directly on the polymer concentration. The rheological properties of the associating polymers are examined under two different conditions, (i) the dilute state, where the concentration of the polymer is below the critical concentration, and (ii) the semi-dilute state, where the concentration of the polymer is above the critical concentration.

2.5.1 Viscosity in Dilute State

The intrinsic viscosity k and Huggin's constant η parameters can be used to study the rheological properties of the associating polymers. These parameters are used to determine the degree of hydrophobic interactions and the molecular weight of the polymer [68].

For the dilute state, according to the Flory-Huggins equation, the reduced viscosity is a linear function of the polymer concentration

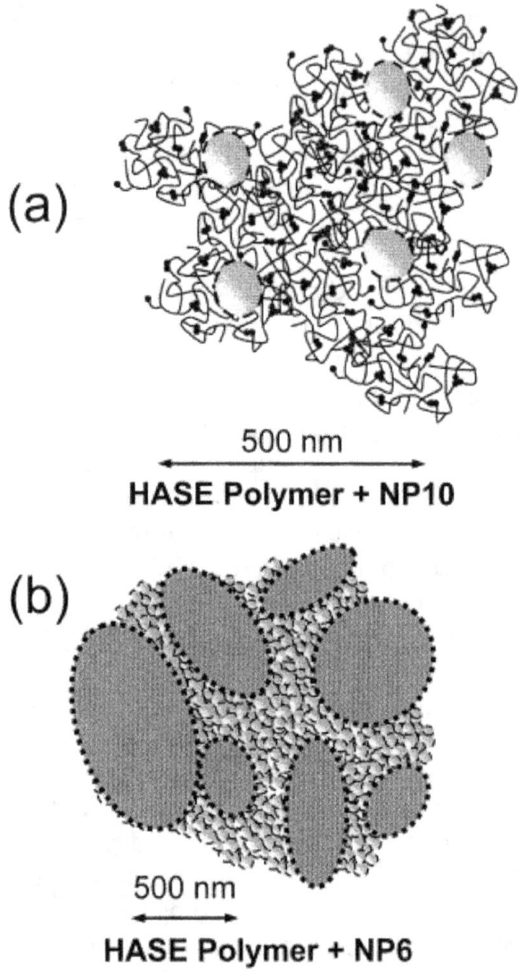

Figure 2.7 Schematic illustrations of (a) HASE polymer with NP10 and (b) HASE polymer with NP6. Reproduced from Reference 66 with permission from American Chemical Society.

$$\frac{\eta - \eta_0}{\eta_0 c} = k[\eta]^2 c$$

where c is the polymer concentration and η_0 is the solvent viscosity [69]. The viscosity measurements of polymer solutions with low polymer concentration can be used to measure the values of η and k.

At a constant molecular weight, the intrinsic viscosity decreases and the Huggins constant η increases as the hydrophobic content increases. The intrinsic viscosity is related to the average molecular weight of the polymer through the equation [70] as

$$[\eta] = K[M_w]^a$$

The increase in the Huggins constant and the decrease in the intrinsic viscosity due to hydrophobic content is mainly due to the polymer chain contraction because of intramolecular association. Hence, the Huggins constant can be used a measurement for hydrophobic interactions in the associating polymers. A value of η greater than 0.8 indicates that there are intramolecular associations in the polymer solution.

Bock *et al.* [71] and Qian *et al.* [72] studied the effect of introducing ionic character to the associating polymers. In two samples of associating polymers with same molecular weight and hydrophobe content, one sample was introduced with ionic character on its backbone. Due to the introduction of the ionic character, an electrostatic repulsion between the negative charges of the carboxylate group was observed. This resulted in the increase of the hydrodynamic volume in the polymer chain. With the increase in the ionic content on the backbone, the intrinsic viscosity of the polymer solution increased and the Huggins constant decreased. The polymer solvent interactions were improved due to the electrostatic repulsion. This resulted in a lower Huggins number, hence, it was clear that the ionic character and the hydrophobic group had opposite effects on the intrinsic viscosity and the Huggins constant.

2.5.2 Viscosity in Semi-Dilute State

As the polymer concentration is above the critical concentration, the polymer-solvent mixture reaches a semi-dilute state. The rheological characteristics of the polymer solution in this state are different than that of the dilute state, as mentioned above. In this state, the increase in the polymer concentration results in intermolecular association and, hence, enhances the viscosity of the polymer solution. For example, if the hydrophobe content is increased from 0.8 mol% to 1 mol%, the viscosity of the solution increases significantly. In addition, at a given concentration of the hydrophobe, increasing the molecular weight of the associating polymer results in increased viscosity of the solution [73,71]. Based on the studies by Bock *et al.* [71] and Grattoni *et al.* [74], the increase in the hydrophobe content resulted in the increase in the viscosity of the solution due to the intermolecular interactions. In addition, it was also observed that the hydrophobes containing phenyl group enhanced the viscosity significantly, especially at high hydrophobe concentrations.

The apparent viscosity of the solution stays constant at concentrations above 3000 ppm. Further, the viscosity increases with the increase in shear, which is called shear thickening. This increase in viscosity stops at a particular point and starts to reduce after a certain period of shear exposure, which is called shear thinning. This phenomenon occurs due to shifting of the intermolecular and intramolecular association with the shear rate. In addition, it can also been seen that the interchain associations are favored by temperature and, hence, the viscosity increases with increase in temperature. However, at very high temperatures, the viscosity decreases due to the degradation of the associating polymer. Figure 2.8 shows the shear thickening behavior of hydrophobically associating acrylamide polymers as a function of hydrophobe level [25].

2.5.3 Chemical Interaction

The analysis of chemical interactions with salts is very critical for the associating polymers because of the conditions in the reservoir. In the presence of a saline environment, the viscosity of associating polymer can increase, this effect is more significant at high polymer concentrations. This phenomenon is similar to that of the micelle formation in the presence of ionic surfactants. In saline environment, the critical micelle concentration (CMC) reduces due to enhanced aggregation. Thus, the enhanced association describes the effect on viscosity of the associating polymer in a saline environment [22,75]. The sensitivity of HPAM to saline environment is considered as a disadvantage at high hydrophobe concentrations.

As discussed in the previous sections, the surfactant concentration also affects the viscosity of the associating polymer systems. Conditions where the surfactant concentration is closer to the CMC, a peak in the viscosity of the associating polymer systems can be observed [76]. Below CMC, the polymer cross-linking efficiency increases due to the presence of more than one type of hydrophobe in the micelles. The cross-linking between polymers, in turn, increases the viscosity of the system. On the other hand, when the surfactant concentration is above the CMC, the number of micelles increases, affecting the cross-linking, thus, resulting in reduced viscosity [77].

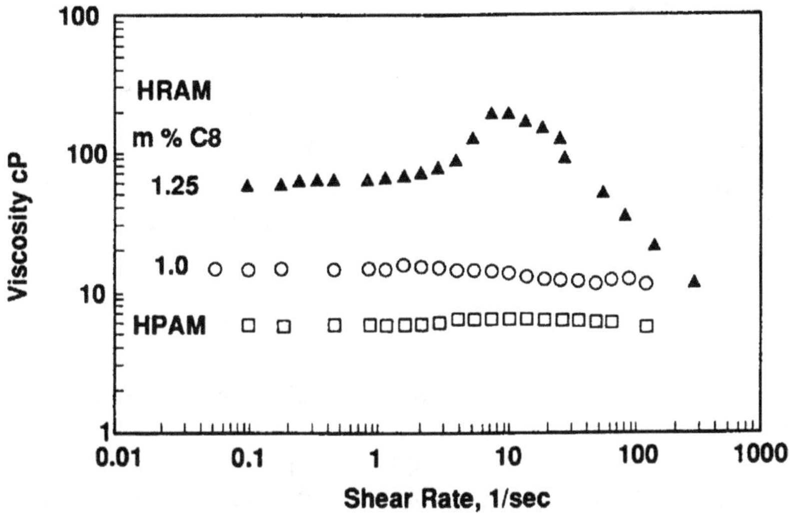

Figure 2.8 Shear thickening behavior of hydrophobically associating acrylamide polymers as a function of hydrophobe level. Reproduced from Reference 25 with permission from American Chemical Society.

2.5.4 Flow of Associating Polymers in Porous Rocks

The study of the flow of the associating polymers in porous media is another factor to be considered for the use of polymers in EOR. The resistance of the associating polymers should be low for the EOR applications. The resistance factor is calculated by Bock *et al.* [62], which is described by the following equation:

$$R = \frac{\Delta P_{polymer}}{\Delta P_{brine}}$$

where ΔP refers to the pressure drop across the core for the respective solution and R represents the resistance factor. The resistance factor can increase if the hydrophobe levels are too high. This is because of the polymer adsorption through the pores, thus, leading to plugging. In order to deal with this issue, modification is performed on the polymer to reduce the level of adsorption. The resistance factor for HPAM peaks at injection velocities at 10 ft/day and drops at higher velocities due to degradation caused by high shear rates [62].

2.6 Conclusion

The primary and secondary oil recovery methods result in 20 to 40% oil recovery from the reservoirs, while adapting to tertiary recovery methods (or EOR) will result in 40 to 60% recovery of reservoir's oil. Though CO_2 injection technique has been used widely across North America and Russia, the applications of associating polymers have been overlooked for the past decade. This review paper summarizes the synthesis, characterization and rheology of associating polymers and their scope in EOR. In addition, this report adds to the previous review on the subject [78] by reviewing the recent developments on the application of associative polymers in EOR.

Micellar polymerization is required to produce hydrophobic associating polymers that are soluble in water. The synthesized associating polymers are then characterized using fluorescence techniques by using pyrene as fluorescent probes. The rheological properties of the polymers depend on various factors like molecular weight, hydrophobe distribution, temperature, etc. An optimum has to be achieved among the molecular weight, ionic characterization, polymer concentration and hydrophobe content.

Polyacrylamide based polymers are used commonly for the synthesis of associating polymers that are used for EOR. Even though other techniques like gas injection and thermal injection are cheaper than the polymer injection technique, the sweep efficiency of the polymer injection technique is significantly higher than other techniques.

This review paper serves as a precursor to the research in the field of synthesis and characterization of associative polymers for achieving enhanced oil recovery.

References

1. Walsh, M. P., and Lake, L. W. (2003) *A Generalized Approach to Primary Hydrocarbon Recovery*, Volume 4, Elsevier Science, The Netherlands.
2. Gozalpour, F., Ren, S., and Tohidi, B. (2005) CO_2 EOR and storage in oil reservoir. *Oil and Gas Science and Technology*, **60**(3), 537-546.

3. Tabary, R., and Bazin, B. (2007) Advances in Chemical Flooding. In: IFP-OAPEC Joint Seminar Improved Oil recovery (IOR) Techniques and Their Role in Boosting the Recovery Factor, France.
4. Pope, G. A. (2007) Overview of Chemical EOR. Online: http://www.uwyo.edu/eori/_files/misc_download/overview%20of%20chemical%20eor.pdf (assessed 31.01.2017).
5. Dake, L. P. (1983) *Fundamentals of Reservoir Engineering*, Volume 8, 1st Edition, Elsevier Science Publishers, The Netherlands.
6. Sastry, N., Dave, P., and Valand, M. (1999) Dilute solution behaviour of polyacrylamides in aqueous media. *European Polymer Journal*, **35**(3), 517-525.
7. Littman, W. (1998) *Polymer Flooding*, Developments in Petroleum Science, Elsevier Science Publishers, The Netherlands.
8. Abidin, A., Puspasari, T., and Nugroho, W. (2012) Polymers for enhanced oil recovery technology. *Procedia Chemistry*, **4**, 11-16.
9. Chauveteau, G., and Sorbie, K. (1991) Mobility control by polymers. In: *Basic Concepts in Oil Recovery Processes*, Elsevier Science Publishers, USA.
10. Liu, D.-X., Zhao, X.-T., and Qiu, G.-M. (2008) Review on the degradation of polyacrylamide used for flooding. *Polymer Materials Science and Engineering*, **9**, 8.
11. Li, J., Yang, X., Yu, X., Xu, L., Kang, W., Yan, W., Gao, H., Liu, Z., and Guo, Y. (2009) Rare earth oxide-doped titania nanocomposites with enhanced photocatalytic activity towards the degradation of partially hydrolysis polyacrylamide. *Applied Surface Science*, **255**(6), 3731-3738.
12. Li, K., Baohui, W., Xuejia, Z., Qin, R., Wei, J., and Xue, Y. (2008) Adsorption of polyacrylamide on soil. *Environmental Pollution and Control*, **1**, 9.
13. Zaitoun, A., and Potie, B. (1983) Limiting Conditions for the Use of Hydrolyzed Polyacrylamides in Brines Containing Divalent Ions. *SPE Oilfield and Geothermal Chemistry Symposium*, Society of Petroleum Engineers, USA, doi:10.2118/11785-MS.
14. Lopez, M., Vargas-Garcia, M., Suarez-Estrella, F., and Moreno, J. (2004) Properties of xanthan obtained from agricultural wastes acid hydrolysates. *Journal of Food Engineering*, **63**(1), 111-115.
15. Garcıa-Ochoa, F., Santos, V., Casas, J., and Gomez, E. (2000) Xanthan gum: production, recovery, and properties. *Biotechnology Advances*, **18**(7), 549-579.
16. Strauss, U. P., and Jackson, E. G. (1951) Polysoaps. I. viscosity and solubilization studies on an n-dodecyl bromide addition compound of poly-2-vinylpyridine. *Journal of Polymer Science*, **6**(5), 649-659.
17. Dubin, P., and Strauss, U. P. (1967) Hydrophobic hypercoiling in

copolymers of maleic acid and alkyl vinyl ethers. *The Journal of Physical Chemistry*, **71**(8), 2757-2759.

18. Beihoffer, T. W., Lundberg, D. J., and Glass, J. E. (1989) Influence of polarity in water- soluble polymer synthesis. In: *Polymers in Aqueous Media*, American Chemical Society, USA, pp. 151-164.
19. Corson, F. P., and Evani, S. (1977) Aqueous Thickening Agents Derived from Vinyl Benzyl Ether Polymers, US Patent 4,008,202.
20. Corson, F. P., and Evani, S. (1977) Vinyl Benzyl Ethers and Nonionic Water Soluble Thickening Agents Prepared Therefrom, US Patent 4,029,873.
21. Wei, B., Romero-Zeron, L., and Rodrigue, D. (2014) Evaluation of two new self-assembly polymeric systems for enhanced heavy oil recovery. *Industrial and Engineering Chemistry Research*, **53** (43), pp 16600-16611.
22. Zou, C., Gu, T., Xiao, P., Ge, T., Wang, M., and Wang, K. (2014) Experimental study of cucurbit[7]uril derivatives modified acrylamide polymer for enhanced oil recovery. *Industrial and Engineering Chemistry Research*, **53**(18), 7570-7578.
23. Emery, O., Lalot, T., Brigodiot, M., and Marechal, E. (1997) Free-radical polymerization of acrylamide by horseradish peroxidase-mediated initiation. *Journal of Polymer Science, Part A: Polymer Chemistry*, **35**(15), 3331-3333.
24. Odian, G. G., and Odian, G. (2004) *Principles of Polymerization*, Wiley-Interscience, USA.
25. Bock, J., Siano, D. B., Valint, P. L., and Pace, S. J. (1989) Structure and properties of hydrophobically associating polymers. In: *Polymers in Aqueous Media: Performance through Association*, Glass, J. E. (ed.), Advances in Chemistry Series 223; American Chemical Society, USA, Chapter 22.
26. Mittal, V., Matsko, N. B., Butte, A., and Morbidelli, M. (2008) Swelling deswelling behavior of PS-PNIPAAM copolymer particles and PNIPAAM brushes grafted from polystyrene particles & monoliths. *Macromolecular Materials and Engineering*, **293**, 491-502.
27. Mittal, V., Matsko, N. B., Butte, A., and Morbidelli, M. (2008) Functionalized polystyrene latex particles as substrates for ATRP: Surface and colloidal characterization. *Polymer*, **48**, 2806-2817.
28. Hawker, C. J. (1997) "Living" free radical polymerization: A unique technique for the preparation of controlled macromolecular architectures. *Accounts of Chemical Research*, **30**(9), 373-382.
29. Greesh, N., Hartmann, P. C., Cloete, V., and Sanderson, R. D. (2008) Impact of the clay organic modifier on the morphology of polymer–clay nanocomposites prepared by in situ free-radical polymerization in emulsion. *Journal of Polymer Science, Part A: Polymer Chemistry*, **46**(11), 3619-3628.
30. Hawker, C. J., Elce, E., Dao, J., Volksen, W., Russell, T. P., and Barclay,

G. G. (1996) Well-defined random copolymers by a "living" free-radical polymerization process. *Macromolecules*, **29**(7), 2686-2688.

31. Wan, D., Satoh, K., Kamigaito, M., and Okamoto, Y. (2005) Xanthate-mediated radical polymerization of n-vinylpyrrolidone in fluoroalcohols for simultaneous control of molecular weight and tacticity. *Macromolecules*, **38**(25), 10397-10405.
32. Ray, B., Kotani, M., and Yamago, S. (2006) Highly controlled synthesis of poly (n-vinylpyrrolidone) and its block copolymers by organostibine-mediated living radical polymerization. *Macromolecules*, **39**(16), 5259-5265.
33. Haddleton, D. M., Kukulj, D., Duncalf, D. J., Heming, A. M., and Shooter, A. J. (1998) Low-temperature living "radical" polymerization (atom transfer polymerization) of methyl methacrylate mediated by copper (i) n-alkyl-2-pyridylmethanimine complexes. *Macromolecules*, **31**(16), 5201-5205.
34. Strehmel, V., Laschewsky, A., Wetzel, H., and Gornitz, E. (2006) Free radical polymerization of n-butyl methacrylate in ionic liquids. *Macromolecules*, **39**(3), 923-930.
35. Ganachaud, F., Monteiro, M. J., Gilbert, R. G., Dourges, M.-A., Thang, S. H., and Rizzardo, E. (2000) Molecular weight characterization of poly (n-isopropylacrylamide) prepared by living free-radical polymerization. *Macromolecules*, **33**(18), 6738-6745.
36. Xiao, D., and Wirth, "M. J. (2002) Kinetics of surface-initiated atom transfer radical polymerization of acrylamide on silica. *Macromolecules*, **35**(8), 2919-2925.
37. Buback, M., Kowollik, C., Kurz, C., and Wahl, A. (2000) Termination kinetics of styrene free-radical polymerization studied by time-resolved pulsed laser experiments. *Macromolecular Chemistry and Physics*, **201**(4), 464-469.
38. Zhang, H., Hong, K., and Mays, J. W. (2002) Synthesis of block copolymers of styrene and methyl methacrylate by conventional free radical polymerization in room temperature ionic liquids. *Macromolecules*, **35**(15), 5738-5741.
39. Jakubowski, W., Min, K., and Matyjaszewski, K. (2006) Activators regenerated by electron transfer for atom transfer radical polymerization of styrene. *Macromolecules*, **39**(1), 39-45.
40. Nakano, T., Takewaki, K., Yade, T., and Okamoto, Y. (2001) Dibenzofulvcnc, a 1, 1- diphenylethylene analogue, gives a π-stacked polymer by anionic, free-radical, and cationic catalysts. *Journal of the American Chemical Society*, **123**(37), 9182-9183.
41. Sumerlin, B. S., Donovan, M. S., Mitsukami, Y., Lowe, A. B., and McCormick, C. L. (2001) Water-soluble polymers. 84. Controlled polymerization in aqueous media of anionic acrylamido monomers

via raft. *Macromolecules*, **34**(19), 6561-6564.

42. Peiffer, D. (1990) Hydrophobically associating polymers and their interactions with rod- like micelles. *Polymer*, **31**(12), 2353-2360.
43. Cook, R. L., King Jr, H., and Peiffer, D. G. (1992) Pressure-induced crossover from good to poor solvent behavior for polyethylene oxide in water. *Physical Review Letters*, **69**, 3072.
44. Uhrich, K. E., Cannizzaro, S. M., Langer, R. S., and Shakesheff, K. M. (1999) Polymeric systems for controlled drug release. *Chemical Reviews*, **99**(11), 3181- 3198.
45. Svec, F., Peters, E. C., Sykora, D., and Frechet, J. M. (2000) Design of the monolithic polymers used in capillary electrochromatography columns. *Journal of Chromatography A*, **887**, 3-29.
46. Rudzinski, W., Dave, A., Vaishnav, U., Kumbar, S., Kulkarni, A., and Aminabhavi, T. (2002) Hydrogels as controlled release devices in agriculture. *Designed Monomers and Polymers*, **5**(1), 39-65.
47. Kickelbick, G. (2003) Concepts for the incorporation of inorganic building blocks into organic polymers on a nanoscale. *Progress in Polymer Science*, **28**(1), 83-114.
48. Fevola, M. J., Hester, R. D., and McCormick, C. L. (2003) Molecular weight control of polyacrylamide with sodium formate as a chain-transfer agent: Characterization via size exclusion chromatography/multi-angle laser light scattering and determination of chain-transfer constant. *Journal of Polymer Science Part A: Polymer Chemistry*, **41**(4), 560-568.
49. Evani, S., and van Phung, K. (1988) Amphiphilic Monomer and Hydrophobe Associative Composition Containing a Polymer of a Water-soluble Monomer and Said Amphiphilic Monomer, US Patent 4,728,696.
50. Evani, S., and van Phung, K. (1990) Hydrophobe Associative Composition Containing a Polymer of a Water-soluble Monomer and an Amphiphilic Monomer, US Patent 4,921,902.
51. Geise, G. M., Lee, H.-S., Miller, D. J., Freeman, B. D., McGrath, J. E., and Paul, D. R. (2010) Water purification by membranes: the role of polymer science. *Journal of Polymer Science, Part B: Polymer Physics*, **48**(15), 1685-1718.
52. Edwards, J. H., Feast, W. J., and Bott, D. C. (1984) New routes to conjugated polymers: 1. a two-step route to polyacetylene. *Polymer*, **25**(3), 395-398.
53. Li, H., Liu, Y., Zhang, Z., Liao, H., Nie, L., and Yao, S. (2005) Separation and purification of chlorogenic acid by molecularly imprinted polymer monolithic stationary phase. *Journal of Chromatography A*, **1098**(1), 66-74.
54. Dualeh, A. J., and Steiner, C. A. (1990) Hydrophobic microphase formation in surfactant solutions containing an amphiphilic graft

copolymer. *Macromolecules*, **23**(1), 251-255.

55. Koussathana, M., Lianos, P., and Staikos, G. (1997) Investigation of hydrophobic interactions in dilute aqueous solutions of hydrogen-bonding interpolymer complexes by steady-state and time-resolved fluorescence measurements. *Macromolecules*, **30**(25), 7798-7802.
56. Jorand, F., Boue-Bigne, F., Block, J., and Urbain, V. (1998) Hydrophobic/hydrophilic properties of activated sludge exopolymeric substances. *Water Science & Technology*, **37**, 307-315.
57. Yessine M.-A., and Leroux J.-C. (2004) Membrane-destabilizing polyanions: interaction with lipid bilayers and endosomal escape of biomacromolecules. *Advanced Drug Delivery Reviews*, **56**(7), 999-1021.
58. Liu, S. Q., Tong, Y., and Yang, Y.-Y. (2005) Incorporation and in-vitro release of doxorubicin in thermally sensitive micelles made from poly isopropylacrylamide n,n-dimethylacrylamide)-poly (d,l-lactide-glycolide) with varying compositions. *Biomaterials*, **26**(24), 5064-5074.
59. Lo, C.-L., Lin, K.-M., and Hsiue, G.-H. (2005) Preparation and characterization of intelligent core-shell nanoparticles based on poly (d,l-lactide)-g-poly (n-isopropyl acrylamide-co-methacrylic acid). *Journal of Controlled Release*, **104**(3), 477-488.
60. Varadaraj, R., Bock, J., Brons, N., and Pace, S. (1993) Probing hydrophobic microdomains of hydrophobically associating acrylamide-n-alkylacrylamide copolymers in solution using a solvatochromic absorption dye probe. *The Journal of Physical Chemistry*, **97**, 12991-12994.
61. Varadaraj, R., Branham, K. D., McCormick, C. L., and Bock, J. (1994) Analysis of hydrophobically associating copolymers utilizing spectroscopic probes and labels. In: *Macro- molecular Complexes in Chemistry and Biology*, Dubin, P., Bock, J., Davis, R., Schulz, D. N., Thies, C. (eds.), Springer-Verlag, Germany, pp. 15-31.
62. Bock, J., Pace, S. J., and Valint, P. L. (1987) Enhanced Oil Recovery with Hydrophobically Associating Polymers Containing Sulfonate Functionality, US Patent 4,702,319.
63. Hickner, M. A., Ghassemi, H., Kim, Y. S., Einsla, B. R., and McGrath, J. E. (2004) Alternative polymer systems for proton exchange membranes (pems). *Chemical Reviews*, **104**(10), 4587-4612.
64. Annable, T., Buscall, R., Ettelaie, R., and Whittlestone, D. (1993) The rheology of solutions of associating polymers: Comparison of experimental behavior with transient network theory. *Journal of Rheology*, **37**(4), 695-726.
65. Aubry, T., and Moan, M. (1994) Rheological behavior of a hydrophobically associating water soluble polymer. *Journal of Rheology*,

38(6), 1681-1692.

66. English, R. J., Laurer, J. H., Spontak, R. J., and Khan, S. A. (2002) Hydrophobically modified associative polymer solutions: Rheology and microstructure in the presence of nonionic surfactants. *Industrial and Engineering Chemistry Research*, **41**(25), 6425-6435.
67. Schulz, D., and Bock, J. (1991) Synthesis and fluid properties of associating polymer systems. *Journal of Macromolecular Science - Chemistry*, **28**(11-12), 1235-1243.
68. Ma, X., and Pawlik, M. (2007) Intrinsic viscosities and Huggins constants of guar gum in alkali metal chloride solutions. *Carbohydrate Polymers*, **70**(1), 15-24.
69. Sakai, T. (1968) Huggins constant k for flexible chain polymers. *Journal of Polymer Science, Part A2: Polymer Physics*, **6**(8), 1535-1549.
70. Bock, J., Valint. Jr., P. L., Pace, S. J., Siano, D. B., Schulz, D. N., and Turner, S. R. (1988) Hydrophobically associating polymers. In: *Water-soluble Polymers for Petroleum Recovery*, Stahl, G. A., Schulz, D. N. (eds.), Springer, USA, pp. 147-160.
71. Bock, J., Siano, D., Valint Jr, P., and Pace, S. (1989) Structure and properties of hydrophobically associating polymers. *Polymers in Aqueous Media: Performance through Association, Advances in Chemistry Series*, **223**, 411-424.
72. Bai, Y., Qian, J., Sun, H., and An, Q. (2006) Dilute solution behavior of partly hydrolyzed poly(vinyl acetate) in selective solvent mixtures and the pervaporation performance of their membranes in benzene/cyclohexane separation. *Journal of Membrane Science*, **279**, 418-423.
73. Rossi, S., Ferrari, F., Bonferoni, M. C., and Caramella, C. (2001) Characterization of chitosan hydrochloride-mucin rheological interaction: influence of polymer concentration and polymer: mucin weight ratio. *European Journal of Pharmaceutical Sciences*, **12**(4), 479-485.
74. Grattoni, C. A., Al-Sharji, H. H., Yang, C., Muggeridge, A. H., and Zimmerman, R. W. (2001) Rheology and permeability of crosslinked polyacrylamide gel. *Journal of Colloid and Interface Science*, **240**(2), 601-607.
75. Chiu, Y.-L., Chen, S.-C., Su, C.-J., Hsiao, C.-W., Chen, Y.-M., Chen, H.-L., and Sung, H.-W. (2009) Ph-triggered injectable hydrogels prepared from aqueous n-palmitoyl chitosan in vitro characteristics and in vivo biocompatibility. *Biomaterials*, **30**(28), 4877-4888.
76. Glass, J., Lundberg, D., Zeying, M., Karunasena, A., and Brown, R. (1990) Viscoelasticity and High Shear Rate Viscosity in Associative Thickener Formulations. *Proceedings of the Water-Borne and Higher-Solids Symposium*, USA, pp. 102-120.

77. Piculell, L., Guillemet, F., Thuresson, K., Shubin, V. and Ericsson, O. (1996) Binding of surfactants to hydrophobically modified polymers. *Advances in Colloid and Interface Science*, **63**, 1-21.
78. Taylor K. C., and Nasr-El-Din, H. A. (1998) Water-soluble hydrophobically associating polymers for improved oil recovery: A literature review. *Journal of Petroleum Science and Engineering*, **19**(3), 265-280.

3

Potential of Ionic Liquids for Gas Separation

3.1 Introduction

Natural gas, which is largely hydrocarbon in nature, is one of the most important energy sources for several decades, and has seen its demand as a fuel to grow gradually worldwide [1]. However, besides the main component methane (CH_4), natural gas is generally contaminated with various impurities like hydrogen sulfide (H_2S) and carbon dioxide (CO_2), commonly known as acid gases. These acid gases need to be separated at the well for preventing the technological issues, which may occur during the transportation as well as the liquefaction of the gas [2]. The presence of acid gases in the natural gas can cause shrinkage in the natural gas heating value and can also lead to corrosion in facilities [3]. Further, H_2S is immensely toxic gas and releases SO_2 into the atmosphere after the combustion process [4]. As per the regulations, the ultimate levels of CO_2 and H_2S in the sales grade natural gas are limited to 2 mol % and 4 ppm respectively [5]. In order to keep the acid gas levels within the limit for meeting the sales gas specifications, different processes have been developed for treating the natural gas streams. Gas sweetening process is the acid gas eradication process where the acid gases such as CO_2, H_2S and other sulfur containing components like mercaptans are removed from the natural gas stream. As the requirement for natural gas has enhanced in recent years, the need for developing highly efficient technologies for acid gas sweetening has emerged. Depending on the various operating conditions and different levels of CO_2 and H_2S present in the natural gas, several technologies such as absorption, adsorption, membranes and cryogenic condensation are used for the acid gas separation process.

Currently, the most widely employed techniques for the acid gas separation are the chemical and physical absorption processes [6]. The benefits of chemical absorption process, utilizing the aqueous amine solutions, include low hydrocarbon loss, high efficiency for acid gas separation at different conditions and low operational cost.

Haleema Saleem and Vikas Mittal, Khalifa University of Science and Technology, Abu Dhabi, UAE

However, this method suffers from various drawbacks like excessive corrosion, amine loss and intrinsically higher regeneration cost [7]. The physical absorption employing physical solvents such as Selexol has some advantages like simple regeneration and lower regeneration energy requirement when compared to the chemical absorption process. Nevertheless, physical absorption process has some drawbacks such as higher affinity of solvents towards the hydrocarbons which can lead to hydrocarbon losses as well as appropriateness for feed streams with adequately higher CO_2 pressure [8]. Hence, it is advantageous to develop new solvents with more favorable characteristics, therefore, substantial research effort has been devoted to develop suitable solvents for capturing the acid gases.

The substitution of conventional alkanolamine solutions with ionic liquids (ILs) for the removal of H_2S and CO_2 is a field of growing research [9]. The ILs are molten salts generated by the combination of various organic cations (e.g. pyrrolidinium, imidazolium, ammonium, pyridinium, phosphonium), polyatomic inorganic anions (e.g. hexafluorophosphate, tetrafluoroborate, chloride) or organic anions (e.g. tri-fluoro-methylsulfonate, bis[(trifluoromethyl)sulfonyl]imide,) [10-12]. For instance, Figure 3.1 depicts the reaction of an ionic liquid with CO_2.

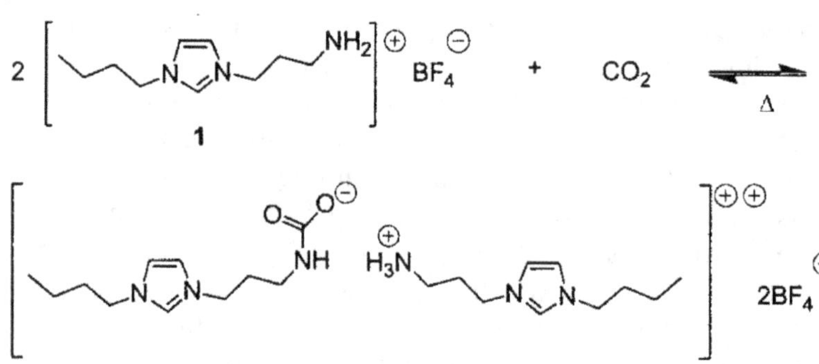

Figure 3.1 Proposed reaction between a task-specific ionic liquid (1) and CO_2. Reproduced from Reference 9 with permission from American Chemical Society.

During the past several years, ILs have acquired boundless attention in many fields due to their interesting features like excellent thermal stability, low volatility, lower regeneration energy, negligible vapor pressure, low corrosivity, fast absorption/desorption kinetics and tunable functionality [13,14]. Hence, these are regarded as environmentally benign solvents, when compared to other volatile organic solvents [15-17]. As these salts are liquids at room temperature, hence they are also known as room temperature ionic liquids (RTILs). The ILs are also called as the "designer" solvents because they can be tailored for different specific applications [18], by the suitable selection of cation, anion and substituents on the cations. Among the various applications of ILs, the two major areas of current research are (1) analyzing the possibility of utilizing ILs for the acid gas (CO_2 and H_2S) separation, and (2) the fixation as well as sequestration of CO_2, which is considered to be an important greenhouse gas [17,19]. Figure 3.2 demonstrates the use of equimolar solutions of primary alkanolamines (i.e., MEA) and secondary alkanolamines (i.e., DEA) with imidazolium based ionic liquids for CO_2 scrubbing [19]. Several theoretical as well as experimental studies have been performed to analyze the properties of various ILs for the

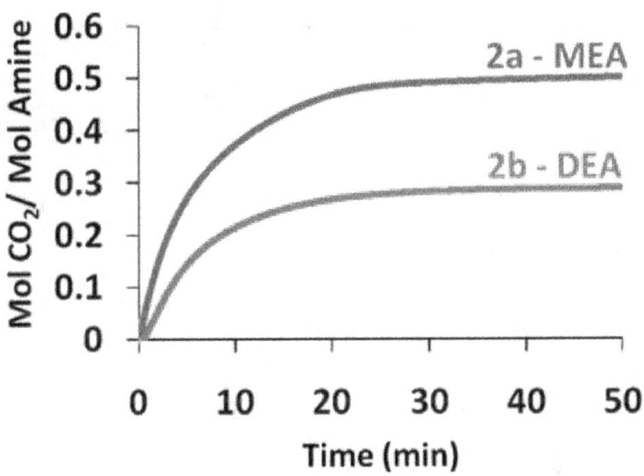

Figure 3.2 CO_2 uptake in equimolar solutions of 2a-MEA and 2b-DEA, where 2a and 2b are imidazolium based ionic liquids. Reproduced from reference 19 with permission from American Chemical Society.

removal of acid gases [20-22]. In a recent study, Jalili *et al.* [22] studied the solubility of hydrogen sulfide in three ionic liquids, 1-butyl-3-methylimidazolium hexafluorophosphate ([bmim][PF_6]), 1-butyl-3-methylimidazolium tetrafluoroborate ([bmim][BF_4]), and 1-butyl-3-methylimidazolium bis(trifluoromethylsulfonyl)imide ([bmim][Tf_2N]), at temperatures ranging from (303.15 to 343.15) K and pressures up to 1 MPa. As observed in Figure 3.3 depicting the comparison of the performance of N-methyldiethanolamine (MDEA) with ILs for the dissolution of H_2S, the low-pressure part with a low slope was attributed to absorption due to stoichiometric reaction between alkanolamine and the acid gas. However, as the overall molality of the sour gas reaches the overall molality of MDEA, the total pressure was observed to increase and additional acid gas must be dissolved physically. Comparing this phenomenon with the curves for ILs used for the study, it was observed that the solubility of H_2S in the ILs was typical of that of physical solvents. In summary, knowledge of different acid gases solubility as well as rate of solubility, i.e., diffusion coefficients of gases at different pressures and temperatures, is critical in the assessment of ILs as a gas dissociation medium and also for the design as well as operation of gas sweetening processes. Further, from the analysis of the solubility of the gases in ILs, beneficial information regarding the interaction of the gases with the ILs can also be obtained [23].

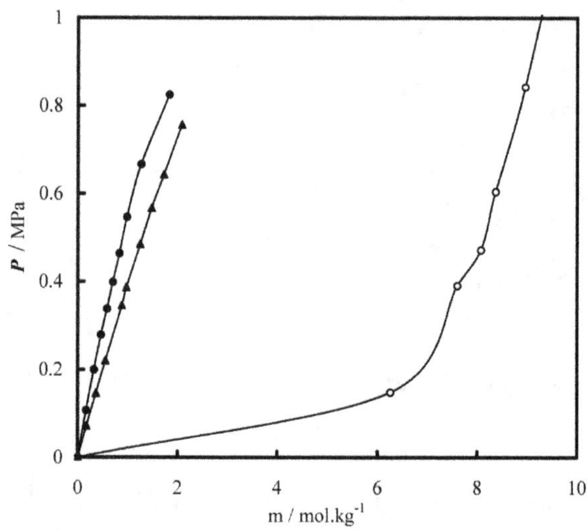

Figure 3.3 Comparison between the solubility of hydrogen sulfide as molality, m, in the ILs and an aqueous solution of N-methyldiethanolamine (MDEA) at 313.15 K. ●, H_2S in [bmim][Tf_2N]; ▲, H_2S in [bmim][BF_4]; O, H_2S in aqueous MDEA solution. Reproduced from reference 22 with permission from American Chemical Society.

In this chapter, the different types of ILs employed for the efficient removal of acid gases during the natural gas sweetening process have been discussed. Various IL techniques such as RTILs, TSILs (task specific ILs), PILs (polymerized ILs), or SILs (supported

ILs) can be employed for the H_2S or CO_2 removal, be employed for the H_2S or CO_2 removal, ranging from low temperature to high temperature applications. In addition to this, the review deals with the analysis of theoretical predictions, experimental results as well as emerging ideas on the various ILs for the acid gas separation, with specific emphasis on CO_2 and H_2S. Several observations in terms of advantages as well as disadvantages of employing various ILs are also detailed. Further, the review sheds light on the influence of different anions, cations, alkyl substituents, polymer backbones of anions and cations, and crosslinking on the acid gas sorption ability and separating effectiveness.

3.2 Different IL Based Techniques

In this section, analysis of different ILs such as pure ILs, polymerized ILs and supported ILs, used for the acid gas separation, is carried out. Among the different supported ILs, ILs supported on membranes, MOFs as well as graphene have been provided more attention due to their potential for large scale applications.

3.2.1 Pure Ionic Liquids

Extensive reviews regarding gas solubility, exclusively CO_2, in ILs and utilization of ILs in H_2S/CO_2 removal from industrial gas streams have been reported in the literature [7,24,25]. In one such review, Bara *et al.* [25] reported an extensive analysis of imidazolium based room temperature ionic liquids for CO_2 separation. Figure 3.4 demonstrates the CO_2 solubility performance of oligo ethylene glycol, nitrile and analogous *n*-alkyl functionalized [Rmim][Tf_2N] RTILs. Overall, for enhancing the CO_2 solubility in ILs, many studies shave investigated the effect of various parameters like alkyl chain size of cations, different cation types, alkyl chains fluorination on the cation, various anion types and degree of fluorination of anions [23,26-28]. The major observation obtained from the various research studies is that the CO_2 solubility in the ILs with anions containing $-CF_3$ groups like $[PF_6]^-$, $[Tf_2N]^-$ and $[BF_4]^-$ is greater because of the CO_2 philic nature of the fluoroalkyl group, i.e., more number of fluorinated anions causes increased CO_2 solubility in ILs.

Imidazolium based ILs are of distinguished interest among the various ILs employed for the acid gas removal. Different studies on the solubility of H_2S and CO_2 in different imidazolium-based ILs, have been performed at varying pressure and temperature [25]. The first study on the H_2S solubility on imidazolium ILs was reported by Jou and Mather [20]. The authors reported the H_2S solubility in the IL 1-butyl 3-methyl imidazolium hexa-fluorophosphate ([C4mim][PF_6]) at pressure till 9.6 MPa and temperatures ranging from 298.15-403.15 K. As mentioned earlier, Jalili *et al.* [22] analyzed the H_2S solubility in three different imidazolium based ILs and concluded that the H_2S solubility in 1-butyl 3-methyl imidazolium based ILs followed the order $[BF_4]^- < [PF_6]^- < [Tf_2N]^-$. In another study by Jalili *et al.* [29], the solubility as well as diffusion coefficient

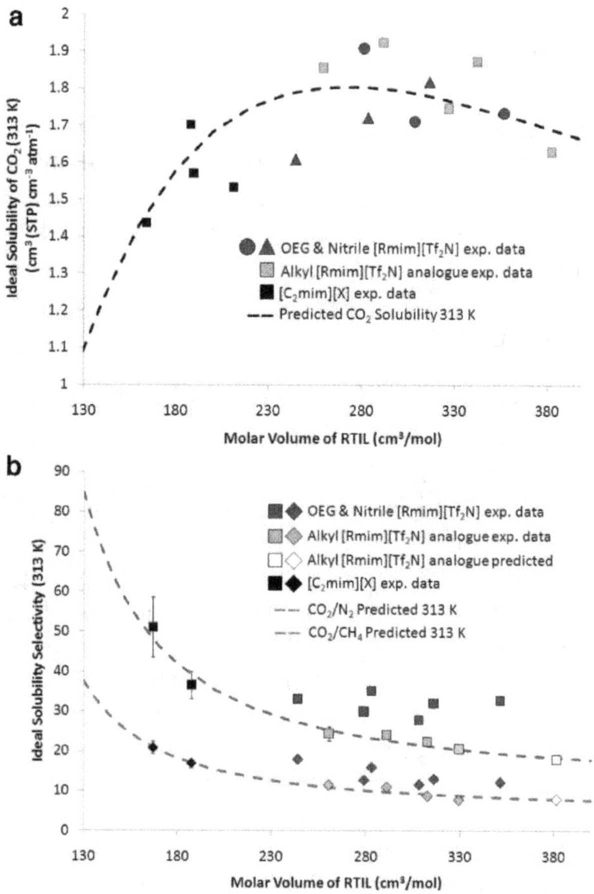

Figure 3.4 (a) Ideal CO_2 solubility as a function of molar volume, for OEG, nitrile and analogous n-alkyl functionalized [Rmim][Tf_2N] RTILs; (b) ideal solubility selectivity as a function of molar volume. Reproduced from Reference 25 with permission from American Chemical Society.

of H_2S and CO_2 gases in the IL 1-ethyl 3-methyl imidazolium ethyl-sulfate ([emim][$EtSO_4$]) were determined, at pressures till 1.6 MPa along with temperatures 303.15-353.15 K [29]. The comparison confirmed that the solubility of H_2S was more than the CO_2 solubility in the IL, and the diffusion coefficient of H_2S was almost twice the magnitude as that of CO_2 in the IL.

Nematpour *et al.* [30] studied the solubility of H_2S and CO_2 gases in the IL 1-ethyl 3-methyl imidazolium tri-fluoromethanesulfonate ([C_2mim][OTf]) at pressures up to 3 MPa along the temperature range 303.15-353.15 K. The results indicated that at same pressure and temperature, the H_2S solubility in [C_2mim][OTf] was more than four times than the CO_2 solubility. The molality of H_2S at 2.0 MPa pressure and 303.15 K temperature was almost 1.3 times and 2.5 times that of [C_2mim][Tf_2N] and [C_2mim][eFAP], respectively. The order of CO_2 solubility in the ILs followed the order [C_2mim][eFAP] > [C_2mim][Tf_2N] > [C_2mim][OTf]. As the temperature was increased to 353.15 K, the H_2S solubility in [C_2mim][OTf] reduced to 1.8 times the solubility in [C_2mim][eFAP]. Thus, as compared to other ILs, [C_2mim][OTf] was confirmed to have better potential for the separation of H_2S and CO_2 gases from each other.

Soriano *et al.* [31] studied the CO_2 solubility in the IL 1-ethyl 3-methyl imidazolium tri-fluoromethanesulfonate [Emim][triflate] till 5.9 MPa pressure in the temperature range of 303.2-343.2 K with the help of a thermogravimetric microbalance. The CO_2 solubility was examined from absorption equilibrium data keeping fixed pressure and temperature. The maximal solubility was observed at 303.2 K. An enhancement in the solubility was observed with increase in pressure, while it reduced as the temperature increased. A two parameter extended Henry's law equation was employed to analyze the experimental results.

In another study by Pomelli *et al.* [21], H_2S solubility in various [C_4mim]$^+$ based ILs with diverse anions and a series of [$Tf_{2\text{-}}N$] ILs having distinct cations was analyzed at 1400 kPa and 298.15 K, with the help of a medium pressure NMR method. From the obtained results, it was concluded that the interaction between the anion of ILs and H_2S was the major factor contributing to the the higher H_2S solubility in the ILs. Safavi *et al.* [32] reported experimental data for the solubility of CO_2 and H_2S in 1-octyl-3methyl imidazolium hexafluorophosphate ([C_8mim][PF_6]) at pressures till 2 MPa in the temperature range from 303.15-353.15 K. It was observed that the solubility of CO_2 and H_2S enhanced with increasing pressure, however, it reduced with the increase in temperature. In this study, the H_2S solubility was about three times than that of CO_2 in the specific IL examined. The authors also examined the influence of cation alkyl chain length on H_2S and CO_2 solubility by comparing the experimental results with the data of previous reports. It was noted that the CO_2 as well as H_2S solubility increased as the cation alkyl chain length increased. In addition to this, the anion effect on the solubility was analyzed by comparing the H_2S as well as CO_2 solubility in [C_8mim][PF_6] with that of [C_8mim][Tf_2N]. It was concluded that the H_2S and CO_2 solubility got increased in the ionic liquid with anion [Tf_2N]$^-$.

Muldoon *et al.* [28] studied the CO_2 solubility in 1-hexyl 3-methyl-1H-imidazol-3-ium tris(1,1,2,2,2-penta-fluoroethyl) trifluorophosphate. It was observed that the ILs with increased fluoroalkyl chains on either anion or cation enhanced the CO_2 solubility, when compared to the lesser fluorinated ILs. Zhang *et al.* [33] examined the CO_2 solubility using ILs having various anions and cations and it was confirmed that the ILs having the [eFAP]$^-$ were better solvents for the removal of CO_2, when compared to other ILs. The authors also conducted molecular simulation analysis to gain further insights about the enhancement of CO_2 solubility in [eFAP]$^-$ based ILs [34].

Shiflett *et al.* [35] analyzed the results obtained by the vapour-liquid-liquid equilibrium (VLLE) measurements of single gases CO_2, H_2S, as well as their binary mixtures by using [C_4mim][PF_6] at pressures till 6.5 MPa and temperatures from 273 K to 342 K. It was observed that the selectivity (CO_2/H_2S) was almost independent of the quantity of IL added, and ranged from 3.2 to 4.0. However, in the case of large CO_2/H_2S mole ratios (9/1), the incorporation of IL enhanced the CO_2/H_2S selectivity from 1.2 to 3.7, at 298.15 K temperature.

In another study by Jalili *et al.* [36], the solubility of CO_2 and H_2S in 1-ethyl-3-methyl imidazolium tris(penta-fluoroethyl)trifluorophosphate ([C_2mim][eFAP]) was analyzed at pressures till 2 MPa and temperature range 303-353 K. The results indicated that H_2S was more soluble in the specific IL than the CO_2 gas. The quantity of H_2S dissolved was greater than twice the CO_2 quantity, at fixed pressure and temperature. Thus, the aforementioned studies suggest that the pure IL solvent can be potentially employed for the separation of H_2S and CO_2 gases during the industrial gas sweetening processes.

3.2.2 Polymeric Ionic liquids

An emerging family of functional polymers, typically called as poly(ionic liquid)s or polymeric ionic liquids (PILs) have exclusive characteristics of ILs as well as macromolecular framework [37]. PILs are polyelectrolytes containing IL species in the form of anions or cations in monomers, which are linked to the polymer chain, thus, forming an extensive molecular design. The PILs can be generated by the monomer polymerization or by the co-polymerization of monomers with IL species being in repeating units. This specific class of materials possess excellent CO_2 sorption, good thermal stability and faster adsorption-desorption rates than the RTILs [38,39]. Due to the weak intermolecular forces, the monomeric IL molecules are liquid at the room temperature. However, as the PILs are macromolecules with strong intermolecular forces between the chains, these materials are solids with greater CO_2 sorption ability than the corresponding monomeric ILs. The PILs also possess increased durability, stability, excellent processability and improved control over their nano- and meso-structures.

Majority of the research studies dealing with PILs have concentrated on the addition of polymerizable species into IL cations and subsequently investigated the CO_2 and H_2S sorption as well as separation capability of PILs [37,40,41]. These studies indicated the capability of PILs as acid gas sorption material, however, more research effort is required to study different polycation-anion pairs that could govern the acid gas sorption capacity of PILs. Different functionalities can be incorporated on the interior or surface of polymeric framework, which could offer remarkable features to the porous polymers like the selectivity and reversibility. The key specifications for ideal CO_2 sorption are low cost, greater CO_2 solubility, lesser energy requirement for regeneration, long term reusability and environmental friendliness. The selection of proper cations is of great significance in the development of a PIL skeleton for the removal of acid gases. The nature of cation exhibits an essential role in describing the features of PIL, which is different than the RTIL, which are anion dependent.

Tang *et al.* [38] examined the potential of three different types of PILs as CO_2 sorption material, namely poly[2-(methacryloyloxy)ethyl-3-butyl imidazolium tetra-fluoroborate] (PIL-3), poly[1-(p-vinylbenzyl)-3-butyl-imidazolium hexa-fluorophosphate] (PIL-2b) and poly[1-(p-vinylbenzyl)3-

butylimidazolium tetra-fluoroborate] (PIL-2a). At 22 °C and 592.3 mmHg CO_2, the CO_2 removal ability of RTIL [bmim][BF4] (1-butyl-3-methyl imidazolium tetra-fluoroborate) was found to be 1.30 mol%. However, in the case of PILs, the CO_2 sorption values of PIL-3, PIL-2a and PIL-2b were analyzed to be 1.78 mol% (1.4 moles), 2.27 mol% (1.7 moles) and 2.8 mol% (2.1 moles), respectively. In this study, the IL monomers exhibited no CO_2 sorption because of their crystalline structures. On the other hand, the PILs attained their 90% CO_2 sorption abilities within 3-4 minutes. In addition to this, the PILs attained entire CO_2 sorption within 30 minutes, however, the RTIL [bmim][BF_4] required more than 400 minutes to obtain the equilibrium [42].

Majority of PILs are based on the aliphatic backbone and are brittle, whereas the PILs based on condensation polymerization possess certain aromaticity in the backbone of main chain [43]. Films of such PILs have the capability to withstand pressure till 10 atm, during the permeation analysis. Tang *et al.* [44] proved that the P[VBTMA] based PILs with distinct anions [BF_4, Tf_2N and PF_6] exhibited significant CO_2 sorption capacity. The PILs can be generated either by IL monomer polymerization [45] or by anion exchange process of a precursor polymer [46]. Tang *et al.* [45,47] observed that the tetra-alkylammonium based PILs displayed reversible sorption/desorption at quicker rates as well as greater CO_2 sorption abilities, almost 6.0-7.6 times, when compared to RTILs. In addition to this, tetra-alkylammonium based PILs containing same anions offered higher CO_2 sorption, when compared to the imidazolium based PILs, which were studied earlier by the authors [42]. The steady interaction of the tetra-alkylammonium cation with the CO_2 is due to the greater positive charge density, when compared to the imidazolium cation where the positive charges are delocalized. Different types of PILs, mainly ammonium and imidazolium based polymers, have been widely prepared and compared to examine their CO_2 sorption abilities. Heintz *et al.* [41] also examined the solubility of CO_2 and N_2/H_2S mixture in a polymeric ammonium polyether-based IL with chloride anion at pressure range 0.23-3 MPa and temperature range 300 K to 500 K.

In a study by Bhavsar *et al.* [40], PILs were generated based on P[VBTMA][Cl] (poly(vinylbenzyltrimethylammonium chloride)) and P[DADMA][Cl] (poly(diallyldimethyl ammonium chloride)) as precursors, using anion exchange technique. Na or Li salts were employed to prepare water insoluble PILs, while the Ag salt of needed anion was employed to generate water soluble PILs with greater than 96% anion exchange in a single step. The PILs containing carboxylate anions exhibited great CO_2 sorption as well as higher selectivity over N_2 and H_2. With increasing order of molar masses, the CO_2 sorption in PILs containing inorganic anions was observed to be enhanced. P[DADMA][Bz] and P[DADMA][Ac] displayed significant CO_2 sorption along with higher CO_2 based selectivities S_{CO2}/S_{N2} of 41.5 and 114.3, respectively. The capability of PIL with [Ac] anion was confirmed further by detecting similar higher CO_2 sorption properties for the PIL based on the cation P[VBTMA]. The CO_2 sorption in a series of PILs with changing anion was found to be enhanced by reducing the PIL density and raising anion basicity. Xiong *et al.* [48] generated imidazolium based ILs with symmetrical ester as well as hydroxyl groups and the equivalent polymer was prepared by melt condensation polymerization. The properties and structure of the generated PILs were characterized by X-ray diffraction (XRD), differential scanning calorimetry (DSC) gel permeation chromatograph (GPC), scanning electron microscopy (SEM) and proton nuclear magnetic resonance. Further, the CO_2 sorption characteristics of the IL monomers and corresponding PILs were examined at a temperature of 25 °C and pressure of 648.4 mmHg CO_2 with the help of a thermogravimetric analyzer. Among the various PILs and IL monomers, the CO_2 sorption ability of 1,3-bis(2-hydroxylethyl)-imidazolium hexa-fluorophosphate ([HHIm]PF_6, 10 mol%) was observed to be highest. In addition to this, the sorption equilibrium of [HHIm]PF_6 was attained in a shorter time-period. Thus, many PILs have been developed as CO_2 sorbents, however, there is a requirement for additional exploration for understating the relationship between gas permeability and macromolecular/molecular composition [49]. The literature on PILs is generally dominated by PF_6, Tf_2N and BF_4 as anions. Several other evaluations have also been carried on anions such as OTf, dca and SbF_6 [50]. In addition to this, the CO_2 sorption abilities of imidazolium and ammonium based PILs displayed higher sorption, when compared to other polymers like polymethacrylates, polyethylenes and polystyrenes/polycarbonates [51]. Among the various cations, the ammonium cation present in simple/porous PILs exhibited greater CO_2 separation ability when compared to other cations.

In another study, Wilke *et al.* [52] reported the synthesis of a mesoporous poly(ionic liquid) network via a hard-templating pathway. The structure was stabilized by achieving a high degree of cross-linking. The synthesized mesoporous PIL exhibited faster CO_2 adsorption than the non-porous counterpart. Figure 3.5 demonstrates the synthetic pathway for the synthesis of mesoporous PIL along with template method for the network generation. The cross-linkable monomer was prepared by reaction of 1- vinylimidazole with 4-vinyl benzyl chloride. The authors observed favorable interactions of the PIL with CO_2 and the presence of large mesopores in the network led to enhanced mass transfer. The PIL also exhibited higher selectivity for CO_2 as compared to N_2.

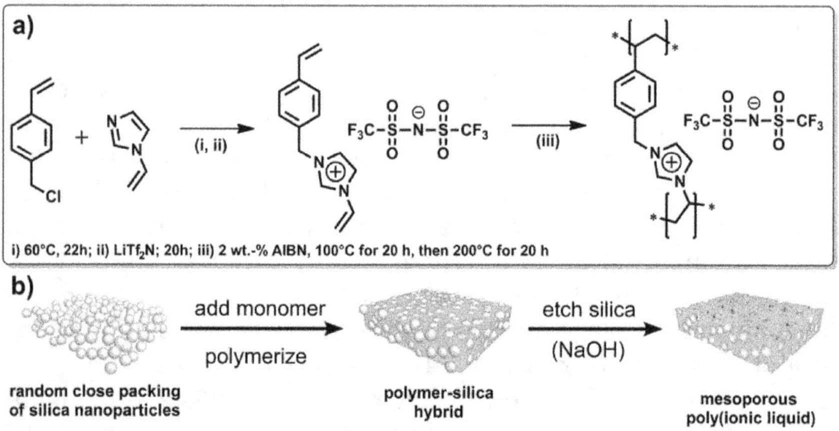

Figure 3.5 (a) Synthetic pathway for the generation of PIL and (b) schematic of the template method used to generate the network. Reproduced from Reference 52 with permission from American Chemical Society.

In a study by Shahrom *et al.* [53], nitrile imidazolium based TSILs having different alkyl chain length and various anions ([SBA], [DDS], [DOSS], [TFMS], [BS]), PILs containing [VBTEA], [VBTMA] and [METMA] cations and [NO_3], [BF_4], [Cl], [TFMS] and [PF_6] anions and amino acid-based polymerized ionic liquids [AAPILs] incorporating [Arg], [Pro] as anions and [VBTMA] as cation were prepared. It was observed that the CO_2 sorption enhanced with the enhancement in alkyl chain length, and the [C_2CNDim][DOSS] displayed the maximum CO_2 sorption (0.55 mol fraction at 298 K and 10 bar) among the different TSILs. In the case of PILs, the CO_2 sorption capacity enhanced in polymeric forms when compared to that of monomers. For further increasing the efficiency of PIL for the CO_2 adsorption, AAPILs were examined, and the results indicated that the AAPILs were able to adsorb more CO_2 than the PILs and TSILs, due to the presence of functionalized amine tethered at the anion in AAPIL. The results also indicated that at 298 K and lower pressure (0.7 bar), the CO_2 sorption of the AAPILs [VBTMA][Pro] and [VBTMA][Arg] were 0.38 and 0.530 mol fractions, respectively. On comparing the PILs and ILs, the CO_2 based selectivities of certain PILs were found to be higher, which clearly indicated that PILs are next generation promising materials for CO_2 sorption.

3.2.3 Supported Ionic Liquids

Despite the fact that ILs are promising candidates for the acid gas capture, the viscosity, energy necessity for recycling and high cost of preparation are the three main drawbacks which limit their practical applications [54]. To overcome these limitations, a new approach known as supported ILs (SILs) was proposed [55], where the ILs are encapsulated into porous solids for bettering the removal efficiency [56]. SILs are, thus, derivatives of ILs, and the IL immobilization processes combine their properties with the advantageous features of the substrates. As a result, the SILs have been explored in nearly all the fields involving ILs and have exhibited features like efficient performance, low cost operation and environmental friendly nature.

Ionic liquids Supported on Membranes

Supported liquid membranes (SLMs), the porous supports where the pores are impregnated using a solvent, have exhibited remarkable potential in various applications [57]. In the recent years, the concept of combining RTILs with membranes has been developed for the CO_2 separation [58,59]. By utilizing the ionic liquid membrane (ILM) technology, the drawbacks of ILs can be overcome. In this system, IL could be stabilized by impregnation inside the support membrane's pores. The ILM consists of feed as well as permeate phases, which are separated by the membrane holding the IL, allowing the concurrent extraction and stripping at each side of ILM. The use of ILM technique offers advantages like the requirement of lesser quantity of IL as carrier and no need of additional steps for IL recycling. Further, this technique also offers some other benefits like simple fabrication of compact and flexible devices, lower capital as well as operating costs, lesser energy requirements, etc. Due to its higher viscosity as well as negligible vapor pressure, the stability of

ILM is higher when compared to the traditional SLM based on different organic solvents [60,61]. Thus, ILMs exhibit encouraging application potential for the removal of CO_2 and other gases. Various types of membrane processes accommodating ILs have been studied, which include supported IL membranes (SILMs), polymer/IL gel membranes, polymerized ionic liquid (PIL) membranes and membrane absorption utilizing membrane contactors (MCs) employing ILs as absorbent.

In a study by Scovazzo *et al.* [62], the selective separation of the gas pairs CO_2/N_2 and CO_2/CH_4 employing continued flows of mixed gases, at different CO_2 concentrations (till 2 bars of CO_2 partial pressure) was carried out. In this system, the ILs [emim][BF_4], [emim+][TfO$^-$] as well as [emim$^+$][dca$^-$] were supported in polyvinylidene fluoride (PVDF) membrane, while the ILs [emim$^+$][BETI$^-$], [hmim+][Tf_2N^-] and [emim$^+$][Tf_2N^-] were supported in a polyethersulfone (PES) membrane. The maximum CO_2/N_2 and CO_2/CH_4 selectivities were 21.2 and 27, by employing the ILs [emim+][Tf_2N^-] and [emim$^+$][BF_4], respectively. The segregation of CO_2, from CH_4 and N_2, was also favorably performed with IL polymer films, which were prepared from IL monomers along with polymerizable groups like acrylate and styrene [63]. The separation of CO_2 from CH_4 was also studied by utilizing task specific ILs like [NH_2pmim$^+$][$CF_3SO_3^-$] and [NH_2pmim$^+$][Tf_2N-], where the functional groups present in the ILs are able to chemically complex with CO_2 [64]. This type of SILMs, having amine terminated ILs, promoted the transportation of CO_2 through the membrane, exhibiting greater selectivities when compared to the IL [bmim$^+$][NTf_2^-] for the removal of CO_2 from the CO_2/CH_4 gas mixture. Effective gas separations by employing SILM have also been reported in other studies such as H_2S/CH_4 [65] and CO_2/He [66]. Park *et al.* [65] studied the multiphase separation method, using RTIL and polymer to develop SILMs, for the separation of acidic gases from the crude natural gas. PVDF material was used for the membrane, while the $BMImBF_4$ was employed as the RTIL. SEM analysis was carried out for the structural analysis of generated SILMs. For the examination of the permeation behavior, the SILMs were tested with H_2S, CO_2 and CH_4 at different operating conditions. Due to the fact that the gases like H_2S and CO_2 have greater affinity towards the RTILs than the CH_4, the permeability coefficients of H_2S and CO_2 were found to be significantly higher at 160-1100 and 30-180 Barrer, respectively. In addition to this, the selectivity of H_2S/CH_4 and CO_2/CH_4 were observed to be 130-260 and 25-45, respectively. The influence of various operating conditions in gas separation has also been examined. Neves *et al.* [67] analyzed the potential of utilizing the SILMs based on 1-n-alkyl 3 methyl imidazolium cation for the CO_2/CH_4 as well as CO_2/N_2 separations. Here, the effect of existence of water vapor in the gas stream on the gas selectivity and permeability was studied. It was found that the water vapor presence in the gas stream enhanced the gas permeability of SILM, however, reduced the CO_2/CH_4 and CO_2/N_2 selectivity remarkably, when compared to the dry gas stream. This reduction in the selectivity was due to the water cluster development within the membrane, and the influence was more evident for the less hydrophobic ILs. In another study by Cserjesi *et al.*, [68] the analysis of permeability for CO_2, N_2, H_2 and CH_4 gases and the selectivity of SILM based on broad variety of ILs at various transmembrane pressures and temperatures was performed. It was observed that the enhancement in temperature increased the permeability, while the enhancement in trans-membrane pressure reduced the permeability. Also, the SILMs exhibited relatively higher long-term stability, as their permeability did not exhibit any remarkable variation in the course of the experiments.

IL based membranes have also been utilized for gas permeation at higher temperatures [66]. In these type of membranes, the diffusivity selectivity benefits the smaller molecules like H_2 or He, while CO_2 is highly benefitted by solubility selectivity. Myers *et al.* [69] synthesized task specific ILs containing functional groups which are able to form complexes with CO_2. This was employed as fiber supported liquid membrane (FSLM) and the CO_2/H_2 selectivity of the FSLM reached till 10-20 at about 85 °C temperature. Albo *et al.* [70] analyzed the CO_2 removal in the IL [emim][$EtSO_4$] (1-ethyl-3-methyl-imidazolium ethylsulfate) using a polypropylene (PP) hollow fiber membrane contactor for simulated flue gas at ambient temperatures (Figure 3.6). Here, a mathematical model considering a parallel flow configuration was employed for a cross flow system, for explaining the mass transfer rate.

In another study by Gomez-Coma *et al.* [71], the analysis of the temperature effect on the CO_2 removal efficiency using two ILs, [emim][$EtSO_4$] (1-Ethyl-3-methyl-imidazolium ethylsulfate) and [emim][Ac] (1-Ethyl-3-methyl-imidazolium acetate) was carried out. In this system, a PP hollow fiber module served as the membrane device where the absorption of CO_2 took place and the temperature ranged from room temperature to 333 K. The CO_2 separation efficiency was attained from the experimental data, and it was observed that only in the case of [emim][Ac], the temperature dependence was observed. Here, the CO_2 removal efficiency doubled from temperature 291 K to 318 K.

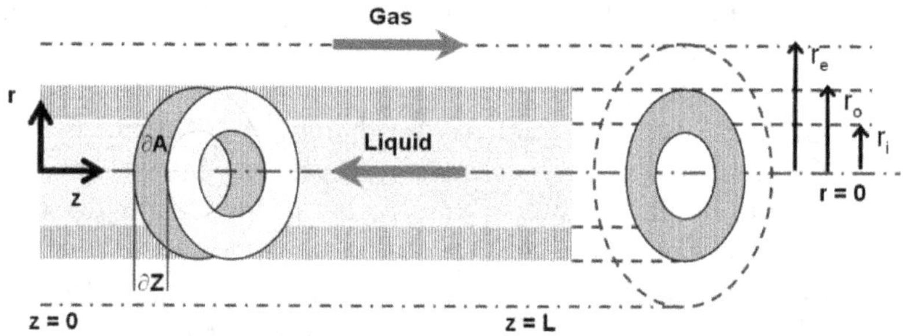

Figure 3.6 Axial and radial coordinates of the fiber. Reproduced from Reference 70 with permission from American Chemical Society.

Mulikutla *et al.* [72] analyzed the CO_2 reactive absorption in the novel non-volatile absorbent containing IL [bmim][DCA] (1-butyl 3-methyl-imidazolium-dicyanamide) which consisted of 20 wt% polyamidoamine (PAMAM) dendrimer Gen 0. Here, a humidified simulated flue gas which contained almost 14% CO_2 was employed, and favorable separation of bulk of CO_2 was observed. The absorption and regeneration was attained by using rectangular cross flow hollow fiber membrane modules which were made of porous PP fibers with porous polymerized fluorosiloxane coating on the outside diameter. Chau *et al.* [73] studied the novel cyclic 5 valve pressure swing membrane absorption process (PSMAB) for the separation of CO_2 from low temperature shifted syngas in hollow fiber membrane contactor using IL absorbent ([bmim][DCA]) at about 100 °C. At higher temperature, the CO_2 removal efficiency was remarkably increased by the addition of PAMAM dendrimer Generation 0 to the IL for a 60% He, 40% CO_2 feed. The absorption behavior in this process depends on the selective sorption, absorbent viscosity and hollow fiber module design.

The IL based supported membranes possess several limitations like poor stability which leads to poor long term performance and operation being achievable only at low pressures [63,74,75]. For overcoming these limitations, the inclusion of IL nature in the polymer backbone is developing as a promising technology. Due to the limitations of PILs based on aliphatic backbone, such as inability to form film and limitation of obtained membranes to be useful only at lower pressure about 2 atm [63], efforts have been taken for transforming these into membranes by utilizing techniques such as copolymerization or crosslinking for efficient flat film formation [76]. The casting of IL monomer on a porous polymer support, followed by UV polymerization, has also been studied [76]. The gas removal behavior of the produced membranes could be analyzed at lower pressure, i.e., till 2 atm upstream pressure, due to its fragile nature. By utilizing the crosslinking methodology, Li *et al.* [74] generated composite membrane of IL and PIL, as shown in Figure 3.7. In

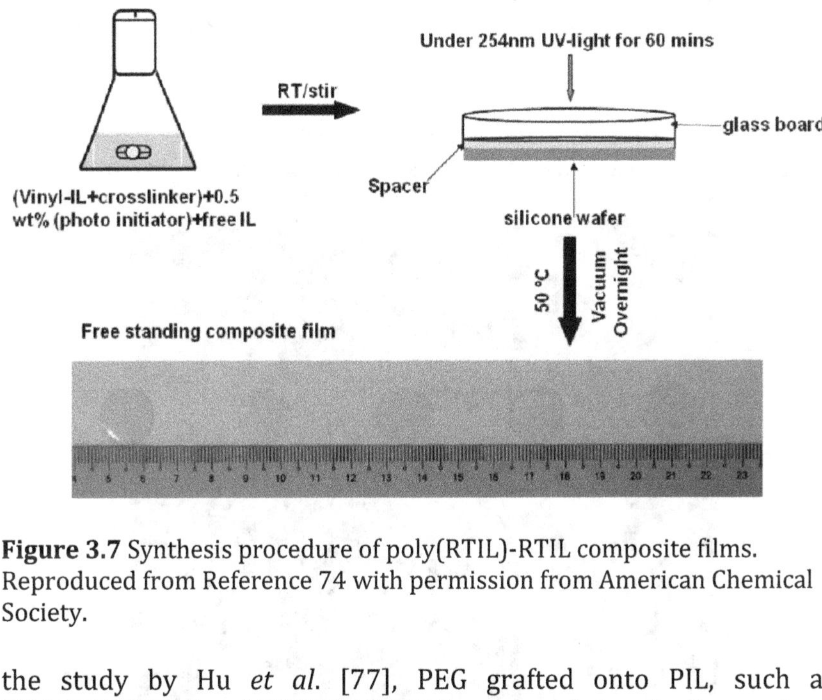

Figure 3.7 Synthesis procedure of poly(RTIL)-RTIL composite films. Reproduced from Reference 74 with permission from American Chemical Society.

the study by Hu *et al.* [77], PEG grafted onto PIL, such as P[MATMA][BF_4] and P[VBTMA][BF_4] generated chemically, mechanically as well as thermally stable CO_2 selective membranes. PEG475 was observed to be less effective than the PEG2000. The membranes generated using grafted polymers were found to be less brittle when compared to the membrane made of pure P[MATMA][BF_4] and (P[VBTMA][BF_4]). The P[VBTMA][BF_4]-g-PEG2000 and P[MATMA][BF_4]-g-PEG2000 exhibited similar CO_2/CH_4 separation, when compared to the polymeric membranes. The selectivity of P[MATMA][BF_4]-g-PEG2000 and P[VBTMA][BF_4]-g-PEG2000 for the CO_2/N_2 and CO_2/CH_4 separation was mainly because of the solubility differences, and not due to the differences in their diffusivities.

Ionic Liquids Supported on Metal Organic Frameworks

In the past several decades, metal organic frameworks (MOFs) have obtained significant attention because of their exclusive properties like large pore volume, higher surface area and tunable structure [17]. Many research studies have demonstrated that the MOFs can be used as a promising porous support for the ILs [78-81]. Jiang *et al.* [78] studied the composite of [BMIM][PF_6] which was supported on a MOF IRMOF-1, for the CO_2/N_2 separation by means of molecular simulations (Figure 3.8). It was confirmed that the selectivity could be increased when the ILs are added into the pores of the IRMOF-1. The IL present in the composite displayed an ordered structure, as seen from the radial distribution functions, because of the confinement effect. In the open pores of IRMOF-1, the large $[BMIM]^+$ cation resided, while the small anion $[PF_6]^-$ was located in

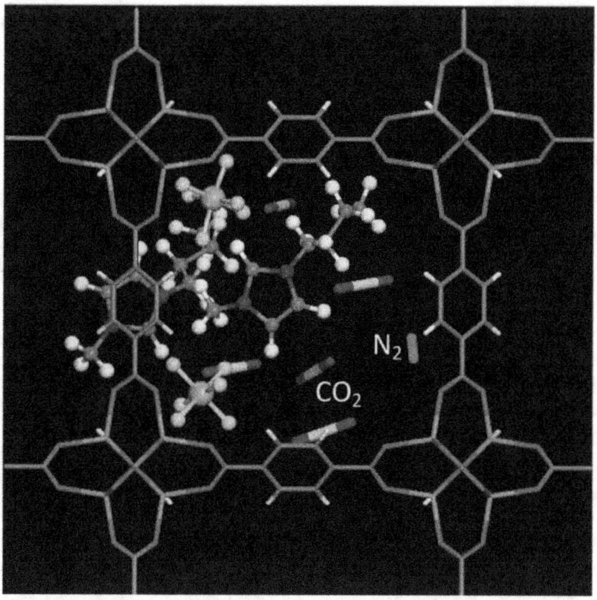

Figure 3.8 Simulation snapshot of CO_2/N_2 mixture in IL/IRMOF-1 at $W_{IL/IRMOF-1} = 0.4$. Reproduced from Reference 78 with permission from American Chemical Society.

the metal cluster corner. In comparison with $[PF_6]^-$, $[BMIM]^+$ illustrated higher mobility. By enhancing the IL ratio in the composite, the confinement effect could be increased, thereby reducing the motility of $[PF_6]^-$ and $[BMIM]^+$. Ions present in the composite offered stronger interaction with CO_2, especially the $[PF_6]^-$ anion was the most favorable site for the adsorption of CO_2. The composite offered selective CO_2 adsorption from the CO_2/N_2mixture, with selectivity remarkably high in comparison with several other supported ILs. In addition, by enhancing the IL ratio in the composite, the selectivity was observed to increase. Subsequently, the authors computationally investigated the efficiency of MOF supported IL membranes for the CO_2/N_2 separation [81]. The IRMOF-1 supported IL membranes were analyzed for the capture of CO_2 by atomistic simulations. The role of hydrophilic/hydrophobic framework and the effect of anions of the ILs was analyzed. In this system, the IL consisted of same cation 1-n-butyl-3-methylimidazolium $[BMIM]^{(+)}$, however, four different types of anions namely tetrafluoroborate $[BF_{(4)}]^{(-)}$, thiocyanate $[SCN]^{(-)}$, hexafluorophosphate $[PF_{(6)}]^{(-)}$ and bis(trifluoromethylsulfonyl)imide $[Tf_{(2)}N]^{(-)}$ were used. The anions possessed a stronger interaction with IRMOF-1 and highly ordered structure in the IRMOF-1, when compared to the cation. The bulky as well as chain like $[Tf_{(2)}N]^{(-)}$ and $[BMIM]^{(+)}$ occupied place near the phenyl ring, whereas the small anions $[SCN]^{(-)}$, $[BF_{(4)}]^{(-)}$ and $[PF_{(6)}]^{(-)}$ were observed to be located near the metal cluster, specifically the quasi-spherical$[BF_{(4)}]^{(-)}$ and $[PF_{(6)}]^{(-)}$. Among the four different anions, $[BMIM]^{(+)}$ exhibited strongest interaction with IRMOF-1, while $[Tf_{(2)}N]^{(-)}$ had the weakest interaction. By enhancing the weight ratio of IL to IRMOF-1, the selectivity of CO_2/N_2 at infinite dilution could be increased. At a specified weight ratio $W_{IL/IRMOF\text{-}1}$, the selectivity decreased in the order $[SCN]^{(-)}$> $[BF_{(4)}]^{(-)}$> $[PF_{(6)}]^{(-)}$> $[Tf(2)N]^{(-)}$. The simulation analysis clearly acknowledged that the anion had substantial influence on the microscopic properties of ILs and also proposed that the MOF supported ILs are capable material for the CO_2 capture.

Vicent-Luna *et al.* [79] investigated different IL/Cu-BTC composites for the CO_2 removal, with special focus on the effect of various anions and the quantity of ILs. A molecular simulation study was carried out to analyze the effect on gas adsorption (CO_2, CH_4, N_2 and their mixtures), when the RTILs are incorporated into the pore of Cu-BTC MOF. It was found that the existence of RTILs in the pores of MOF increases remarkably the adsorption of CO_2 at lower pressure, whereas the N_2 and CH_4 adsorption was unaffected (Figure 3.9). In another study conducted by Tzialla *et al.* [80], it was observed that the permeability and CO_2 selectivity of a [omim][TCM]/ZIF 69 composite membrane were greater when compared to the bulk IL and pure Zeolitic imidazolate framework ZIF-69 membrane (Figure 3.10). The ZIF-69 membrane was developed on porous α-alumina substrates by seeded secondary growth and was subsequently functionalized using a CO_2-selective alkylmethyl imidazolium cation/tricyanomethanide anion based IL. In this system, the ZIF intergrain boundaries as well as the defects were fixed by a medium offering higher CO_2 selectivity. Thus, the selectivity of the hybrid membrane was remarkably greater than that of as grown ZIF membranes, due to the presence of ZIF channels, and the permeability was greater when compared to bulk IL. It was observed that the CO_2 permeation

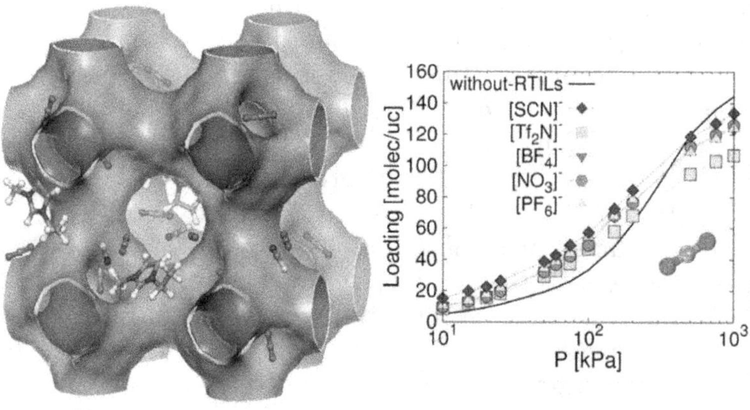

Figure 3.9 Comparison of various anions for CO_2 removal. Reproduced from Reference 79 with permission from American Chemical Society.

was 20 times faster than the N_2 through the undamaged ZIF pores, and 65 times faster than through the IL phase. The developed membranes under a 2 bar transmembrane pressure and at room temperature displayed CO_2 permeance of 3.7×10^{-11} mol $m^{-2}s^{-1}Pa^{-1}$ and 5.6×10^{-11} $molm^{-2}s^{-1}Pa^{-1}$, and real CO_2/N_2 selectivities of 64 and 44 for CO_2/N_2 mixtures containing 75% and 44% (v/v) CO_2, respectively. Thus, the study presented a promising solution for the issues related to the defect formation during the growth of ZIFs, zeolite and inorganic membranes used for the removal of CO_2.

Xue *et al.* [82] conducted a computational study to analyze the dispersion characteristics of ILs in MOFs and covalent organic frameworks (COFs). In addition, the separation behavior of the developed composites for the CO_2/N_2 and CO_2/CH_4 mixtures was also examined. For the IL 1-n-butyl-3-methyl imidazolium thiocyanate [BMIM][SCN], eight COFs and five MOFs, with various pore structure as well as chemical properties, were picked as the supports. The results indicated that the stronger Coulombic interactions provided by the MOF framework caused the better dispersion of IL molecules in their pores, when compared to COFs. The gas removal ability was remarkably increased by the addition of [BMIM][SCN] into COFs and MOFs. The superior dispersion of the IL in the support provided improved removal efficiency of the composite, and this phenomenon

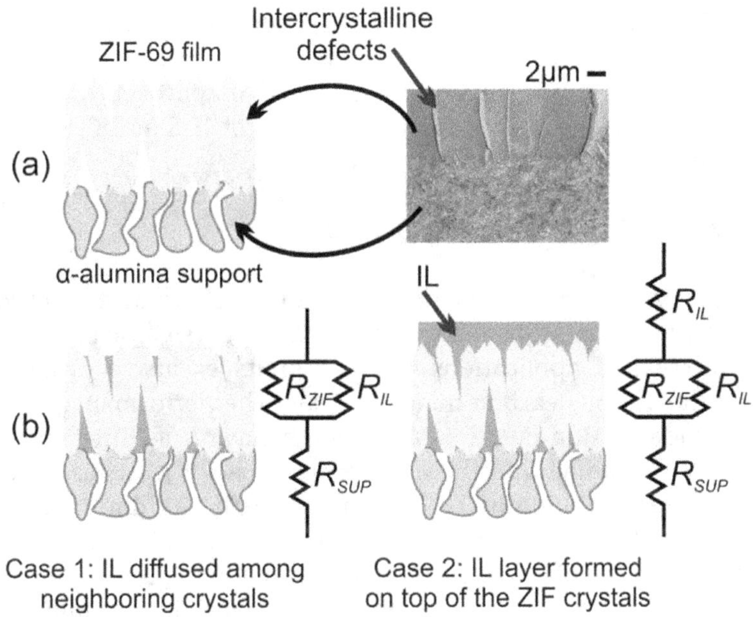

Figure 3.10 Schematic illustration of ZIF-69 membrane morphologies. Reproduced from Reference 80 with permission from American Chemical Society.

was more visible for the CO_2/CH_4 mixture, when compared to the CO_2/N_2 mixture. The study also clearly mentioned that employing IL supports with strong adsorption sites such as co-ordinatively unsaturated metal sites cannot accomplish remarkable increase in the gas removal ability of the composites. In another study, Li *et al.* [83] performed molecular simulations to analyze the performance of IL/Cu-TDPAT composites for the separation of the H_2S/CH_4 mixture. In this study, the ILs consisted of four types of anions ($[PF_6]^-$, $[Cl]^-$, $[BF_4]^-$, $[Tf_2N]^-$]) and identical cation $[BMIM]^+$. The results indicated that the H_2S adsorption ability could be remarkably increased by the addition of IL into the pores of Cu-TDPAT, and the composite with $[Cl]^-$ anion exhibited the highest heat of adsorption. Within the pressure range analyzed, the H_2S/CH_4 adsorption selectivities of each composite were remarkably greater than that of the pure Cu-TDPAT, and the selectivity mainly exhibited an enhancement with increasing the IL loading. The study also confirmed that the [BMIM][Cl]/Cu-TDPAT composite attained the best separation performance in both PSA and VSA processes. Thus, the observations made in the study provided beneficial information for the development of new assuring IL/MOF composites for H_2S capture from the natural gas stream. Overall, it was also confirmed that the application of MOFs as the IL support is an effective approach to prepare new promising adsorbents for the removal of H_2S and CO_2 during the natural gas purification.

Ionic Liquids Supported on Graphene

Graphene, a material with two-dimensional (2D) lattice of carbon atoms having one atom thickness, has attained significance in adsorption related applications because of its exclusive properties among the various carbon nanomaterials. The performance of graphene oxide (GO) in the CO_2 adsorption behavior of chitosan based aerogel was studied and the adsorption capacity was observed to become almost double by the addition of 20% GO, due to the enhanced specific surface area [84]. Further, it was also reported that the hydrogen facilitated exfoliation-co-reduction of graphite oxide, which resulted in the formation of highly wrinkled graphene, offered greater affinity sites for holding the CO_2 molecules [85]. In a study by Ghosh *et al.* [86], the CO_2 adsorption on graphene at 195 K temperature and 100 kPa was examined. The analysis of CO_2 adsorption behavior of graphene confirmed that the adsorption energy of chemisorption was in the range of 297.9-301.7 $kJ.mol^{-1}$, while that of physisorption was in the range 8.8-13.8 $kJ.mol^{-1}$, depending on the CO_2 molecule orientation [87]. As the adsorption behavior depends on the surface anchoring sites as well as the lateral surface properties (porosity and specific surface area), the task specific functionalization of graphene with an appropriate moiety present the potential to tune the adsorption behavior of graphene [88]. It was also observed that the CO_2 solubility in ILs can be increased through supporting ILs using graphene based substrate.

Tamilarasan *et al.* [89] studied the integration of IL or PIL with graphene to examine the increase in CO_2 adsorption properties. Here, graphene was subjected to non-covalent functionalization by IL or PIL, followed by determination of CO_2 adsorption as well as desorption behavior at lower pressures (<100 kPa). The IL uniformly enveloped the surface of graphene, whereas the PIL formed greatly distributed porous nanoparticles. The IL functionalization enhanced the CO_2 adsorption capacity by 2% than graphene, whereas the PIL functionalized graphene exhibited 22% greater adsorption capacity. Thus, the study demonstrated the benefit of polymerizing the IL for the CO_2 adsorption process. Further, it was also noted that the adsorption behavior of the integrated system was greater, when compared to the single constituents (either PIL or IL or graphene). The authors also observed that the PIL functionalization provided more adsorption favorability, with a greater adsorption energy. Due to the ease of regeneration of adsorbent, the isosteric heat of adsorption was determined to be in the range of 18-28 $kJ.mol^{-1}$. These observations underline the potential of PIL integration with greater surface area nanostructures for the enhancement in the adsorption ability. In another study by the authors, the preparation and application of adsorptivity based amine rich IL (ARIL) i.e., 3,5-diamino-1-methyl-1,2,4-triazolium tetra-fluoroborate grafted graphene (HEG/ARIL) for CO_2 adsorption was studied [90]. At standard pressure and temperature, the IL appeared in solid state, and formed solid like short range ordering on the graphitic substrate. It was noted from the molecular vibrational spectroscopy that the ARIL molecules were physically adsorbed on the graphene surface, with no identifiable chemical bonding. By grafting graphene with ARIL, the adsorbate confinement of graphene got significantly increased. Further, on the HEG/ARIL, the entropy and isosteric heat of CO_2 adsorption was observed to be lower than that of pure HEG. Thus, it was concluded that the ARIL functionalization of graphene enhanced the CO_2 adsorption behavior and reduced the interaction between graphene and CO_2, due to the fact that it increased the density of lesser affinity adsorption sites.

Bian *et al.* [91] proposed a new method of IL- assisted growth of $Cu_3(BTC)_2$ on GO sheets and used it for increasing the CO_2 adsorption ability as well as adsorption rate. To analyze the influence of anions and cations of ILs on the GO-IL/MOF composite structure, three ILs, namely, 1-butyl-3methyl-imidazolium tetra-fluoroborate ([Bmim]BF_4), triethylene tetramine tetrafluoroborate (TETA-BF_4)

and triethylene tetramine acetate (TETA-Ac) were examined. With imidazole or amine cations adsorbed at the GO surface and the closely attached anions, the GO-ILs can contribute a number of active sites for the Cu^{2+} cations absorption through coordination. Among the different GO-IL/MOF composites, GO-TAc/MOF-60 exhibited a superimposed structure, which could result in more adsorption activity sites and lessen the transfer distance. Further, the GO sheets present in the GO-IL/MOF composites contributed channels for the quicker transfer and exhibited a greater CO_2 adsorption ability of 5.62 mmol/g at 100 kPa and 25 °C. The material also exhibited greater CO_2 kinetic separation behavior. Thus, the composite was observed to have a better cyclic adsorption/desorption stability. The relation between the CO_2 adsorption ability and the composite specific structures was experimentally determined to obtain a suitable method for designing and preparing hierarchical MOF composites.

3.3 Summary and Outlook

The increasing level of CO_2 in the atmosphere plays a significant role towards the global warming as well as climate change and can cause substantial damage to the environment. In addition, CO_2 and H_2S are present in the natural gas streams and are required to be removed for enhancing the usefulness of natural gas as well as to safeguard the facilities agonist corrosion and other harmful effects. H_2S is specifically highly toxic and releases SO_2 into the atmosphere following combustion. To reduce the levels of these gases, ionic liquids offer an exceptionally versatile as well as tunable platform to generate an extensive variety of sorbents for the H_2S and CO_2 capture. This review has highlighted the current developments in the area of ILs as sorbents for H_2S and/or CO_2 separation, and has also enlisted the challenges impeding the increased efficiency of these materials. In this category, various IL bases techniques such as RTIL, PIL and SILs have been extensively reviewed. This brief scrutiny on the present trends on the IL mediated CO_2 and H_2S capture suggests that these acid gas capture by ILs is a feasible practice. From the analysis of different literature studies, it was observed that the parameters like anions types, cation types, alkyl chain size of cations, alkyl chains fluorination on the cation, pressure, degree of fluorination of anions, temperature and the anion interaction with acid gases have a major role in determining the gas solubility of ILs. Further, it was observed that the conversion of ILs into the macromolecules (PILs) can remarkably increase the CO_2 or H_2S sorption behavior. The proper selection of cations is of considerable significance in the development of a PIL skeleton for the removal of CO_2. Further, the benefits of using supported ILs such as ILs supported on membranes, MOFs and graphene were discussed. Supported ionic liquid membranes combine the benefits of IL with the solid supports and it can be regarded as excellent materials for gas separation because of the higher selectivity and satisfactory permeability. MOFs have been proved to be promising porous supports for the ILs. It was also seen that the integration of IL or PIL with graphene enhanced its CO_2 adsorption properties. Even though the unavailability of low cost and diverse ILs is the major concern hindering the large scale application of IL systems for CO_2 and H_2S separation, however, developments in the synthesis methods would result in large scale production to become feasible in near future. Thus, with optimized choice of the IL system components, low coat and efficient performance, the IL based technologies represent strong potential to replace other absorption and adsorption based methods for gas separation.

References

1. Berg, S. V. (1998) Lessons in electricity market reform: regulatory processes and performance. *The Electricity Journal*, **11**(5), 13-20.
2. Kohl, A. L., and Nielsen, R. B. (1997) *Gas Purification*, 5th edition, Gulf Publishing Company, USA.
3. Rufford, T. E., Smart, S., Watson, G. C. Y., Graham, B. F., Boxall, J., Diniz da Costa, J. C., and May, E. F. (2012) The removal of CO_2 and N_2 from natural gas: a review of conventional and emerging process technologies. *Journal of Petroleum Science and Engineering*, **94**, 123-154.
4. Koech, P. K., Rainbolt, J. E., Bearden, M. D., Zheng, F., and Heldebrant, D. J. (2011) Chemically selective gas sweetening without thermal-swing regeneration. *Energy and Environmental Science*, **4**, 1385-1390.
5. Abdulrahman, R. K., and Sebastine, I. M. (2013) Natural gas sweetening process simulation and optimization: a case study of Khurmala field in Iraqi Kurdistan region. *Journal of Natural Gas Science and Engineering*, **14**, 116-120.
6. Bhide, B. D., Voskericyan, A., and Stern, S. A. (1998) Hybrid processes for the removal of acid gases from natural gas. *Journal of Membrane Science*, **140**, 27-49.

7. Karadas, F., Atilhan, M., and Aparicio, S. (2010) Review on the use of ionic liquids (ILs) as alternative fluids for CO_2 capture and natural gas sweetening. *Energy & Fuels*, **24**, 5817-5828.
8. Revelli, A.-L., Mutelet, F., Jaubert, J.-N. (2010) High carbon dioxide solubilities in imidazolium-based ionic liquids and in poly(ethylene glycol) dimethyl ether. *The Journal of Physical Chemistry B*, **114**, 12908-12913.
9. Bates, E. D., Mayton, R. D., Ntai, I., and Davis, J. H. (2002) CO_2 capture by a task-specific ionic liquid. *Journal of the American Chemical Society*, **124**, 926-927.
10. Rahmati-Rostami, M., Ghotbi, C., Hosseini-Jenab, M., Ahmadi, A. N., and Jalili, A. H. (2009) Solubility of H_2S in ionic liquids [hmim][PF_6], [hmim][BF4], and [hmim][Tf_2N]. *The Journal of Chemical Thermodynamics*, **41**, 1052-1055.
11. Plechkova, N. V., and Seddon, K. R. (2008) Applications of ionic liquids in the chemical industry. *Chemical Society Reviews*, **37**, 123-150.
12. Wang, J., Luo, J., Feng, S., Li, H., Wan, Y., and Zhang, X. (2016) Recent development of ionic liquid membranes. *Green Energy & Environment*, **1**(1), 43-61.
13. Zhang, X., Zhang, X., Dong, H., Zhao, Z., Zhang, S., and Huang, Y. (2012) Carbon capture with ionic liquids: Overview and progress. *Energy and Environmental Science*, **5**, 6668-6681.
14. Ju, Y.-J., Lien, C.-H., Chang, K.-H., Hu, C.-C., and Wong, D. S.-H. (2012) Deep eutectic solvent-based ionic liquid electrolytes for electrical double-layer capacitors. *Journal of the Chinese Chemical Society*, **59**, 1280-1287.
15. Seddon, K. (2002) Ionic liquids: designer solvents for green synthesis. *Chemical Engineer*, **730**, 33-35.
16. Duchet, L., Legeay, J. C., Carrie, D., Paquin, L., Vanden Eynde, J. J., and Bazureau, J. P. (2010) Synthesis of 3,5-disubstituted 1,2,4-oxadiazoles using ionic liquid-phase organic synthesis (IoLiPOS) methodology. *Tetrahedron*, **66**, 986-994.
17. Gurdal, Y., and Keskin, S. (2013) Predicting noble gas separation performance of metal organic frameworks using theoretical correlations. *The Journal of Physical Chemistry C*, **117**, 5229-5241.
18. Rogers, R. K., and Seddon, K. R. (2002) *Ionic Liquids: Industrial Applications to Green Chemistry*, Oxford University Press, USA.
19. Camper, D., Bara, J. E., Gin, D. L., and Noble, R. D. (2008) Room-temperature ionic liquid– amine solutions: Tunable solvents for efficient and reversible capture of CO_2. *Industrial and Engineering Chemistry Research*, **47**, 8496-8498.
20. Jou, F.-Y., and Mather, A. E. (2007) Solubility of hydrogen sulfide in [bmim][PF6]. *International Journal of Thermophysics*, **28**, 490-495.
21. Pomelli, C. S., Chiappe, C., Vidis, A., Laurenczy, G., and Dyson, P. J.

(2007) Influence of the interaction between hydrogen sulfide and ionic liquids on solubility: Experimental and theoretical investigation. *The Journal of Physical Chemistry B*, **111**, 13014-13019.
22. Jalili, A. H., Rahmati-Rostami, M., Ghotbi, C., Hosseini-Jenab, M., and Ahmadi, A. N. (2009) Solubility of H_2S in ionic liquids [bmim][PF_6], [bmim][BF_4], and [bmim][Tf_2N]. *Journal of Chemical & Engineering Data*, **54**, 1844-1849.
23. Anthony, J. L., Maginn, E. J., and Brennecke, J. F. (2002) Solubilities and thermodynamic properties of gases in the ionic liquid 1-n-butyl-3-methylimidazolium hexafluorophosphate. *The Journal of Physical Chemistry B*, **106**, 7315-7320.
24. Lei, Z., Dai, C., and Chen, B. (2014) Gas solubility in ionic liquids. *Chemical Reviews*, 114, 1289-1326.
25. Bara, J. E., Carlisle, T. K., Gabriel, C. J., Camper, D., Finotello, A., Gin, D. L., and Noble, R. D. (2009) Guide to CO_2 separations in imidazolium-based room-temperature ionic liquids. *Industrial & Engineering Chemistry Research*, **48**, 2739-2751.
26. Aki, S. N. V. K., Mellein, B. R., Saurer, E. M., and Brennecke, J. F. (2004) High-pressure phase behavior of carbon dioxide with imidazolium-based ionic liquids. *The Journal of Physical Chemistry B*, **108**, 20355-20365.
27. Aziz, N., Yusoff, R., and Aroua, M. K. (2012) Absorption of CO_2 in aqueous mixtures of N-methyldiethanolamine and guanidinium tris (pentafluoroethyl) trifluorophosphate ionic liquid at high-pressure. *Fluid Phase Equilibria*, **323**, 120-125.
28. Muldoon, M. J., Aki, S. N. V. K., Anderson, J. L., Dixon, J. K., and Brennecke, J. F. (2007) Improving carbon dioxide solubility in ionic liquids. *The Journal of Physical Chemistry B*, **111**, 9001-9009.
29. Jalili, A. H., Mehdizadeh, A., Shokouhi, M., Ahmadi, A. N., Hosseini-Jenab, M., and Fateminassab, F. (2010) Solubility and diffusion of CO_2 and H_2S in the ionic liquid 1-ethyl-3-methylimidazolium ethylsulfate. *The Journal of Chemical Thermodynamics*, **42**, 1298-1303.
30. Nematpour, M., Jalili, A. H., Ghotbi, C., and Rashtchian, D. (2016) Solubility of CO_2 and H_2S in the ionic liquid 1-ethyl-3-methylimidazolium trifluoromethanesulfonate. *Journal of Natural Gas Science and Engineering*, **30**, 583-591.
31. Soriano, A. N., Doma, Jr., B. T., and Li, M.-H. (2009) Carbon dioxide solubility in 1-ethyl-3-methylimidazolium trifluoromethanesulfonate. *The Journal of Chemical Thermodynamics*, **41**(4), 525-529.
32. Safavi, M., Ghotbi, C., Taghikhani, V., Jalili, A. H., and Mehdizadeh, A. (2013) Study of the solubility of CO_2, H_2S and their mixture in the ionic liquid 1-octyl-3-methylimidazolium hexafluorophosphate: Experimental and modelling. *The Journal of Chemical Thermodynamics*, **65**, 220-232.
33. Zhang, X., Liu, Z., and Wang, W. (2008) Screening of ionic liquids to

capture CO_2 by COSMO-RS and experiments. *AIChE Journal*, **54**, 2717-2728.
34. Zhang, X., Huo, F., Lio, Z., Wang, W., Shi, W., and Maginn, E. J. (2009) Absorption of CO_2 in the ionic Liquid 1-n-Hexyl-3-methylimidazolium tris (pentafluoroethyl) trifluorophosphate ([hmim][FEP]): A molecular view by computer simulations. *The Journal of Physical Chemistry B*, **113**, 7591-7598.
35. Shiflett, M. B., and Yokozeki, A. (2010) Separation of CO_2 and H_2S using room-temperature ionic liquid [bmim][PF_6]. *Fluid Phase Equilibria*, **294**, 105-113.
36. Jalili, A. H., Shokouhi, M., Maurer, G., and Hosseini-Jenab, M. (2013) Solubility of CO_2 and H_2S in the ionic liquid 1-ethyl-3-methylimidazolium tris(pentafluoroethyl)trifluorophosphate. *The Journal of Chemical Thermodynamics*, **67**, 55-62.
37. Mecerreyes, D. (2011) Polymeric ionic liquids: Broadening the properties and applications of polyelectrolytes. *Progress in Polymer Science*, **36**, 1629-1648.
38. Tang, J., Sun, W., Tang, H., Radosz, M., and Shen, Y. (2005) Enhanced CO_2 absorption of poly(ionic liquid)s. *Macromolecules*, **38**, 2037-2039.
39. Yu, G., Man, Z., Li, Q., Li, N., Wu, X., Asumana, C., and Chen, X. (2013) New crosslinked-porous poly-ammonium microparticles as CO_2 adsorbents, *Reactive and Functional Polymers*, **73**, 1058-1064.
40. Bhavsar, R. S., Kumbharkar, S. C., and Kharul, U. K. (2012) Polymeric ionic liquids (PILs): Effect of anion variation on their CO_2 sorption. *Journal of Membrane Science*, **389**, 305-315.
41. Heintz, Y. J., Sehabiague, L., Morsi, B. I., Jones, K. L., Luebke, D. R., and Pennline, H. W. (2009) Hydrogen sulfide and carbon dioxide removal from dry fuel gas streams using an ionic liquid as a physical solvent. *Energy & Fuels*, **23**, 4822-4830.
42. Tang, J., Tang, H., Sun, W., Radosz, M., and Shen, Y. (2005) Poly (ionic liquid)s as new materials for CO_2 absorption. *Journal of Polymer Science, Part A: Polymer Chemistry*, **43**, 5477-5489.
43. Li, P., Zhao, Q., Anderson, J. L., Varanasi, S., and Coleman, M. R. (2010) Synthesis of copolyimides based on room temperature ionic liquid diamines. *Journal of Polymer Science, Part A: Polymer Chemistry*, **48**, 4036-4046.
44. Tang, J., Shen, Y., Radosz, M., and Sun, W. (2009) Isothermal carbon dioxide sorption in poly(ionic liquid)s. *Industrial & Engineering Chemistry Research*, **48**, 9113-9118.
45. Tang, J., Tang, H., Sun, W., Plancher, H., Radosz, M., and Shen, Y. (2005) Poly(ionic liquid)s: a new material with enhanced and fast CO_2 absorption. *Chemical Communications*, 3325-3327.
46. Marcilla, R., Blazquez, J. A., Rodriguez, J., Pomposo, J. A., and Mecerreyes, D. (2004) Tuning the solubility of polymerized ionic liquids

by simple anion-exchange reactions. *Journal of Polymer Science, Part A: Polymer Chemistry*, 42, 208-212.

47. Tang, J., Tang, H., Sun, W., Radosz, M., and Shen, Y. (2005) Low-pressure CO_2 sorption in ammonium-based poly(ionic liquid)s. *Polymer*, **46**, 12460-12467.
48. Xiong, Y. B., Wang, H., Wang, Y. J., and Wang, R. M. (2011) Novel imidazolium-based poly(ionic liquid)s: preparation, characterization, and absorption of CO_2. *Polymers for Advanced Technologies*, **23**(5), 835-840.
49. Green, O., Grubjesic, S., Lee, S., and Firestone, M. A. (2009) The design of polymeric ionic liquids for the preparation of functional materials. *Journal of Macromolecular Science C: Polymer Reviews*, **49**, 339-360.
50. Bara, J. E., Gin, D. L., and Noble, R. D. (2008) Effect of anion on gas separation performance of polymer-room-temperature ionic liquid composite membranes. *Industrial & Engineering Chemistry Research*, **47**, 9919-9924.
51. Kato, S., Tsujita, Y., Yoshimizu, H., Kinoshita, T., and Higgins, J. S. (1997) Characterization and CO_2 sorption behaviour of polystyrene/polycarbonate blend system. *Polymer*, **38**, 2807-2811.
52. Wilke, A., Yuan, J., Antonietti, M., and Weber, J. (2012) Enhanced carbon dioxide adsorption by a mesoporous poly(ionic liquid). *ACS Macro Letters*, **1**, 1028-1031.
53. Shahrom, M. S. R., Wilfred, C. R., and Taha, A. B. K. Z. (2016) CO_2 capture by task specific ionic liquids (TSILs) and polymerized ionic liquids (PILs and AAPILs). *Journal of Molecular Liquids*, **219**, 306-312.
54. Zheng, J., Li, S., Wang, Y., Li, L., Su, C., Liu, H., Zhu, F., Jiang, R., and Ouyang, G. (2014) In situ growth of IRMOF-3 combined with ionic liquids to prepare solid-phase microextraction fibers. *Analytica Chimica Acta*, **829**, 22-27.
55. Selvam, T., Machoke, A., and Schwieger, W. (2012) Supported ionic liquids on non-porous and porous inorganic materials - A topical review. *Applied Catalysis A*, **445-446**, 92-101.
56. Yu, Y., Mai, J., Wang, L., Li, X., Jiang, Z., and Wang, F. (2014) Ship-in-a-bottle synthesis of amine-functionalized ionic liquids in NaY zeolite for CO_2 capture. *Scientific Reports*, **4**, 5997-6005.
57. Muthuramam, G., and Palanivelu, K. (2006) Transport of textile dye in vegetable oils based supported liquid membrane. *Dyes and Pigments*, **70**, 99-104.
58. Scovazzo, P., Visser, A. E., Davis, J. H., Rogers, R. D., Koval, C. A., DuBois, D. L., and Noble, R. D. (2002) Supported ionic liquid membranes and facilitated ionic liquid membranes. *Ionic Liquids*, **818**, 69-87.
59. Scovazzo, P., Kieft, J., Finan, D. A., Koval, C., DuBois, D., and Noble, R.

(2004) Gas separations using non-hexafluorophosphate $[PF_6]^-$ anion supported ionic liquid membranes. *Journal of Membrane Science*, **238**(1-2), 57-63.

60. Luis, P., Van Gerven, T., and van der Bruggen, B. (2012) Recent developments in membrane-based technologies for CO_2 capture. *Progress in Energy and Combustion Science*, **38**(3), 419-448.
61. Al Marzouqi, M. H., Abdulkarim, M. A., Marzouk, S. A., El-Naas, M. H., and Hasanain, H. M. (2005) Facilitated transport of CO_2 through immobilized liquid membrane. *Industrial & Engineering Chemistry Research*, **44**, 9273-9278.
62. Scovazzo, P., Havard, D., McShea, M., Mixon, S., and Morgan, D. (2009) Long-term, continuous mixed-gas dry fed CO_2/CH_4 and CO_2/N_2 separation performance and selectivities for room temperature ionic liquid membranes. *Journal of Membrane Science*, **327**, 41-48.
63. Bara, J. E., Lessmann, S., Gabriel, C. J., Hatakeyama, E. S., Noble, R. D., and Gin, D. L. (2007) Synthesis and performance of polymerizable room-temperature ionic liquids as gas separation membranes. *Industrial & Engineering Chemistry Research*, **46**, 5397-5404.
64. Hanioka, S., Maruyama, T., Sotani, T., Teramoto, M., Matsuyama, H., Nakashima, K., Hanaki, M., Kubota, F., and Goto, M. (2008) CO_2 separation facilitated by taskspecific ionic liquids using a supported liquid membrane. *Journal of Membrane Science*, **314**, 1-4.
65. Park, Y.-I., Kim, B.-S., Byun, Y.-H., Lee, S.-H., Lee, E.-W., and Lee, J.-M. (2009) Preparation of supported ionic liquid membranes (SILMs) for the removal of acidic gases from crude natural gas. *Desalination*, **236**, 342-348.
66. Ilconich, J., Myers, C., Pennline, H., and Luebke, D. (2007) Experimental investigation of the permeability and selectivity of supported ionic liquid membranes for CO_2/He separation at temperatures up to 125 °C. *Journal of Membrane Science*, **298**, 41-47.
67. Neves, L. A., Crespo, J. G., and Coelhoso, I. M. (2010) Gas permeation studies in supported ionic liquid membranes. *Journal of Membrane Science*, **357**, 160-170.
68. Cserjesi, P., Nemestothy, N., and Belafi-Bako, K. (2010) Gas separation properties of supported liquid membranes prepared with unconventional ionic liquids. *Journal of Membrane Science*, **349**, 6-11.
69. Myers, C., Pennline, H., Luebke, D., Ilconich, J., Dixon, J. K., Maginn, E. J., and Brennecke, J. F. (2008) High temperature separation of carbon dioxide/hydrogen mixtures using facilitated supported ionic liquid membrane. *Journal of Membrane Science*, **322**, 28-31.
70. Albo, J., Luis, P., and Irabien, A. (2010) Carbon dioxide capture from flue gases using a cross-flow membrane contactor and the ionic liquid 1-ethyl-3-methylimidazolium ethyl sulfate. *Industrial & Engineering Chemistry Research*, **49**, 11045-11051.

71. Gomez-Coma, L., Garea, A., and Irabien, A. (2014) Non-dispersive absorption of CO_2 in [emim][EtSO4] and [emim][Ac]: temperature influence. *Separation and Purification Technology*, **132**, 120-125.
72. Mulukutla, T., Obuskovic, G., and Sirkar, K. K. (2014) Novel scrubbing system for post-combustion CO_2 capture and recovery: experimental studies. *Journal of Membrane Science*, **471**, 16-26.
73. Chau, J., Jie, X., and Sirkar, K. K. (2016) Polyamidoamine-facilitated poly(ethylene glycol)/ionic liquid based pressure swing membrane absorption process for CO_2 removal from shifted syngas. *Chemical Engineering Journal*, **305**, 212-220.
74. Li, P., Pramoda, K. P., and Chung, T. S. (2011) CO_2 separation from flue gas using polyvinyl-(room temperature ionic liquid)-room temperature ionic liquid composite membranes. *Industrial & Engineering Chemistry Research*, **50**, 9344-9353.
75. Yuan, J., Mecerreyes, D., and Antonietti, M. (2013) Poly(ionic liquid)s: an update. *Progress in Polymer Science*, **38**, 1009-1036.
76. Bara, J. E., Hatakeyama, E. S., Gabriel, C. J., Zeng, X., Lessmann, S., Gin, D. L., and Noble, R. D. (2008) Synthesis and light gas separations in cross-linked gemini room temperature ionic liquid polymer membranes. *Journal of Membrane Science, 316*, 186-191.
77. Hu, X., Tang, J., Blasig, A., Shen, Y., and Radosz, M. (2006) CO_2 permeability, diffusivity and solubility in polyethylene glycol-grafted polyionic membranes and their CO_2 selectivity relative to methane and nitrogen. *Journal of Membrane Science*, **281**, 130-138.
78. Chen, Y., Hu, Z., Gupta, K. M., and Jiang, J. (2011) Ionic liquid/metal-organic frameworks composite for CO_2 capture: A computational investigation. *The Journal of Physical Chemistry C*, **115**, 21736-21742.
79. Vicent-Luna, J. M., Gutierrez-Sevillano, J. J., Anta, J. A., and Calero, S. (2013) Effect of room-temperature ionic liquids on CO_2 separation by a Cu-BTC metal-organic framework. *The Journal of Physical Chemistry C*, **117**, 20762-20768.
80. Tzialla, O., Veziri, C., Papatryfon, X., Beltsios, K. G., Labropoulos, A., Iliev, B., Adamova, G., Schubert, T. J. S., Kroon, M. C., Francisco, M., Zubeir, L. F., Romanos, G. E., and Karanikolos, G. N. (2013) Zeolite imidazolate framework-ionic liquid hybrid membranes for highly selective CO_2 separation. *The Journal of Physical Chemistry C*, **117**, 18434-18440.
81. Gupta, K. M., Chen, Y., Hu, Z., and Jiang, J. (2012) Metal-organic frameworks supported ionic liquid membranes for CO_2 capture: Anion effects. *Physical Chemistry and Chemical Physics*, **14**, 5785-5794.
82. Xue, W., Li, Z., Huang, H., Yang, Q., Liu, D., Xu, Q., and Zhong, C. (2016) Effects of ionic liquid dispersion in metal-organic frameworks and covalent organic frameworks on CO_2 capture: A compu-

tational study. *Chemical Engineering Science*, **140**, 1-9.

83. Li, Z., Xiao, Y., Xue, W., Yang, Q., and Zhong, C. (2015) Ionic liquid/metal-organic framework composites for H_2S removal from natural gas: A computational exploration. *The Journal of Physical Chemistry C*, **119**, 3674-3683.
84. Alhwaige, A. A., Agag, T., Ishida, H., and Qutubuddin, S. (2013) Biobased chitosan hybrid aerogels with superior adsorption: Role of graphene oxide in CO_2 capture. *RSC Advances*, **3**, 16011-16020.
85. Bienfait, M., Zeppenfeld, P., Dupont-Pavlovsky, N., Muris, M., Johnson, M. R., Wilson, T., DePies, M., and Vilches, O. E. (2004) Thermodynamics and structure of hydrogen, methane, argon, oxygen, and carbon dioxide adsorbed on single-wall carbon nanotube bundles. *Physical Reviews B*, 70, 035410.
86. Ghosh, A., Subrahmanyam, K. S., Krishna, K. S., Datta, S., Govindaraj, A., Pati, S. K., and Rao, C. N. R. (2008) Uptake of H_2 and CO_2 by graphene. *The Journal of Physical Chemistry C*, **112**, 15704-15707.
87. Lee, K.-J., and Kim, S.-J. (2013) Theoretical investigation of CO_2 adsorption on graphene. *Bulletin of Korean Chemical Society*, **34**, 3022-3026.
88. Mishra, A. K., and Ramaprabhu, S. (2012) Nanostructured polyaniline decorated graphene sheets for reversible CO_2 capture. *Journal of Materials Chemistry*, **22**, 3708-3712.
89. Tamilarasan, P., and Ramaprabhu, S. (2015) Integration of polymerized ionic liquid with graphene for enhanced CO_2 adsorption. *Journal of Materials Chemistry A*, **3**, 101-108.
90. Tamilarasan, P., and Ramaprabhu, S. (2016) Amine-rich ionic liquid grafted graphene for sub-ambient carbon dioxide adsorption. *RSC Advances*, **6**, 3032-3040.
91. Bian, Z., Zhu, X., Jin, T., Gao, J., Hu, J., and Liu, H. (2014) Ionic liquid-assisted growth of $Cu_3(BTC)_2$ nanocrystals on graphene oxide sheets: Towards both high capacity and high rate for CO_2 adsorption. *Microporous and Mesoporous Materials*, **200**, 159-164.

4

Xanthan Gum as Advanced Polymer System for Enhanced Oil Recovery

4.1 Introduction

Gum is the terminology used for describing the hydro-colloidal gels, which have water affinity and show binding characteristics with water as well as with other organic/inorganic materials. Chemically, the gums are polysaccharides or carbohydrate polymers, and exist in entire life forms. The first microbial polysaccharide used commercially was dextran, which was developed in early 1940 [1]. Xanthan, a natural polysaccharide, is the second microbial polysaccharide utilized for the commercial purposes. Xanthan gum or polysaccharide B-1459 is an extracellular high molecular hetero-polysaccharide, and is prepared by the xanthomonas campestris NRRL B-1459 bacterium. It was initially developed at the Northern Regional Research Center of United States Agriculture Department in the year 1963 [2]. Substantial production of xanthan gum began in 1964 and numerous developments in the manufacturing processes of xanthan gum have been made since then. Nowadays, xanthan gum is regarded as the most critical microbial polysaccharide commercially. It has been widely studied due of its characteristics that permit it to supplement several other natural as well as synthetic water soluble gums. Due to exclusive rheological properties of xanthan gum, it is employed as stabilizer for emulsions as well as suspensions, and acts as a rheological control agent in the aqueous systems [3]. Xanthan gum solutions are very pseudo-plastic and display excellent suspending properties [4]. A beneficial characteristic of xanthan gum is its capability to generate highly viscous aqueous solutions as well as dispersions, which can retain their viscosity in the occupancy of various salts. The xanthan gum solutions are non-Newtonian in nature and its apparent viscosity varies remarkably when the diversified shear stresses are enforced. The viscosity of xanthan gums decreases at greater shear [5]. The safety as well as toxicological properties of this material for the pharmaceutical and

Haleema Saleem and Vikas Mittal, Khalifa University of Science and Technology, Abu Dhabi, UAE

food applications have also been widely studied. As a result, the commercialization of xanthan gum as stabilizer and thickener has continuously developed at an annual rate of 5–10% due to its superior physico-chemical characteristics, when compared to several other commercial polysaccharides.

In the petroleum industry, biopolymer xanthan gum is used for several applications such as oil drilling, pipeline cleaning, fracturing as well as workover and completion. Xanthan gum has good compatibility with salt and good resistance against thermal degradation, thus, leading to its application in the drilling fluids [6]. High viscosity of xanthan gum at low shear helps to generate drilling fluids with low concentration of suspended solids. Xanthan gum can retain this characteristic even at extreme conditions like high concentration alkali, acid and salt in the solution as well as high temperature. This characteristic is very important in the application of xanthan gum in the offshore drilling operations or other severe conditions. The pseudoplasticity of the xanthan gum solutions exhibits lower viscosity at the drill bit, where higher shear rate occurs, whereas high viscosity is generated in the annulus region, where the shear is depressed. Hence, the biopolymer helps in multi-purpose mode by allowing rapid penetration at the bit along with suspending cuttings in the annulus region.

For each barrel of the oil produced, on an average, almost two barrels remain in the ground. Hence, enhanced oil recovery (EOR) has become a substantial application of xanthan gum in the recent years. The fundamental principle practiced is to enhance the separation of water and oil, thus, increasing the oil recovery. Nevertheless, the quality of xanthan gum is an important consideration, as many impurities present might enhance the complication during the oil refining process. During the utilization of xanthan gum in the tertiary oil recovery process using micellar-polymer flooding, the polymer thickened brine solution is employed to drive the surfactant slug through the porous reservoir rock for mobilizing the residual oil. Polymer restricts the bypassing of water through the surfactant as well as guarantees good area sweeping [7]. Xanthan gum is compatible with most of the surfactants as well as several additional injection fluid additives, which are employed in the tertiary oil recovery formulations [8]. For both the aforementioned applications, the objective of the polymer is to lower the motility of the injected water by viscosity enhancement. The thermal resistance property of xanthan gum makes it an ideal mobile control agent as well as a reliable displacing agent. Other different specialty usages for xanthan gum include rust removal, wet slag, welding rods and cleaning waste from the gas pipelines.

In this review chapter, application of the biopolymer xanthan gum in the oil and gas industries is focused. The structure, properties and production method of xanthan gum are also briefly outlined. An up-to-date review of the application of xanthan gum in the EOR process, oil drilling operation (as dispersant and tackifier) and in fracturing fluids is carried out. In addition to this, applications of xanthan gum for the chemical absorption of carbon dioxide and corrosion inhibition in the mild steel are also mentioned.

4.2 Structure

Xanthan gum has a complex molecular structure, with the primary structure comprising of replicated pentasaccharide units which are made up of two units of glucose, two units of mannose and one unit of glucuronic acid, present in the molar ratio 2.8:2.0:2.0. The main chain contains β-D-glucose units, which are connected at the positions 1 and 4. The primary chain chemical structure is identical to the cellulose structure. The tri-saccharide side chains consist of D-glucuronic acid units between the two D-mannose units, which are connected at the O-3 position of intermittent glucose residue present in the primary chain [9,10]. Almost 50% of the terminal D-mannose consists of pyruvic acid, which are connected by the keto group to the 4 and 6 positions. The D-mannose units are connected to the primary chain containing an acetyl group at O-6 position. The existence of pyruvic as well as acetic acids generates an anionic polysaccharide type [11].

The trisaccharide units are jointly aligned with the polymeric backbone, as a result, a stiff chain is generated which might exist as a triple, double or single helix [12,13]. The chain interacts with other polymer chains, thereby leading to the formation of a complex. Molecular weight distribution range of xanthan gum is 2×10^6 - 20×10^6 Da. This depends on the organization between the chains to form diverse distribution of individual chains. Molecular weight of xanthan is greatly influenced by the changes in the fermentation conditions utilized in the production process. Xanthan solution generated by the dissolution process at moderate temperatures exhibits extreme viscosity. Thus, the overall viscosity is greatly influenced by the dissolution temperature, molecular conformations as well as the emergence of ordered structures [14]. The structure of xanthan gum is also depicted in Figure 4.1 [15].

The preparation of xanthan gum is considered to be identical to that of the preparation of exopolysaccharide by other gram-negative bacteria. The preparation process can be split into three sections (1) uptake of simple sugars, subsequently converting to nucleotidal derivatives, (2) aggregation of pentasaccharide units that are connected to iso-pentyl pyrophosphate carrier, and (3) the polymerization of pentasaccharide repeating units, followed by secretion [16].

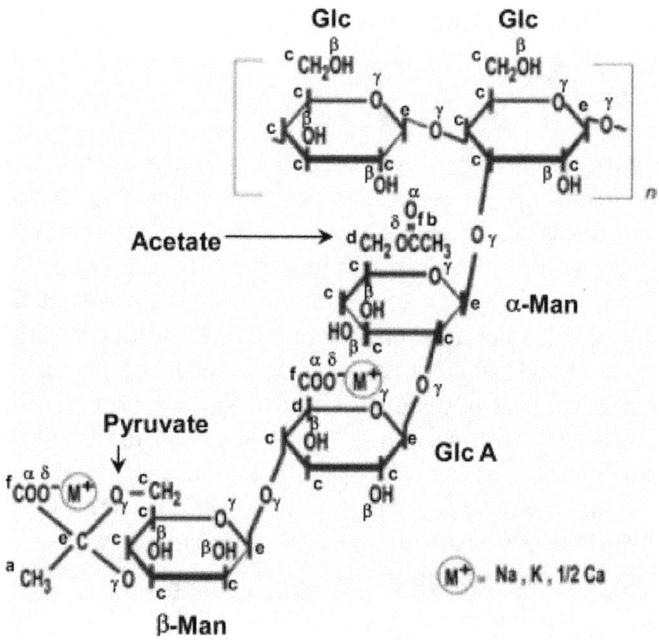

Figure 4.1 Structure of xanthan gum. Reproduced from Reference 15 with permission from American Chemical Society.

4.3 Production

Most of the commercial xanthan gum production processes employ glucose or invert sugars, and the majority of the industries prefer the batch production processes rather than the continuous processes [17]. Different alternate substrates like hydrolyzed rice, acid whey, sucrose, barley, coconut juice, sugar cane molasses, sugar cane etc., are available, however, glucose is still regarded as the best in terms of supply, product quality and product yield [18,19]. It was confirmed that the production as well as properties of the biopolymer xanthan gum is greatly influenced by culture medium [20], bacterial strain [21], pH [22], temperature [23], agitation rate [24] and time of fermentation [25]. For the generation of xanthan gum, the microbial strain is initially conserved for attaining long term storage by utilizing established methods for maintaining the appropriate characteristics [14]. Minute quantity of preserved culture is developed by growth on the surface of solid or in liquid media for attaining the inoculum for large bioreactors. The microorganism growth and the generation of xanthan gum is affected by various factors like the mode of operation (continuous or batch), type of bioreactor employed, medium composition as well as culture conditions (concentration of dissolved oxygen, pH and temperature). At the end of the fermentation process, the broth consists of bacterial cells, xanthan and several other chemicals. For the recovery of xanthan, cells are normally separated first, by either centrifugation or filtration [26]. The purification might include precipitation employing water-miscible non-solvents (acetone, ethanol, isopropanol etc.), incorporation of several salts as well as pH adjustments [26]. Followed by the precipitation process, the product is dewatered and dried.

4.4 Properties and Surface Modifications

Xanthan is a white to cream colored free-flowing powder and tasteless biopolymer. The biopolymer is soluble in cold as well as hot water, however, the material is insoluble in most of the organic solvents. The industrial significance of this material is dependent on its capability to govern the rheology of water based systems. Xanthan is utilized as a thickener in various industries, stabilizing the emulsions and suspensions in textile and paper mill industries [27] and in enhanced oil recovery processes. Xanthan is also commonly used in the food industry due to its higher viscosity at lower concentrations, solubility in cold or hot water, good stability in acid systems and very less change in viscosity with temperature variation. In addition to this, the material also displays good suspending properties due to its higher yield value and capacity to contribute great freeze-thaw stability [28,29]. The solution of xanthan has good stability over a wide range of temperatures (till 90 °C), salt concentrations (till 150 g/l NaCl) and pH (2-11) [30]. The rheological properties of xanthan solutions change with the nature of polymer, i.e., average molecular weight, and pyruvate as well as acetate contents [31,32]. In the molecule, the levels of acetal and pyruvic acid substitutions may change with operational conditions [33], bacterial development [34], and type of *Xanthomonas sp.* utilized [35]. Xanthan gum with a greater level of pyruvilation and acetylation degree enhances the viscosity of its water solutions due to intermolecular associations [36]. Nevertheless, pyruvate free xanthan is considered to be more appropriate for the generation of mobility control solutions, which are utilized in the enhanced oil recovery processes [37]. It was proved that by the incorporation of salt (monovalent or divalent), the xanthan gum chains were subjected to a co-operative conformational transition from a disarranged conformation to an arranged and highly rigid structure [38,39].

At different pH values, xanthan gum exists in an ordered conformation, where the acetal links are secured from the chemical attack. Normally, xanthan gum of 1000 ppm concentration range are utilized for the EOR application [40]. At these concentrations, xanthan gum does not degrade by the application of typical shear forces existing in the petroleum reservoirs. Further, this material is comparatively insensitive to the greater electrolyte concentrations and does not precipitate in the presence of divalent ions [41]. Dissolving xanthan gum in solution is a two-stage slow process. During the first stage, a gel is generated when solvent molecules begin to dissolve and connect with the polymer chains. Second stage greatly depends on the type of mixing. A good dispersion is necessary for preventing minute pore plugging during the field usage [42]. Xanthan solution properties are influenced by the measurement and dissolution temperature, and additional non-xanthan polymers.

A drawback of xanthan gum is its sensitivity to the bacterial degradation. It was established that the salt tolerant anaerobic as well as aerobic microorganisms can cause the degradation of xanthan gum chains, thereby, leading to a loss in the solution viscosity [43]. Biocides are employed for suppressing the growth of microorganisms that degrade xanthan gum. In majority of the cases, formaldehyde is used as an efficient biocide. Nevertheless, the application of biocides for protecting xanthan gum makes the lesser environmental impact of xanthan gum polymer somewhat controversial.

Roy *et al.* [44] reported modified amphiphilic xanthan having hydrophobic octyl moieties onto the carboxylic acid functional groups using a grafting reaction. In the course of the reaction, the reagents (octylamine, carbodiimide (EDAC) and N-hydroxysuccinimide

(NHS)) were incorporated in the same stoichiometry with reference to the carboxylic functional groups. The derivative xanthan gum followed similar organization as the precursor, however, the intermolecular hydrophobic interactions greatly affected the transient network lifetime. The rheological as well as structural characteristics of these derivatives were also analyzed employing the rheological and fluorescence analyses. It was observed that an enhancement in the fluorescence intensity occurred for both xanthan samples as the concentration of polymer increased. This event was, however, more distinct for the modified xanthan as compared to the precursor. Figure 4.2 also demonstrates the rheological performance of the precursor and modified xanthan gum sample.

In other study describing the physical modification of xanthan gum, Wang *et al.* [45] synthesized aerogel composites based on sodium montmorillonite clay (Na+-MMT) and xanthan gum employing an environmental benign freeze-drying technique. For enhancing the characteristics of XG/clay aerogels, a bio-based polysaccharide was utilized. Fourier transform infrared (FTIR) spectroscopy confirmed the molecular interactions between biopolymer and clay in the aerogel composite. The thermo-gravimetric analysis (TGA) confirmed that the clay enhanced the thermal stability of the aerogels.

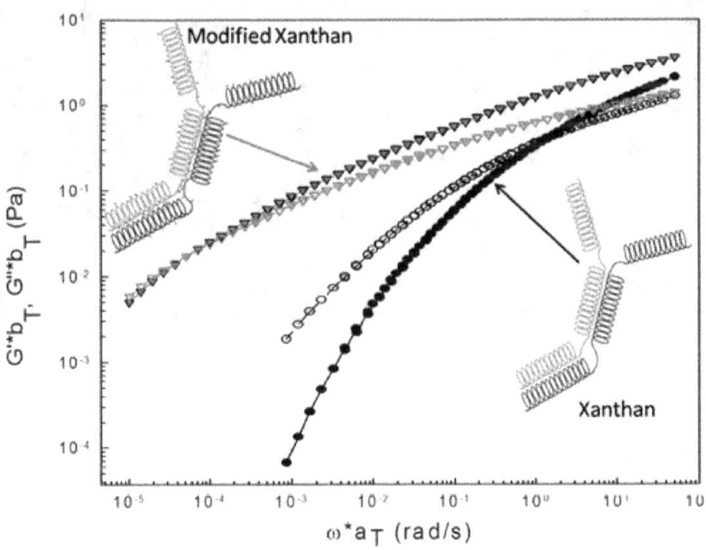

Figure 4.2 Rheological properties of the two xanthan gum samples as a function of angular viscosity. Reproduced from reference 44 with permission from American Chemical Society.

The bio-based aerogels exhibited identical degradation pattern with two major stages: bound water loss as well as aerogel matrix degradation. It was further concluded that the incorporated glycerol was removed between the temperatures 180 and 250 °C, which overlapped with the onset degradation temperatures of xanthan at approximately 235 °C and 240 °C, respectively. Figure 4.3 also shows the scanning electron micrographs of the various aerogels at high magnification [45].

Jampala *et al.* [15] also investigated the structure and rheological properties of xanthan gum modified in a cold plasma environment. The modification was carried out using dichlorosilane (DS)-plasma conditions and subsequently *in-situ* amination by ethylenediamine (ED). Network strength of the polymer was observed to increase after plasma treatment. Optimization of plasma treatment time and polymer concentration was performed to obtain stable gels of xanthan gum. Figure 4.4 shows the high-resolution electron spectroscopy for chemical analysis (ESCA) spectra of samples treated with DS-plasma for 15 min, followed by *in-situ* ED-grafted.

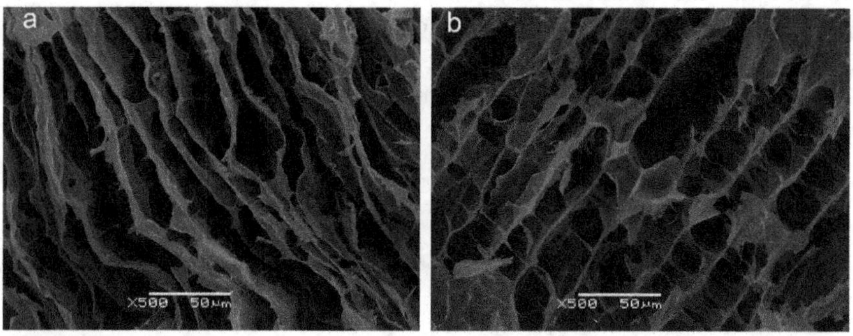

Figure 4.3 SEM micrographs of two different xanthan gum based aerogels. Reproduced from Reference 45 with permission from American Chemical Society.

4.5 Applications

4.5.1 Enhanced Oil Recovery Process

Almost 60-70% of the original oil in place (OOIP) is generally not generated by the conventional methods, which depends on hydrocarbon as well as reservoir drive. The secondary recovery is generally water flooding which recovers approximately 20-50% of the OOIP. The EOR normally involves the recovery beyond the secondary stage. The EOR methods achieved particular attention due to depleting resources of crude oil worldwide. The different EOR processes are chemical and gas flooding, combustion, steam and electric heating. The chemical flooding methods are recognized by the particular chemical, which is injected. Chemical EOR processes are one of the predominantly used methods for the petroleum production because of feasible additional production costs and easy application. The chemical EOR techniques employ interfacial tension minimization and/or the mobility ratio improvement for mobilizing the oil [46]. These EOR processes are contingent on the polymer flooding (PF) or the alkali-surfactant-polymer (ASP) flooding [47]. The existence of the surfactant decreases the interfacial tension between the hydrocarbon and aqueous phases. The PF aqueous solutions have been regarded as an interesting alternative to the conventional water flooding due to its ability for the recovery of extra oil from the reservoirs [48,49]. The application of aqueous solutions in the EOR stage is very useful for keeping the rate of crude oil production at the necessary economical levels, after the conventional methods have been disabled. In the PF process, when the water soluble

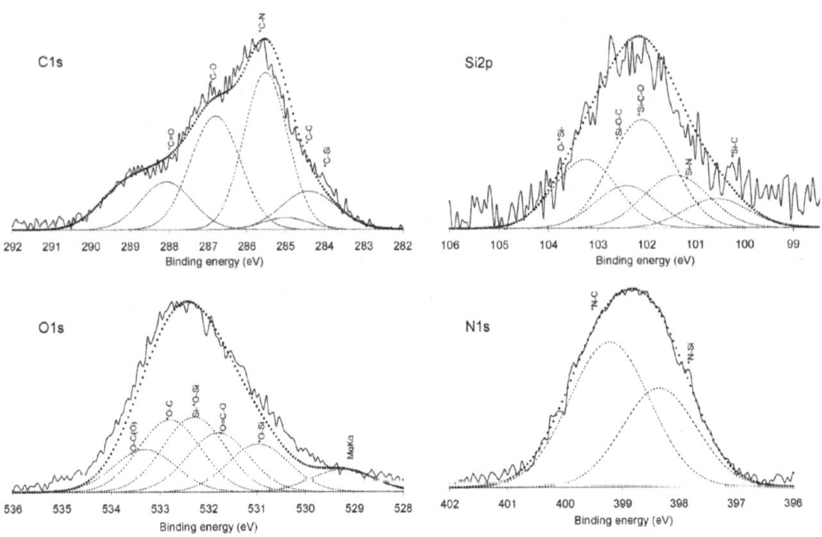

Figure 4.4 High-resolution ESCA spectra of 15-min DS-plasma-treated and subsequently in situ ED-grafted. Reproduced from Reference 15 with permission from American Chemical Society.

polymers are incorporated into the injection water, it can enhance the water phase viscosity along with improvement in the mobility ratio. It can also increase the swept volume, thereby, increasing the efficiency of oil recovery [49, 50]. The custom made EOR injection fluids generally consist of water soluble polymers like polyacrylamides. These polymers lower permeability contrast and enhance the aqueous phase viscosity. The operational hazard related with PF involves polymer degradation, loss of polymer to porous media and loss of injectivity.

In EOR applications, the most commonly employed synthetic water soluble polymer is partially hydrolyzed polyacrylamide (HPAM). Though the polymer has been successfully used in the recovery of tertiary oil [49,50], the EOR polymers based on polyacrylamide have some disadvantages. Due to the presence of recurrent electric charges on the backbone, the HPAM collaborates with the charged species existing in the solution. HPAM is also very sensitive towards the existence of surfactants and hydrocarbons in the reservoir. From an environmental outlook, the application of synthetic polymer in EOR is troublesome. In most cases, after the injection, HPAM will remain in the reservoir. Due to the environmental regulations, the industrial processes are required to minimize the spillage, and the use of non-biodegradable polymer can lead to bio-accumulation during accidental environmental release. Hence, the biopolymer based EOR technology will be more promising in the future, especially if the existing limitations are taken care of. During the chemical EOR processes, other environmentally beneficial materials like several polysaccharides and nano-cellulose have the potential to fully or partially substitute HPAM in the coming years. The important physical properties of biopolymer materials are lesser sensitivity towards the shear degradation and greater viscosifying efficiency at lesser additive concentration. Xanthan gum is regarded as one of important biopolymers which can be practiced commercially in the petroleum reservoirs [51]. For the effective replacement of EOR polymers like HPAM, the biopolymer materials should have properties such as greater salinity tolerance, higher shear and mechanical stability, good adsorption and rheological properties, and excellent viscosifying efficacy as well as thermal tolerance. For instance, despite the fact that xanthan gum is high-priced and more susceptible to biodegradation, it provides excellent viscosity generating capacity and very good stability under the mechanical shear and high salinity conditions, when compared to HPAM [52]. However, in the presence of oxidizing agents, xanthan gum is exposed to the oxidative degradation process. This type of degradation has a vital role in the viscosity loss of the xanthan gum solutions [53]. In addition to this, various bonds present in the xanthan gum molecule, like the acetyl groups [54], pyruvate-ketals and glycosidic linkages [55], are exposed to cleavage by the hydrolysis process, which causes remarkable viscosity loss. Jang *et al.* studied the promising application of xanthan gum and HPAM as the PF agents for the purpose of heavy oil recovery under a range of saline conditions [56]. The rheological analysis was performed to analyze the variation in shear viscosity, when the polymer was subjected to a range of reservoir conditions. As the xanthan gum concentration was increased, the shear viscosity was also observed to increase. For the polymer flooding in EOR, the enhanced shear viscosity of xanthan gum can raise the sweep efficiency as well as the residual resistance factor [57]. The results indicated that the Newtonian viscosity area observed at lower shear rates were not seen at the higher concentrations. On the other hand, remarkable shear thinning property was observed, which was due to disentanglement and alignment of the xanthan gum chains along the direction of flow. Overall, the rheological test analysis revealed that the HPAM solution was more sensitive to higher temperature and salinity than the xanthan gum solution. The xanthan gum structure was reported to have helix as well as random coil conformations, which greatly depends on the temperature of dissolution [16]. A flood test was also carried out with xanthan gum solution at 3 wt% salinity, and it was observed that the recovery of oil was 7.2% greater than HPAM solution. The solution of xanthan gum at 3 wt% and 10 wt% salinity exhibited no remarkable difference in the oil recovery. As the main objective of employing the polymers in EOR operation is to enhance the solution viscosity and lower the rock permeability, and both of these changes help to enhance the sweep efficiency by lowering the mobility of displacing fluid [58]. Thus, the results obtained in this analysis confirmed the superiority of xanthan gum over HPAM for polymer flooding applications for enhanced oil recovery, especially for the applicability of xanthan gum at high salinity reservoir conditions. Thus, the use of biopolymers like xanthan gum for EOR applications has the potential to provide superior performance in comparison with the synthetic polymers.

The mixing of xanthan gum with surfactants was also studied by Taugbol *et al.* [59] and the combination was proved to be very useful for EOR. The authors demonstrated that by using xanthan gum along with the surfactant alkyl propoxy-ethoxy sulfate ($C_{12-15}(PO)_4$-$(EO)_2$-OSO_3-Na^+), greater than 50% of residual oil could be recovered. However, it was noted that the recovery of residual oil by utilizing only a surfactant solution was lesser. In addition to this, it was also observed that the consolidation of xanthan gum and dodecyl-*o*-xylene sulfonate surfactant recovered even lesser residual oil when compared to the case where only surfactant solution was utilized [60]. The authors suggested the explanation for this phenomenon as the development of large micellar aggregates that exhibit an adverse effect on the flow behavior of surfactant through the porous media. In a study by Wei *et al.* [61], the chemical formulation as well as the rheological properties of a self-assembling biopolymer (SAP) system were studied. The aim was to ascertain the suitability of the system as a substitute in the polymer flooding process for oil recovery, specifically under extreme reservoir conditions of salinity, temperature and hardness. The viscoelastic SAP system was generated by the non-bonding associations of xanthan gum side branches with the EOR surfactant b-cyclodextrin inclusion complex in aqueous solutions. It has been reported that the water soluble polymers can link with the ionic surfactants through the non-bonding interactions, which can enhance the rheological and physico-chemical properties of the basic polymer [62,63]. Thus, the improved viscoelastic system displays higher thermal and mechanical stability, along with increased hardness and resistance to brine salinity because of the interlocking effect. From the sand pack flood tests performed at the reservoir conditions, it was observed that this SAP system generated remarkably greater resistance factors than xanthan gum during the flow in porous media. This clearly indicated its efficiency in governing the motility of the displacing fluid. Further, the SAP system displayed good capability as the *in-situ* permeability modifier, thus, confirming the superior performance.

Sveistrup *et al.* [64] examined the polysaccharides starch, xanthan gum and scleroglucan, and confirmed these materials to be environmentally friendly EOR agents. In this study, the viscosifying properties of these polysaccharide based biopolymers were examined, especially viscometric and interfacial tension assessments. The polysaccharides exhibited robust shear tolerance, supportive viscosifying performance, electrolyte tolerance as well as temperate interactions with the surfactants. These polysaccharides were also observed to exhibit encouraging viscosifying accomplishment in semi-dilute regime, due to the rigid helical structure of the chains, which are characterized by the extensive persistence lengths. Further, both xanthan and scleroglucan displayed sufficient tolerance towards the electrolyte, particularly at moderate and greater shear rates, where the biopolymers remained in shear aligned states characterized by the scattered inter-chain entanglements. The xanthan gum, which is a polyelectrolyte, has side chains which are connected with backbone under deserted electrostatic conditions, which results in marginal reduction in the hydrodynamic volumes and viscosities during the complete range of measured shear rates. At low and moderate shear rates, the polyelectrolyte backbone recantation occurs because of the existence of electrolyte. However, at the higher rates of shear, the shear alignment synthetically expands the polyelectrolyte chain, contrasting the coil recantation effects.

The biopolymer-surfactant interactions are generally much more complicated, compared to the biopolymer-electrolyte interactions. For xanthan gum, there exists only minute effect of extra anionic surfactant at greater salt content, when compared to the surfactant-free brine. The viscosity of xanthan gum is clearly dependent on temperature, with a double helical configuration at lesser temperatures to a disarranged coil structure at greater temperatures [65]. The xanthan gum viscosities are observed to be reduced in the presence of surfactant. The noticed viscosity variations are attributable to the polymer–surfactant interactions that affect the biopolymer's secondary helical structure. In the case of xanthan gum, unbound as well as bound anionic surfactants result in electrostatic repulsive interplay on the polyelectrolyte backbone, which results in the polymer recantation and lower inter-chain entanglements. Non-electrostatic interactions are also observed between the surfactants and xanthan gum. At moderate and higher shear rates, viscosity has been observed to be reduced, as the non-electrostatic surfactant-polymer interactions operate mainly on the xanthan gum's inherent backbone, oppositely affecting the side-chain association property. These non-electrostatic interactions are capable to promote a structural transition to a weaker as well as less arranged conformation, from a rigid ordered double helix. The interfacial tension analysis has also been observed to confirm the binding of anionic surfactant to the backbone of xanthan gum. Overall, xanthan gum has proved to an excellent environmental friendly EOR agent, which is appropriate for oil recovery particularly under extreme reservoir conditions of temperature, hardness and salinity. Consequently, xanthan gum continues to draw research interest for enhanced utilization in the EOR processes.

4.5.2 Dispersant for Drilling

The demand for oil and gas has increased manifold in the past several decades because of industrialization, increased population and enhanced living standards. However, numerous producing oilfields have turned into mature fields, which causes a loss in the production. The development of new oilfields is also being undertaken, even in challenging reservoir environments, like deep offshore regions or high pressure-high temperature (HPHT) reservoir conditions. Thus, it has become very challenging to design the drilling fluids. In accordance with the American Petroleum Institute (API), the drilling fluid is described as a circulating fluid which is employed in the rotary drilling for performing any or all of the different functions necessary during the drilling operations. Several kinds of drilling fluids like oil based drilling mud (OBM), water based drilling mud (WBM) and their derivatives are being utilized, along with appropriate clay as well as polymer additives. The important functions of the drilling fluids include (1) regulation of the downhole formation pressures, (2) preventing damage of producing formation, (3) cooling and lubricating drill bit, (4) overcoming the formation fluid pressure, and (5) removal of cuttings, developed by drill bit from bore hole [66]. For this, an effective drilling fluid should possess certain characteristics like desired rheological behavior, temperature stability, prevention of fluid loss, stability under different pressure and operating conditions [67].

The classification of drilling muds as either WBMs or OBMs, depends on the nature of continuous phase of the mud [42]. OBMs normally employ hydrocarbon oil as the major liquid component along with alternate materials like clays or colloidal asphalts, which are incorporated to contribute the required viscosity together with polymers, emulsifiers and additives such as weighting agents. WBMs generally consists of viscosifiers, weighting agents, fluid loss control agents, emulsifiers, lubricants, salts, pH control agents and corrosion inhibitors. For the rheology control function, various polymers like xanthan gum and partially hydrolyzed polyacrylamide (PHPA) are generally employed. Sea water mud is a type of WBM which is designed for the offshore drilling application, where make-up water is collected from ocean. In the salt water muds, dissolved sodium chloride (NaCl) acts as the main component. Some of the most commonly used type of sea water mud additives are bentonite clay, lignite, lignosulfonate and caustic soda. Biopolymer xanthan gum can be utilized instead of bentonite. In order to eliminate the problems related to the reduction of viscosity in polymer-based aqueous fluids, formates like sodium formate and potassium formate are generally incorporated in the fluids for increasing their thermal stability. Nevertheless, the technology of utilizing formates leads to high price. The stability of wellbore treatment can be preserved at temperatures till 135-160°C (275-325 °F), by adding different polysaccharides to the fluids.

Nano-fluids demonstrate potential usage in upstream oil and gas industry for increasing the performance of different processes like exploration, drilling as well as completion, production and EOR operations [68, 69]. Some of the current research studies addressed the application of nanoparticles for solving drilling related issues such as lost circulation, pipe sticking, torque and drag [70]. The advancement of appropriate nano-fluids for the WBM needs an acceptable dispersion medium for the nanoparticles, which should have good compatibility with the oilfield chemicals. Biopolymers like xanthan gum and guar gum are well known for their usage in the oil and gas industries as a drilling fluid additive, fluid loss controlling agent, rheological modifier and emulsion stabilizer. Xanthan gum generates exclusive rheological properties like high low shear viscosity as well as strong shear thinning behavior, due to the existence of a complicated network in the solution originated from the stiff polymer chains.

In a recent study, the ability of xanthan gum and guar gum for generating the stabilized iron-based nano-fluids was confirmed [71]. The study aimed to amend the guar gum solutions with small quantity of xanthan gum in order to remarkably enhance the biopolymer capability for stabilizing highly concentrated iron nano- and micro-particle suspensions. The cooperative effect between xanthan gum and guar gum developed a viscoelastic gel which had the ability to maintain 20 g/L iron particles suspension for over 24 h. The xanthan gum/guar gum viscoelastic gel was characterized by a marked shear thinning property. This behavior, along with lower biopolymer concentration, regulated lower viscosity values at greater shear rates, thereby, expediting the injection in porous media. In addition to this, the thermo-sensitivity of soft elastic polymeric network stimulates greater stability and prolonged storage times at low temperatures and sudden decline in the viscosity at greater temperatures. This property can be utilized for improving the flowability and delivery of suspensions to the target, as well as to efficiently tune and control the release of iron particles. In another recent study by Ponmani *et al.* [72], zinc oxide (ZnO) and copper oxide (CuO) nano-fluids were generated, using the two-step method in a 0.4 wt% xanthan gum aqueous solution as a base. The CuO and ZnO nano-fluids were characterized for dispersion and stability, utilizing the dynamic light scattering and scanning electron microscopy. It was observed that the stability of ZnO nano-fluid was lower as compared to the CuO nano-fluid in the xanthan gum aqueous solution. Xanthan gum formed a jelly structure over the nanoparticles, keeping the particles suspended in solution. The study also confirmed that the electrical and thermal conductivities of these nanofluids enhanced as the nanoparticle concentration was increased. The thermal and the electrical conductivities increased by almost 25% and 50%, for the ZnO and CuO nano-fluid, respectively. This signifies the advancements in the nano-fluid based technology with xanthan gum as the primary additive in upstream oil industries.

Karen *et al.* [73] generated nano-fluid enhanced WBM (NWBM) by employing the nano-fluids of CuO and ZnO in the xanthan gum aqueous solution as the base fluid. The concentration of nanoparticles in the nanofluids were 0.1 wt%, 0.3 wt% and 0.5 wt% [72]. The generated nano-fluids were added to WBM as an additive (1% by volume). An enhancement in the concentration of nanoparticles increased the thermal conductivity in both the nano-fluid based drilling fluid. The improvement in the thermal conductivity of ZnO based nano-fluids was observed to be from 12% to 23%. However, in the case of CuO based nano-fluids, greater enhancement in the thermal conductivity was observed, i.e., from 28% to 53%. In the case of ZnO nano-fluid systems with WBM, an improvement in the thermal conductivity from 17% to 34% was seen. For the CuO nano-fluid system with WBM, the thermal conductivity enhancement was from 20% to 38%. This increased thermal conductivity of the fluids was a clear indication of the capability of mud to cool down faster as it proceeds up to surface. Further, high pressure rheological analysis was performed on NWBM to determine the influence of nanofluids on the rheological behavior at different pressures (0.1 and 10 MPa) and temperatures (25 °C, 70 °C, 90 °C and 110 °C). The remarkable performance at high pressures and temperatures revealed high potential of NWBM of oil drilling applications.

4.5.3 Tackifier for Drilling

Both xanthan gum and polyacrylamide gum are very common polyelectrolyte polymers used in various industries [74]. The anionic polyacrylamide (HPAM) is commonly used to treat industrial and municipal wastewater due to its low cost and effective flocculence. Large number of polar groups are present in PAM molecular chain that can adsorb the solid particles which are suspended in the drilling fluid. PAM is also commonly used as an industrial oil displacement agent due to its water solubility, ionic properties and several other functions [75]. Interaction of xanthan gum with other polymers like PAM, the performance can even improve further. For instance, PAM/xanthan gum solution has excellent salt resistant characteristic, and hence can be employed as a better salt resistant thickening agent as well as flocculant. As HPAM encounters xanthan gum, the dipole reaction of the hydrogen groups and hydrophobic groups induces molecular chain association, thus causing the formation of macromolecular structure.

During the drilling of the natural gas hydrate (NGH) core samples, the temperature of the drilling fluid in the bore circulation should be lowered below the freezing point to make sure that the hydrate core is preserved in stable and balanced condition. In a study performed by Zhao *et al.* [76], the influence of K^+, Na^+, Mg^{2+} and Ca^{2+}cations, at different concentrations, on the rheological behavior of PAM containing cationic and anionic PAMs (CPAM and HPAM) as well as xanthan gum solution was studied. The analysis was performed after the macromolecules were generated due to the interaction between the xanthan gum (tackifier) and polyacrylamide (flocculant). Further, the crosslinking reaction mechanism of polyacrylamide/xanthan gum macromolecular system and the cation disruption mechanism of crosslinking process were also examined. It was observed that the CPAM/xanthan gum solution exhibited ehnaced cation resistant property than the HPAM/xanthan gum solution. On comapring the viscosity values of two different types of PAM/xanthan gum solutions in varying proportions (f_{PAM}), at a total concentration of 0.02%, it was noted that the viscosity of the solution increased as the PAM content developed. This was due to the fact that the PAM is a flexible chain molecule, while xanthan gum is a rigid chain molecule. The variation in conformation was not difficult for flexible chain, where the inherent tangle was collapsed through the segmental motion, thereby, reducing the flow resistance.

Thus, xanthan gum, in interaction with other polymers, has a strong potential for use as tackifier for drilling operations. Especially, the useful performance is achieved at low concentrations.

4.5.4 Corrosion Inhibitor for Mild Steel

Mild steel (MS) has a wide variety of industrial applications, hence, the corrosion stability of MS in acidic media is an important concern for the material technologists and corrosion scientists [77]. Metals as well as alloys are exposed to immensely corrosive acids during industrial processes like acid cleaning, acid descaling, oil well acidizing and acid pickling [78]. Such extreme conditions lead to accelerated corrosion rates and material degradation, resulting in large economic losses. Natural polymers which show a high affinity for the metal surfaces and exhibit excellent corrosion inhibition effectiveness have been focus of research due to these properties, in addition to the environmental friendly nature [79]. In addition, grafted or modified polysaccharides have also the capability to function as corrosion inhibitors as these polymeric materials possess numerous adsorption sites, which can adsorb more efficiently to the surfaces, thus, helping to achieve effective corrosion inhibition.

In a study by Biswas *et al.* [80], xanthan gum and xanthan gum-grafted-polyacrylamide (XG-*g*-PAM) were analyzed as corrosion inhibitors for mild steel in 15% HCl medium. Different characterization like potentiodynamic polarization, gravimetric analysis and electrochemical impedance spectroscopic analysis were performed for ascertaining the effectiveness of the aforementioned inhibitors. It was revealed that both of these inhibitors followed the Langmuir adsorption isotherm. It was observed that the inhibition efficiency (η%) enhanced with the increase in the concentration of inhibitor. At 0.4 g/L concentration, xanthan gum exhibited 90.2% inhibition efficiency, whereas, the efficiency of XG-g-PAM was 92.7%. Further, there was no remarkable change in η% as the concentration was increased. The superior efficiency of XG-g-PAM was because of the fact that it enclosed the surface of metal more effectively than the xanthan gum, thereby, retarding the dissolution of the mild steel. The SEM analysis indicated the protection of surface of metal by xanthan gum and XG-g-PAM. The microscopy images also revealed that xanthan gum as well as XG-g-PAM formed a barrier film on the surface of metal, which prevented the spread of corrosion. Overall, XG-g-PAM was concluded to demonstrate a superior corrosion inhibition ability than xanthan gum for mild steel protection in the acidic conditions.

Thus, in addition to the large variety of applications of xanthan gum in oil and gas industry, it along with its derivatives have also strong potential of large scale application for the corrosion inhibition of metallic surfaces under extreme corrosion conditions.

4.5.5 Chemical Absorption of Carbon Dioxide

The gas-liquid mass transfer in the non-Newtonian liquid is a relevant example of absorption of gas in the pseudo-plastic flow of the industrial operations like fluidized bed, fermentation broth and slurry [81,82]. Park *et al.* [83] measured the absorption rate of carbon dioxide (CO_2) in the aqueous solution of xanthan gum in the range 0-0.15 wt%, which consisted of NaOH concentration of 0-2.0 kmol/m^3 in a flat-stirred vessel with an agitation speed of 50 rpm at 0.101 MPa and temperature of 25 °C. Volumetric liquid side mass transfer coefficient of CO_2, which was connected to the viscosity and the elastic behavior of xanthan gum solution, was employed to determine the CO_2 chemical absorption rate (R_A). The aqueous solution of xanthan gum with the elastic property of the non-Newtonian liquid caused increase in R_A, when compared to the Newtonian liquid based on the same viscosity of the solution.

4.5.6 Fracturing Fluids

Fracturing is considered to be a hydraulic process, in which overpressure is practiced for cracking the rock and for enhancing the flow of oil. In the interest of avoiding the collapse of the stone, the proppants are settled in these fractures, which perform as spacers. As a consequence, the viscous properties of the fracturing fluid are generally considered to be the most significant factor. The ideal fracturing fluid is required to be efficient with lesser fluid loss. This can be generally achieved by mixing the high fluid viscosity with the fluid loss additives. The most commonly used types of fracturing fluids are water based fluids and oil based fluids. The linear fracturing fluids are water based fracturing fluids with no chemically cross-linked structures. Xanthan gum is one of the most commonly used types of linear gels employed as the fracturing fluids. Due to the high low-shear viscosity of xanthan, it is very productive to maintain the proppants in the suspension and to transport them as far as possible

[84]. Xanthan gum is a vicsosifier, which can be employed as a linear gel system as well as a crosslinked fluid. The main application of xanthan gum in the stimulation has been as a thickener for the hydrochloric acid (HCl). The application of this biopolymer is restricted to the acid concentrations till 15% and temperature of 200 °F (93 °C). A breaker is defined as an additive which facilitates a viscous fracturing fluid to be degraded manageably to a thin fluid which could be generated back out of the fracture. Xanthan can act as an enzyme gel breaker that can cause degradation of the polymer chains by breaking the backbone structure of the thickeners and ultimately of the fluid loss additive. Due to their infinite polymer-degrading activity and inherent specificity, the enzyme breakers have numerous benefits in several other breaker systems. Xanthan gum can also perform as stabilizer for foaming treatments, which are incorporated into the fracturing fluid, to enhance the foam half-life, specifically at the elevated temperature.

4.5.7 Removal of Dyes from Aqueous Solutions

Xantan gum can also be used as an effective adsorbent for the removal of dyes commonly present in aqueous wastewater generated in oi land gas industries. In a recent study, Ghorai *et al.* [85] reported novel nanocomposite consisting of hydrolyzed polyacrylamide (PAM) grafted onto XG along with incorporated nano-silica (h-XG/SiO_2). The composite was observed to be a promising adsorbent for the separation of toxic methyl violet (MV) and methylene blue (MB) from the aqueous solution. The preparation process involved the saponification of grafted PAM and the *in-situ* generation of nanoscale SiO_2 by the sol-gel process. The kinetics of MV and MB adsorption from the aqueous solution confirmed that the adsorption of dyes took place very quickly, in accordance with Langmuir adsorption isotherm and pseudo-second-order kinetics. The process was also observed to be entropy driven. Figure 4.5 presents the adsorption mechanism of MB and MV using the nanocomposite. The significant adsorption ability of dyes on the nanocomposite (99.4% efficiency of MB removal and 99.1% efficiency of MV removal) was justified on the basis of hydrogen bonding interactions as well as the electrostatic and dipole-dipole interactions between the cationic dye and anionic adsorbent. Due to the superior regeneration ability of the nanocomposite, such bio-based nanocomposites have the potential of large scale usage for the adsorption of toxic dyes.

4.6 Summary and Outlook

Xanthan is an outstanding commercially produced industrial gum, obtained by the fermentation process. Compared to carboxymethylcellulose, polyacrylamide, modified starch as well as some plant polysaccharides (guar gum), xanthan gum has fair technical advantage in the oilfield development. Overall, the extensive range of applications for xanthan gum is due to its remarkable properties such as (1) greater viscosity at lower concentrations (600-2000 ppm), (2) non-Newtonian property, (3) mechanical degradation resistance, (4) low sensitivity of viscosity to the variation in salinity, (5) environmental friendly material as it is a biopolymer, and (6) good temperature stability (till 90 °C). Xanthan gum is considered as the one of the few successful biopolymers which can be practiced commercially in the petroleum reservoirs. During the application of xanthan gum in beach, ocean, permafrost drilling and halogen layer, this biopolymer has apparent effect on the blowout prevention. Even lower concentration of solution of xanthan gum can retain the viscosity of water based drilling fluids. The polymer offers good suspension performance, to enable it to restrict the collapse of borehole, prevent the formation blowout and expedite discharging the cutting gravel out of the wells.

The observations made in this review clearly indicate that a large number of studies have been undertaken on establishing the application of xanthan gum in the drilling and EOR applications. Xanthan gum has been proved to be an excellent environmental friendly EOR agent suitable for effective oil recovery, particularly under extreme reservoir conditions of temperature, hardness and salinity. It provides excellent viscosity generating capacity and good stability under mechanical shear and high salinity conditions. Xanthan gum has also been reported to have application in the nano-fluids, which demonstrate potential usage in upstream oil and gas industry for enhancing the performance of different processes like exploration, drilling as well as completion, production and EOR operations. Further, the polymer has been observed to exhibit corrosion inhibiting ability for mild steel in acidic conditions. It is also considered as one of the most commonly used types of linear gels used as fracturing fluids. In summary, there is a significant anticipation that xanthan gum can play a vital role in increasing the production of oil since the utilization of polymer flooding has displayed recovery of more than 20% additional oil from the OOIP.

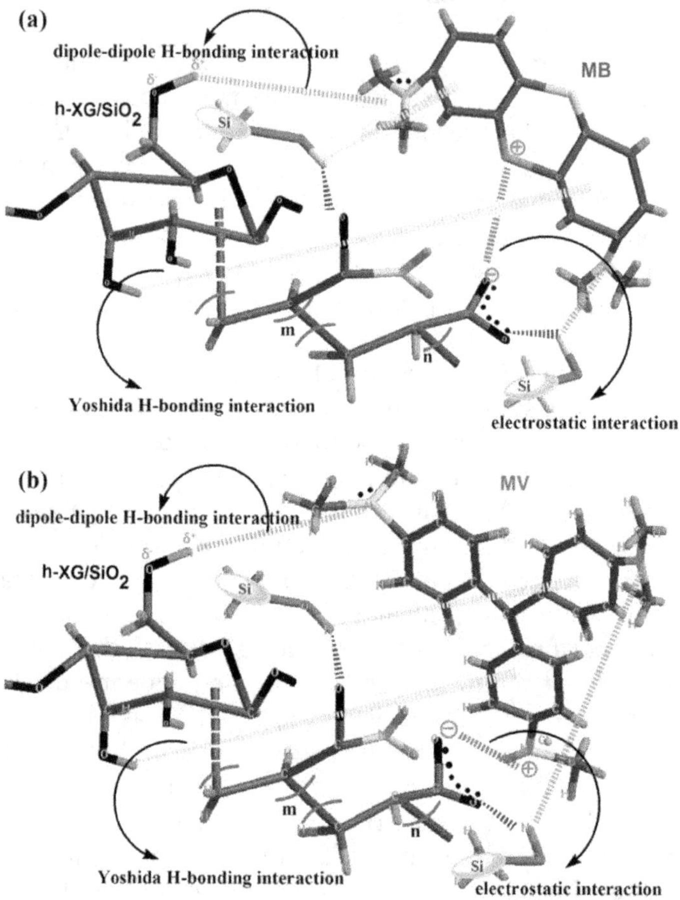

Figure 4.5 Adsorption mechanism of (a) MB and (b) MV using xanthan gum-silica nanocomposites. Reproduced from Reference 85 with permission from American Chemical Society.

References

1. Sarwat, F., Ul Qader, S. A., Aman, A., and Ahmed, N. (2008) Production and characterization of a unique dextran from an indigenous Leuconostoc mesenteroides CMG713. *International Journal of Biological Sciences*, **4**, 379-386.
2. Margaritis, A., and Zajic, J. E. (1978) Biotechnology review: mixing mass transfer and scale-up of polysaccharide fermentations. *Biotechnology and Bioengineering*, **20**, 939-1001.

3. W. H. Mcneely (1967) Biosynthetic polysaccharides. In: *Microbial Technology*, Peppler, H. J. (ed.), Reinhold Publishing Corp., USA, p. 381.
4. Milas, M., Reed, W. F., and Printz, S. (1996) Conformations and flexibility of native and renatured xanthan in aqueous solutions. *International Journal of Biological Macromolecules*, **18**, 211-221.
5. Katzbauer, B. (1998) Properties and applications of xanthan gum. *Polymer Degradation and Stability*, **59**, 81-84.
6. Rosalam, S., and England, R. (2006) Review of xanthan gum production from unmodified starches by Xanthomonas compestris sp. *Enzyme and Microbial Technology*, **39**(2), 197-207.
7. Lee, B. H. (2014) *Fundamentals of Food Biotechnology*, 2nd edition, Wiley-Blackwell, USA.
8. Abidin, A. Z., Puspasari, T., and Nugroho, W. A. (2012) Polymers for enhanced oil recovery technology. *Procedia Chemistry*, **4**, 11-16.
9. Melton, L. D., Mindt, L., Rees, D. A., and Sanderson, G. R. (1976) Covalent structure of the polysaccharide from Xanthomonas campestris: evidence from partial hydrolysis studies. *Carbohydrate Research*, **46**, 245-257.
10. Jansson, P. E., Kenne, L., and Lindberg, B. (1975) Structure of the exocellular polysaccharide from Xanthomonas campestris. *Carbohydrate Research*, **45**, 275-282.
11. Sandford, P. A., and Baird, J. (1983) Industrial utilization of polysaccharides. In: *The Polysaccharides*, Aspinall, G. O. (ed.), volume 2, Academia Press, USA, pp. 411-490.
12. Milas, M., and Rinaudo, M. (1979) Conformational investigation on the bacterial polysaccharide xanthan. *Carbohydrate Research*, **76**, 189-196.
13. Morris, E. R. (1977) Molecular origin of xanthan solution properties. In: *Extracellular Microbial Polysaccharides*, ACS Symposium Series, volume 45, USA, pp. 81-89.
14. Garcia-Ochoa, F., Santos, V. E., Casas, J. A., and Gomez, E. (2000) Xanthan gum: production, recovery, and properties. *Biotechnology Advances*, **18**, 549-579.
15. Jampala, S. N., Manolache, S., Gunasekaran, S., and Denes, F. S. (2005) Plasma-enhanced modification of xanthan gum and its effect on rheological properties. *Journal of Agricultural and Food Chemistry*, 53, 3618-3625.
16. Palaniraj, A., and Jayaraman, V. (2011) Production, recovery and applications of xanthan gum by Xanthomonas campestris. *Journal of Food Engineering*, **106**(1), 1-12.
17. Letisse, F., Chevallereau, P., Simon, J. L., and Lindley, N. D. (2001) *Applied Microbiology and Biotechnology*, **55**, 417-422.
18. Rosalam, S., and England, R. (2006) Review of xanthan gum production from unmodified starches by Xanthomonas camprestris

sp. *Enzyme and Microbial Technology*, **39**, 197-207.

19. El-Salam, M. H. A., Fadel, M. A., and Murad, H. A. (1994) Bioconversion of sugarcane molasses into xanthan gum. *Journal of Biotechnology*, **33**, 103-106.
20. Amanullah, A. S., Satti, S., and Nienow, A. W. (1998) Enhancing xanthan fermentations by different modes of glucose feeding. *Biotechnology Progress*, **14**, 265-269.
21. Rodriguez, H., and Aguilar, L. (1997) Detection of Xanthomonas campestris mutants with increased xanthan production. *Journal of Industrial Microbiology and Biotechnology*, 18(4), 232-234.
22. Esgalhado, M. E., Roseiro, J. C., and Amaral, C. M. T. (1995) Interactive effects of pH and temperature on cell growth and polymer production by Xanthomonas campestris. *Process Biochemistry*, **30**(7), 667-671.
23. Shu, C. H., and Yang, S. T. (1991) Kinetics and modeling of temperature effects on batch xanthan gum fermentation. *Biotechnology and Bioengineering*, **37**, 567-574.
24. Amanullah, A., Tuttiett, B., and Nienow, A. W. (1998) Agitator speed and dissolved oxygen effects in xanthan fermentations. *Biotechnology and Bioengineering*, **57**(2), 198-210.
25. Cacik, F., Dondo, R. G., and Marques, D. (2001) Optimal control of a batch bioreactor for the production of xanthan gum. *Computers and Chemical Engineering*, **25**, 409-418.
26. Flores Candia, J.-L., and Deckwer, W.-D. (1999) Xanthan gum. In: *Encyclopedia of Bioprocess Technology: Fermentation, Biocatalysis, and Bioseparation*, Flickinger, M. C., and Drew, S. W. (eds.), volume 5, Wiley, USA, pp. 2695-2711.
27. Kennedy, J. F., and Bradshaw, I. J. (1984) Production, properties and applications of xanthan. *Progress in Industrial Microbiology*, **19**, pp. 319-371.
28. Rinaudo, M., and Milas, M. (1978) Polyelectrolite behaviour of a bacterial polysaccharide from Xanthomonas campestris: comparison with CMC. *Biopolymers*, **17**, pp. 2663-2678.
29. Margaritis, A., and Zajic, J. E. (1978) Biotechnology review: mixing, mass transfer and scale-up of polysaccharide fermentation. *Biotechnology and Bioengineering*, **20**, pp. 939-1001.
30. Lee, B. H. (1996) *Fundamentals of Food Biotechnology*, VCH Publishers Inc., USA.
31. Moraes, I. C. F., Fasolin, L. H., Cunha, R. L., and Menegalli, F. C. (2011) Dynamic and steady-shear rheological properties of xanthan and guar gums dispersed in yellow passion fruit pulp. Brazilian Journal of Chemical Engineering, 28(3), doi: 10.1590/S0104-66322011000300014.
32. Kang, F. S., and Pettit, D. J. (1993) Xanthan, gellan, welan, and rhamsan. In: *Industrial Gums. Polysaccharides and their Derivatives*,

Whistler, R. L., and BeMiller, J. N. (eds.), 3rd edition, Academic Press, USA, pp. 341-399.
33. Trilsbach, G. F., Pielken, P., Hamacher, K., and Sahm, H. (1984) Xanthan formation by X. campestris under different culture conditions. *Third European Congress on Biotechnology*, Verlag Chemie. Germany, pp. 65-70.
34. Souw, P., and Demain, A. (1979) Nutritional studies on xanthan production by Xanthomonas campestris NRL B-1459. *Applied Environmental Microbiology*, **37**, 1186-1192.
35. Tait, M. I., Sutherland, I. W., and Sturman, C. (1986) Effect of growth conditions on the production, composition and viscosity of Xanthomonas campestris exopolysaccharide. *Journal of General Microbiology*, **132**, 1483-1492.
36. Tako, M., and Nakamura, S. (1984) Rheological properties of deacetylated xanthan in aqueous media. *Agricultural and Biological Chemistry*, **48**, 2987-2993.
37. Holzwarth, G., and Ogletree, J. (1979) Pyruvate-free xanthan. *Carbohydrate Research*, **76**(1), 277-280.
38. Morris, E. R., Rees, D. A., Young, G., Walkinshaw, M. D., and Darke, A. (1977) Order–disorder transition for a bacterial polysaccharide in solution-role for polysaccharide conformation in recognition between Xanthomonas pathogen and its plant host. *Journal of Molecular Biology*, **110**, 1-16.
39. De Mello Luvielmo, M., Borges, C. D., de Oliveira Toyama, D., Vendruscolo, C. T., Scamparini, A. R. P. (2016) Structure of xanthan gum and cell ultrastructure at different times of alkali stress. *Brazilian Journal of Microbiology*, **47**(1), doi: 10.1016/j.bjm.2015.11.006.
40. Chatterji, J., and Borchardt, J. K. (1981) Application of water soluble polymers in the oil field. *Journal of Petroleum Technology*, **33**(11), 2042-2056.
41. Taylor, K. C., and Nasr-El-Din, H. A. (1998) Water-soluble hydrophobically associating polymers for improved oil recovery: A literature review. *Journal of Petroleum Science and Engineering*, **19**, 265-280.
42. Nashawi, I. S. (1991) Laboratory Investigation of the Effect of Brine Composition on Polymer solution – Part 2 Xanthan gum (XG) Case. Society of Petroleum Engineers. Online: https://www.onepetro.org/download/general/SPE-23534-MS?id=general%2FSPE-23534-MS (assessed 29th January 2017).
43. Cadmus, M. C., Jackson, L. K., Burton, K. A., Plattner, R. D., and Slodki, M. E. (1982) Biodegradation of xanthan gum by bacillus sp. *Applied and Environmental Microbiology*, **44**(1), 5-11.
44. Roy, A., Comesse, S., Grisel, M., Hucher, N., Souguir, Z., and Renou, F. (2014) Hydrophobically modified xanthan: An amphiphilic but not

associative polymer. *Biomacromolecules*, **15**, 1160-1170.

45. Wang, l., Schiraldi, D. A., and Sanchez-Soto, M. (2014) Foamlike xanthan gum/clay aerogel composites and tailoring properties by blending with agar. *Industrial and Engineering Chemistry Research*, **53**(18), 7680-7687.
46. Thomas, S. (2008) Enhanced oil recovery - An overview. *Oil and Gas Science and Technology – Rev. IFP*, **63**(1), pp. 9-19.
47. Samanta, A., Ojha, K., and Mandal, A. (2011) Interactions between acidic crude oil and alkali and their effects on enhanced oil recovery, *Energy and Fuels*, **25**(4), 1642-1649.
48. Sorbie, K. S. (1991) *Polymer-Improved Oil Recovery*, Springer, USA.
49. Chelaru, C., Diaconu, I., and Simionescu, I. (1998) Polyacrylamide obtained by plasma-induced polymerization for a possible application in enhanced oil recovery. *Polymer Bulletin*, **40**, 757-764.
50. Wang, W., Liu, Y., and Gu, Y. (2003) Application of a novel polymer system in chemical enhanced oil recovery (EOR). *Colloid and Polymer Science*, **281**, 1046-1054.
51. Standnes, D. C., and Skjevrak, I. (2014) Literature review of implemented polymer field projects. *Journal of Petroleum Science and Engineering*, **122**, 761-775.
52. Guo, X. H., Li, D. W., Tian, J., and Liu, Y. Z. (1999) Pilot Test of Xanthan Gum Flooding in Shengli Oilfield. SPE 57294, *SPE Asia Pacific Improved Oil Recovery Conference*, Malaysia.
53. Harris, M. J., Herp, A., and Pigman, W. (1971) Depolymerization of polysaccharides through the generation of free radicals at a platinum surface: a novel procedure for the controlled production of free-radical oxidations. *Archives of Biochemistry and Biophysics*, **142**(2), 615-622.
54. Ash, S. G., Clarke-Sturman, A. J., Calvert, R. and Nisbet T. M. (1983) Chemical Stability of Biopolymer Solutions. SPE 12085, *58th Annual Technical Conference and Exhibition*, USA.
55. Seright, R. S., and Henrici, B. J. (1990) Xanthan stability at elevated temperatures. *SPE Reservoir Engineering*, **5**(1), 52-60.
56. Jang, H. Y., Zhang, K., Chon, B. H., and Choi, H. J. (2015) Enhanced oil recovery performance and viscosity characteristics of polysaccharide xanthan gum solution. *Journal of Industrial and Engineering Chemistry*, **21**, 741-745.
57. Xu, L., Xu, G., Liu, T., Chen, Y., and Gong, H. (2013) The comparison of rheological properties of aqueous welan gum and xanthan gum solutions. *Carbohydrate Polymers*, **92** (2013), 516-522.
58. Quy, N. M., Ranjith, P. G., Choi, S. K., Giao, P. H., and Jasinge, D. (2009) Analytical assessment of horizontal well efficiency with reference to improved oil recovery of the south-east Dragon oil field southern offshore of Vietnam. *Journal of Petroleum Science and Engineering*, **66**(3-4), 75-82.

59. Taugbol, K., Ly, T. V., and Austad, T. (1995) Chemical flooding of oil-reservoirs. 3. Dissociative surfactant–polymer interaction with a positive effect on oil-recovery. *Colloids and Surfaces A*, **103**, 83-90.
60. Austad, T., and Taugbol, K. (1995) Chemical flooding of oil-reservoirs. 2. Dissociative surfactant–polymer interaction with a negative effect on oil-recovery. *Colloids and Surfaces A*, **103**, 73-81.
61. Wei, B., Romero-Zerón, L., and Rodrigue, D. (2015) Improved viscoelasticity of xanthan gum through self-association with surfactant: β-cyclodextrin inclusion complexes for applications in enhanced oil recovery. *Polymer Engineering and Science*, **55**, 523-532.
62. Ritacco, H., Albouy, P. A., Bhattacharyya, A., and Langevin, D. (2000) Influence of the polymer backbone rigidity on polymer-surfactant complexes at the air/water interface. *Physical Chemistry and Chemical Physics*, **2**, 5243-5251.
63. Thuresson, K., Nystroem, B., Wang, G., and Lindman, B. (1995) Effect of surfactant on structural and thermodynamic properties of aqueous-solutions of hydrophobically-modified ethyl(hydroxyethyl)cellulose. *Langmuir*, **11**(10), 3730-3736.
64. Sveistrup, M., van Mastrigt, F., Norrman, J., Picchioni, F., and Paso, K. (2016) Viability of biopolymers for enhanced oil recovery. *Journal of Dispersion Science and Technology*, **37**(8), 1160-1169.
65. Choppe, E., Puaud, F., Nicolai, T., and Benyahia, L. (2010) Rheology of xanthan solutions as a function of temperature, concentration and ionic strength. *Carbohydrate Polymers*, **82**(4), 1228-1235.
66. Brazzel, R. L. (2009) Multi-component Drilling Fluid Additive, and Drilling Fluid System Incorporating the Additive, US patent 7635667.
67. Melbouci, M., and Sau, A. C. (2008) Water-based Drilling Fluids, US patent 7384892.
68. Kong, X., Ohadi, M. M. (2010) Applications of Micro and Nano Technologies in the Oil and Gas Industry - An Overview of the Recent Progress. *Proceedings of the Abu Dhabi International Petroleum Exhibition Conference*, UAE, pp. 1-4.
69. Friedheim, J., Young, S., Stefano, G., Lee, J., and Guo, Q. (2012) Nanotechnology for Oilfield Applications - Hype or Reality? SPE 157032, *SPE International Oilfield Nanotechnology Conference*, The Netherlands. Online: http://www.nioclibrary.ir/Reports/9140.pdf (assessed 31st January 2017).
70. Amani, M., Al-Jubouri, M., and Shadravan, A. (2012) Comparative study of using oil-based mud versus water based mud in HPHT fields. *Advances in Petroleum Exploration and Development*, **4**(2), 18-27.
71. Xue, D., and Sethi, R. (2012) Viscoelastic gels of guar and xanthan

gum mixtures provide long-term stabilization of iron micro- and nanoparticles. *Journal of Nanoparticle Research*, **14**, 1239.

72. Ponmani, S., William, J. K. M., Samuel, R., Nagarajan, R., Sangwai, J. S. (2014) Formation and characterization of thermal and electrical properties of CuO and ZnO nanofuids in xanthan gum. *Colloids and Surfaces A: Physicochemical and Engineering Aspects*, **443**, 37-43.
73. William, J. K. M., Ponmani, S., Samuel, R., Nagarajan, R., Sangwai, J. S. (2014) Effect of CuO and ZnO nanofluids in xanthan gum on thermal, electrical and high pressure rheology of water-based drilling fluids. *Journal of Petroleum Science and Engineering*, **117**, 15-27.
74. Kennedy, J. R. M., Kent, K. E., and Brown, J. R. (2015) Rheology of dispersions of xanthan gum, locust bean gum and mixed biopolymer gel with silicon dioxide nanoparticles. *Materials Science and Engineering C*, **48**, 347-353.
75. Zhang, R., Shi, W., Yu, S., Wang, W., Zhang, Z., Zhang, B., Li, L., and Bao, X. (2015) Influence of salts, anion polyacrylamide and crude oil on nanofiltration membrane fouling during desalination process of polymer flooding produced water. *Desalination*, **373**, 27-37.
76. Zhao, D., Liu, H., Guo, W., Qu, L. and Li, C. (2016) Effect of inorganic cations on the rheological properties of polyacrylamide/xanthan gum solution. *Journal of Natural Gas Science and Engineering*, **31**, 283-292.
77. Qiu, L. G., Wu, Y., Wang, Y. M., and Jiang, X. (2008) Synergistic effect between cationic gemini surfactant and chloride ion for the corrosion inhibition of steel in sulphuric acid. *Corrosion Science*, **50**, 576-582.
78. Abiola, O. K., and James, A. O. (2010) The effects of aloe-vera extract on corrosion and kinetics of corrosion process of zinc in HCl solution. *Corrosion Science*, **52**, 661-664.
79. de Souza, F. S., and Spinelli, A. (2009) Caffeic acid as a green corrosion inhibitor for mild steel. *Corrosion Science*, **51**, 642-649.
80. Biswas, A., Pal, S., and Udayabhanu, G. (2015) Experimental and theoretical studies of xanthan gum and its graft co-polymer as corrosion inhibitor for mild steel in 15% HCl. *Applied Surface Science*, **353**, 173-183.
81. Ranade, V. R., and Ulbrecht, J. J. (1978) Influence of polymer additives on the gas-liquid mass transfer in stirred tanks. *AIChE Journal*, **24**(5), 796-803.
82. Nakanoh, M., and Yoshida, F. (1980) Gas absorption by Newtonian and non-Newtonian liquids in a bubble column. *Industrial and Engineering Chemistry Process Design and Development*, **19**, 190-195.
83. Park, S.-W., Choi, B.-S., and Lee, J.-W. (2008) Chemical absorption of carbon dioxide into aqueous elastic xanthan gum solution containing NaOH. *Journal of Industrial and Engineering Chemistry*,

14(3), 303-307.

84. Hublik, G. (2012) Xanthan. In: *Polymer Science: A Comprehensive Reference*, volume 10, Matyjaszewski, K., and Moller, M. (eds.), Elsevier, USA, pp. 221-229.
85. Ghorai, S., Sarkar, A., Raoufi, M., Panda, A. B., Schonherr, H., and Pal, S. (2014) Enhanced removal of methylene blue and methyl violet dyes from aqueous solution using a nanocomposite of hydrolyzed polyacrylamide grafted xanthan gum and incorporated nanosilica. *ACS Applied Materials and Interfaces*, **6**(7), 4766-4777.

5

Polymer Nanocomposites for Gas Barrier and Packaging Applications

5.1 Introduction

Over the last 20 years, the usage of metal and glass is progressively being substituted in the packaging industries by plastics for applications where high barrier performance is needed to counteract gases like carbon dioxide (CO_2), oxygen(O_2) and water vapor from permeating through packaging materials. The packaging materials contribute physical stability and generate suitable physicochemical environment for products which are essential for attaining an appropriate shelf life. Packaging system with proper barrier, mechanical and thermal properties avoids the deterioration of product because of biological or physicochemical factors and preserves the comprehensive quality at the time of storage as well as handling. The polymers have replaced the traditional materials in the packaging utilizations due to their low cost, lightweight, functionality and ease of processing. The most commonly used polymeric materials for different packaging applications are polypropylene (PP), polyethylene terephthalate (PET), polyethylene (PE), polyvinyl chloride (PVC) and polystyrene (PS). However, the increasing demands in the market have put considerable pressure on the material developers to follow new ways to produce cost-effective packaging materials/films with significantly improved properties [1]. In spite of their greater flexibility, a restricting characteristic of polymer materials in packaging application is their basic penetrability to vapors as well as gases, inclusive of CO_2, O_2 and organic vapors. The gas penetration through the polymeric films contributes a crucial impact on the service accomplishment. However, the polymer barrier performance could be remarkably improved by the incorporation of impermeable lamellar fillers, like graphene and clay, having adequate aspect ratio for altering the diffusion pathway of the gaseous-penetrant molecules. The aforementioned fillers enable the diffusing gas molecules follow more complicated as well as longer paths while

Haleema Saleem and Vikas Mittal, Khalifa University of Science and Technology, Abu Dhabi, UAE

passing through the nanocomposite film [2]. Subsequently, a remarkable reduction in the gas permeability through the material is observed [3].

In the functional materials field, the advancement of polymer nanocomposites (PNCs) is quickly developing as an interdisciplinary research activity whose outcomes can expand the uses of polymers for the immense advantage of a wide range of industries. PNCs are polymers (elastomers, thermosets or thermoplastics) which are reinforced using limited amount (under 5 wt%) of nanosized particles with greater aspect ratios. In the last decade, the PNCs have captivated remarkable research interest worldwide, mainly because of their new and improved thermal, mechanical, optical and electrical characteristics, when compared to their micro and macro correspondents. Mostly, the PNCs are generated by the dispersion of organic or inorganic nanoparticles into either a thermosetting or a thermoplastic polymer matrix. The nanoparticles could be one-dimensional (1-D) disc-shaped (e.g., clay platelets), two-dimensional (2-D) nanofibers (e.g., whiskers or nanotubes) or three-dimensional (3-D) polyhedral and spherical nanoparticles (e.g. colloidal silica). Compared to the conventional micro- or macro-particles (e.g., carbon fibers, glass, talc, etc.), the nanoparticles contribute significant benefits because of the better adhesion between polymer and filler phases, greater surface area as well as aspect ratio and lesser volume loading for obtaining similar properties [4]. Apart from the enhanced properties, it has to be noted that these improvements are achieved without the increase in polymer density and any reduction in optical characteristics. These materials are also noted to have superior barrier properties against the gases (e.g., CO_2, N_2 and O_2), water as well as hydrocarbons. The gas barrier efficiency of the nanocomposites depends mainly on three factors: the intrinsic barrier feature of the polymeric matrix, properties of filler (volume fraction, gas diffusion resistance and aspect ratio) and dispersion quality of the particles [5]. Most importantly, the successful advancement of PNCs is related to the degree of nanofiller exfoliation in the polymer matrix.

The most appropriate polymer properties for the packaging utilization are barrier and thermal properties, morphology, mechanical and optical properties. These properties greatly affect various essential features of the finished packaging material like transparency, strength and the capability to seal off O_2 and water vapor. The advancement of PNCs signifies an advanced opportunity for improving the polymer barrier properties. If the nanoparticles are effectively exfoliated within the polymer lattice, besides the enhanced barrier features, the PNCs typically display improved mechanical properties as well as enhanced thermal stability [4,6], which add additional value for the PNCs utilized as packaging materials. The pioneer commercial focus of the PNCs is for high barrier packaging applications, like food packaging [7]. Thus, the barrier properties of PNCs have been broadly researched in light of an extensive variety of polymer matrices, inclusive of polyethylene [8,9], polyurethane [10], polyethylene terephthalate [11], epoxy [12,13], poly(methyl methacrylate) (PMMA) [14], polyamide [15] and poly(ε-caprolactone) [16], for various intended applications. Besides the investigation of the permeability of CO_2 [9,17], O_2 [9,15, 18-22], N_2 [9,17,23], moisture [9,20,21] (Figure 5.1) and liquid water [24,25], for the coating [25] and food packaging [7] applications, different permeability analyses for other species like helium [13,15], ethanol [26], acetone [27], hydrogen (H_2) [15], methanol [28,29] and toluene [26] were also carried out for the fuel cell [28] and hydrogen storage [13] applications. Remarkable enhancement in the barrier properties by almost two to three orders of magnitude were also stated by many studies [22,30]. Majority of the research studies on the barrier performance of PNCs are based on the tortuous pathway approach, in which the nanofiller phase is considered to be impermeable for liquid as well as gas molecules. The synergy between polymer matrices and nanoplatelets could influence the orientation of molecules, crystallinity and molecular packing near nanoplatelets [31]. Change in the packing of molecules around nanoplatelets might enhance the PNC barrier properties.

In this chapter, gas barrier and packaging applications of different PNCs based on the nanoscale fillers, such as graphene, carbon nanotubes (CNT), montmorillonite (MMT) clay, silicon dioxide (SiO_2) and zinc oxide (ZnO), are analyzed. Recent studies on the gas barrier performance of polymeric nanocomposites based on different polymers like polypropylene (PP), polyethylene terephthalate, polyethylene, polyacrylonitrile (PAN), epoxy, poly(lactic acid) (PLA), poly(methyl methacrylate), polyimide (PI), poly (vinyl alcohol) (PVA) and polycaprolactone (PCL) have been focused on. The impact of the intrinsic properties of different fillers and their dispersion level in the polymer materials on the gas barrier properties of the resulting polymeric nanocomposites has also been analyzed in detail.

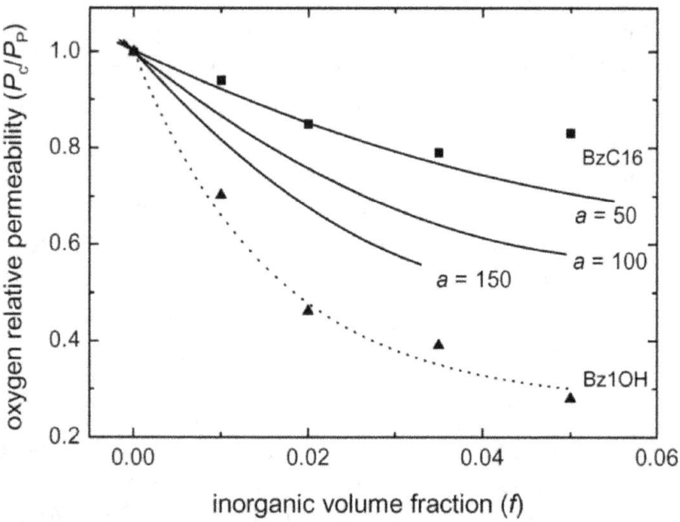

Figure 5.1 Oxygen relative permeability of Bz1OH- and BzC16-epoxy nanocomposites as a function of the inorganic volume fraction. The solid lines are numerical predictions for disk-shaped inclusions with aspect ratios 50, 100, and 150 (diameter/thickness), while the dotted line is simply a guide for the eye. Reproduced from Reference 21 with permission from American Chemical Society.

5.2 Gas Barrier and Packaging Applications

5.2.1 Graphene/Polymer Nanocomposites

Graphene constitutes planar sheets which are one atom thick and has attracted immense research interest in the recent past. It consists of sp^2-bonded carbon atoms which are tightly packed in a hexagonal honeycomb crystal lattice framework. The single-crystalline, defect-free graphene has good mechanical characteristics and excellent transparency, along with gas impermeability [2]. Several attempts have been made to utilize graphene as well as its derivatives as inorganic nanofillers for improving the physical characteristics of PNCs and also to contribute advanced gas barrier features, due to their exclusive properties and better dispersion in typical solvents [32]. The comparatively greater aspect ratio of the graphene-based 2-D materials can surely permit lengthy paths for the gaseous penetrant molecules when compared to alternative nanofillers, provided these materials are effectively dispersed and entirely exfoliated in the polymer matrix [33].

Filler, which has better compatibility with the polymer, generally reduces the permeability, primarily due to the alteration in the transport cross-section as well as the development of the complicated pathways for the gaseous molecules. It is notable that majority of the polymers exhibit inadequate compatibility with graphene, thus, the polymer chains cannot firmly attach to the graphene nanosheets. Consequently, a confined gap will be shaped encompassing the graphene nanosheets. In addition, the graphene nanosheets accumulate as a result of inadequate compatibility with the polymer matrix. The gaseous molecules progress across the confined gap or the permeable polymer matrix, instead of graphene. In this way, these PNCs show a lesser gas barrier effectiveness with non-optimal reduction in the diffusional paths. The graphene surface modification with organic modifiers is an appropriate choice for resolving this problem. Functional groups existing on the surface of graphene nanosheets can enhance the graphene compatibility with the polymer matrices, thereby, contributing improved dispersion and exfoliation of graphene in the PNCs. It is considered that the graphene, without any defect, is not permeable, even to helium [34]. Due to the greater aspect ratio as well as better electron density of carbon rings, this nanofiller is able to repel gaseous molecules and, hence, has a lesser solubility in gases.

The gas barrier performance of graphene based PNCs is certainly influenced by the polymer matrix crystallinity, graphene nanosheets dispersion, orientation and aspect ratio [34]. These factors can be optimized by selecting the appropriate processing technique. Kalaitzidou *et al.* [34] studied the impact of exfoliated graphite nanoplatelets (xGnPs) on the permeability of PP composite. The PNCs were generated by melt blending technique utilizing a twin-screw extruder and injection molding, as a function of aspect ratio of xGnP and its concentration. For studying the influence of the aspect ratio of nanofiller on the xGnP-PP nanocomposite properties, two categories of the nanofillers were utilized. These nanofillers were xGnP-15, with a diameter of 15 μm and a thickness of approximately 10 nm and xGnP-1 with similar thickness and almost 1 μm average diameter. It was noted that the xGnPs' substantial aspect ratio improved the PP O_2 barrier performance by 20%. Further, it was confirmed that carbon black (CB) did not enhance the O_2 barrier performance of PP because of the agglomerates of irregular shape and greater porosity degree inside the aggregates, which did not prohibit the gas molecule diffusion. Both xGnP-1 and vapor grown carbon fiber (VGCF) contributed an identical reduction in the permeability of O_2. The influence of aspect ratio was very noticeable for xGnP-15 which brought about the greatest enhancement in the O_2 barrier performance. Kim and Macosko [35] generated the PNCs filled with graphite nanoplatelets (GnPs) as well as functionalized graphite sheets (FGS) developed by the graphite oxide fractional pyrolysis. Using the melt mixing technique, the dispersion of FGS as well as unexfoliated graphite in poly(ethylene-2,6-naphthalate) (PEN) was studied. It was noted that the H_2 permeability of PEN having 4 wt% FGS reduced by 60%, whereas the similar amount of graphite decreased the H_2 permeability by 25%. Figure 5.2 presents the H_2 permeability at 35 °C temperature through 70 μm thick film specimens. Overall, barrier performance was observed to be decreased by the defect vacancies in FGS sheets, generated through the super-heating preparation method. Additionally, it was also noted that the graphitic flakes were not flawlessly aligned within the film plane, against the composite theory assumption.

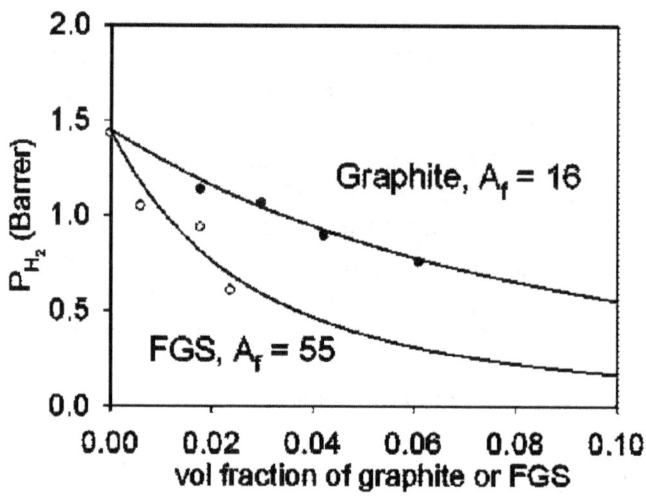

Figure 5.2 H_2 permeability of FGS (open symbols) and graphite (closed symbols) composites at temperature 35 °C. Reproduced from Reference 35 with permission from American Chemical Society.

Expanded graphite (EG) is a layered material, generated when bulk graphite is subjected to solid oxidizers like nitric acid, sulphuric acid, etc. The EG's improved processability enables it to be incorporated into the polymer frameworks, and the low cost as well as scalability of this technique makes EG appealing for a variety of industrial applications. In a study carried out by Prusty and Swain [36], EG was exfoliated in PAN matrix by means of *in-situ* emulsion sonication method. The O_2 transmission rate (OTR) for the EG/PAN nanocomposites was noted to be lower, when compared to pristine PAN till 2.5 psi. Also, it was observed that the permeability of EG/PAN composites remarkably decreased when the content of EG was enhanced from 1% to 4%. For instance, permeability of the nanocomposites was reduced by 13 times with increment in the content of EG. The aforementioned decrease is due to the fact that the graphite nanoparticles perform as a physical impediment, thus, decelerating the gas movement. In another study, Minelli *et al.* [37] illustrated hypothetically that the aligned nanofillers can enhance the gas barrier features significantly. In reality, it is still a crucial challenge for regulating nanoplatelet orientation inside the structural composites.

Jiao *et al.* [38] generated well-ordered magnetic GnP (m-GnP)/epoxy matrix (EP) composites by covering the GnPs with magnetic nanoparticles (Fe_3O_4) and altering GnP alignment in the polymer matrix utilizing a magnetic field. The Fe_3O_4 nanoparticles could be fastened onto the GnP surface by wet chemical co-precipitation technique. From the scanning electron microscopic (SEM) analysis of nanocomposites, it was noted that m-GnPs in the polymer matrix were aligned in the same direction of the magnetic field. In the case of ordered EP/m-GnP nanocomposites generated using 1 wt% m-GnPs, the gas barrier performance enhanced by an order of magnitude than the blank EP composite and 65% greater as compared to the randomly distributed EP/GnP nanocomposites. In accordance with the solubility/diffusion mechanism, two reasons were suggested for the improvement in gas barrier performance of EP/m-GnP composites. To begin with, the m-GnP distributed in the polymer network can unequivocally limit the movement of polymer chains, most likely diminishing the diffusion coefficient of gaseous molecules. Secondly, the m-GnPs with substantial aspect ratio can generate a greatly convoluted pathway for the gaseous molecules. This convoluted way enhances the gaseous molecule diffusion length by re-routing its way through the polymer, at right angles to the direction of thickness, bringing about a lesser rate of transmission across the polymer. Further, the experimental results confirmed that the gas barrier performance diminished remarkably with enhancing gas pressure for the EP/m-GnP nanocomposites, relative to the EP/GnP composites. The aforementioned phenomenon could be attributed to the fragile interface of EP/m-GNP nanocomposite. Fe_3O_4 nanoparticles on the m-GNP surface decreased the interfacial bonding between EP polymer matrix and GNPs, which prompted the initiation of micro-cracks as well as extension under elevated gas pressure at interface.

It is considered that the molecular orientation as well as crystallization are two ways to deal with gas barrier performance improvement. The crystallization diminishes the volume fraction of the penetrable amorphous phase accessible for gaseous transportation, thereby, decreasing the permeability by reducing the total free volume within the polymer matrix [39]. Likewise, the molecular orientation, even with no crystallization, develops the amorphous phase density, thus, diminishing the permeability. However, PET is an exclusion to the guideline in that its amorphous phase may densify at the time of crystallization, subsequently enhancing the amorphous phase permeability and diminishing to certain degree the advantageous effects of crystallization as well as orientation [40]. The PET copolymerization contributes in a different way towards increasing the barrier performance. It is regarded that by the incorporation of GnP, the O_2 barrier performance of PET enhances significantly. The capability of the GnP/PET nanocomposites for O_2 barrier applications to outplay pristine PET is recognizably obvious and in the finest case, for 1.5 wt% GnP, the permeation of O_2 was diminished by over 99%. The process that creates this impact is related to two factors: first one is the constitutional barrier characteristic of the distributed GnP, whose presence within the polymer matrix develops the pathway for the molecular diffusion, and the second one is the greatest level of crystallinity nucleated by GnPs within the polymer matrix. It was also found that the processing conditions regulating the dispersion quality and polymer crystallinity as well as morphology ought to be viewed in combination for optimizing the properties of final materials. The enhanced barrier features of the PNCs because of GnP are accompanied by a greater dimensional strength and enhanced thermal stability. However, the incorporation of GnPs within the polymer also brings about lesser strengths and higher brittleness.

Biopolymers contribute environmental benefits as compared to the petroleum originated polymers because of the fact that these are generated from renewable resources and are biodegradable. Out of the numerous choices, poly(lactic acid) (PLA) is an assuring substitute for the petroleum originated polymers, particularly in food packaging application. In order to use graphene as a barrier, it has been chemically processed for generating graphene oxide (GO) as well as reduced graphene oxide (r-GO) sheets [41]. In the recent past, several research studies have stated the utilization of r-GO as a protective coating and an impenetrable barrier film [42]. r-GO film exhibited extraordinary barrier properties towards water vapor, H_2, NaCl salt, He and HF acid. From different studies, it was confirmed that the GO as well as r-GO sheets exhibited almost identical barrier properties as the defect-free graphene sheets and could be utilized in different food packaging applications [43]. Goh *et al.* [42] developed graphene/PLA nanocomposite films consisting of r-GO or GO films as the core barrier materials for food packaging. An isolated GO film was prepared by means of a filtration procedure prior to reduction of hydroiodic (HI) acid to obtain a r-GO film. The authors took advantage of a sandwiched structure for the graphene/PLA nanocomposite films in which the isolated r-GO or GO film was embedded in the middle of two PLA films to perform as an efficient barrier for food packaging. The water vapor barrier characteristics of the graphene/PLA nanocomposite films were examined at 90% relative humidity (RH) and 38 °C, for analyzing the efficiency of the sandwiched r-GO film as a core barrier. Figure 5.3a presents the water vapor permeability (WVP) results of the materials. Blank PLA exhibited a WVP of 72 $g \cdot mm \cdot m^{-2} \cdot day^{-1} \cdot atm^{-1}$. With the incorporation of the isolated r-GO and GO films as core barriers, the graphene/PLA nanocomposite films demonstrated a more evident water vapor barrier performance with WVP of PLA/r-GO about 8.9 $g \cdot mm \cdot m^{-2} \cdot day^{-1} \cdot atm^{-1}$ and PLA/GO almost 29.6 $g \cdot mm \cdot m^{-2} \cdot day^{-1} \cdot atm^{-1}$. Under both dry (0% RH) and humid (50% RH) environments, the O_2 barrier performance of the films was examined at 23 °C. The O_2 permeability (OP) results are outlined in Figure 5.3b. At RH 0%, the blank PLA demonstrated a OP value of 22.00 $cc \cdot mm \cdot m^{-2} \cdot day^{-1} \cdot atm^{-1}$. A significant decrease of about 99.0% and 97.2% in the oxygen permeability were respectively noted when the isolated r-GO and GO films were inserted as the core barriers. Thus, it was confirmed that the sandwich-type graphene/PLA nanocomposite films with r-GO and GO core barriers demonstrated effective barrier properties towards O_2 and water vapor. Overall, the WVP essentially declined by 87.6% whereas a decrease in OP by twice orders of magnitude was seen with r-GO. Under both moist and dry environments, the lesser OP was also noted to be sustainable. The aforementioned outcomes resulted due to the compact lamellar framework and improved hydrophobicity of sandwiched r-GO film. Also, the limited inter-layer distance between the r-GO sheets and large r-GO lateral dimension generated a broad and convoluted diffusion path, subsequently hindering diffusivity and improving the barrier properties of the nanocomposites.

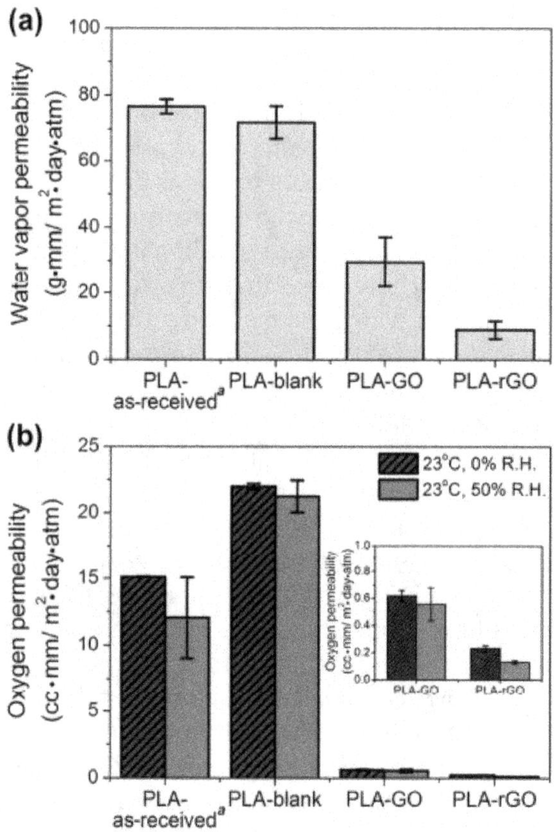

Figure 5.3 (a) WVP and (b) OP of PLA-as-received, PLA-blank and the graphene/PLA nanocomposite films. Reproduced from Reference 42 with permission from American Chemical Society.

GO has the advantageous qualities of an aqueous solution credited to the oxygen consisting functional groups on the basal planes as well as edges of graphene sheets [44]. These functional groups contribute towards homogenous dispersion and complete exfoliation of GO sheets in the polar polymer frameworks and enhance the interfacial bonding remarkably. Zhu *et al.* [45] generated GO from graphite utilizing a modified Hummers' technique and later employed GO as a nanofiller to prepare GO/polyimide (PI) composites using *in-situ* polymerization method. It was noted that the OTR of the PI/r-GO nanocomposite reduced continuously with increase in the loading of GO. It remarkably decreased from 377.78 for pristine PI to 26.07 $cm^3 \cdot m^{-2} \cdot 24\ h^{-1} \cdot atm^{-1}$ for 30 wt% GO loaded nanocomposite, thus, exhibiting a decrease of 93% as compared to pure PI. Morimune *et al.* [44] utilized an ecologically amicable procedure for preparing GO/PMMA nanocomposites, where PMMA was polymerized using emulsion polymerization and integrated with GO utilizing water as the processing medium. Incorporation of 1% w/w of GO to PMMA framework diminished 50% permeability and the nanocomposite having 10% w/w of GO was observed to be totally impermeable. Chen *et al.* [46] prepared a clear gas barrier film made out of GO and poly(vinyl alcohol) (PVA) by consolidated techniques of isothermal re-crystallization and solution blending. Though O_2 molecules passed through the GO/PVA nanocomposite, these adhered to a convoluted diffusion pathway due to the fact that GO sheets are gas impermeable. This resulted in the decrease of the PVA permeability. Moreover, isothermal re-crystallization was utilized to improve the barrier properties. On allowing O_2 to flow through the GO/PVA crystal hybrid framework, it was compelled to take a lengthier diffusional path, generating an extremely low penetrability of the composite. The recrystallized GO/PVA film with 0.07 vol % GO demonstrated an oxygen permeability of $< 5.0 \times 10^{-20}\ cm^3 \cdot cm \cdot cm^{-2} \cdot Pa^{-1} \cdot s^{-1}$. Excellent O_2 barrier performance was, thus, associated with the exclusive hybrid of GO sheets and PVA crystals. Huang *et al.* [3] generated nanocomposites of GO nanosheets (GONS)/PLA. The GONS were observed to be completely exfoliated and randomly dispersed in PLA. At 1.37 vol % GONS loading, the CO_2 and O_2 permeability coefficients of PLA specimens were diminished by around 68% and 45%, respectively. The improved gas barrier efficiency was attributed to the impermeable feature of GONS and the firm PLA-GONS interfacial bonding. In another study, Kim *et al.* [47] utilized distinct graphene layers exfoliated from graphite oxide by means of two procedures, thermal exfoliation and chemical modification, to generate the graphene/thermoplastic polyurethane (TPU) nanocomposites by three unique strategies of dispersion: melt mixing, solution blending and *in-situ* polymerization. The results obtained from the wide-angle X-ray diffraction (WAXRD) and transmission electron microscopy (TEM) confirmed that the solution blending method efficiently dispersed the thin exfoliated layers of graphene in the TPU matrix, as compared to the melt mixing technique. The authors observed that *in-situ* polymerization was not as efficient as solution-mixing in diminishing the penetrability of gas through the composites. The permeability of N_2, $P_{nitrogen}$, of the TPU/graphitic platelets nanocomposite films at 35 °C is presented in Figure 5.4. The permeation of nitrogen (N_2) was observed to be significantly diminished illustrating that the exfoliated sheets of carbon are useful as diffusion barriers in polymer membranes. The graphene dispersion state in the polymer matrix could also be effectively revealed from the reduction in the gas permeability of the nanocomposites.

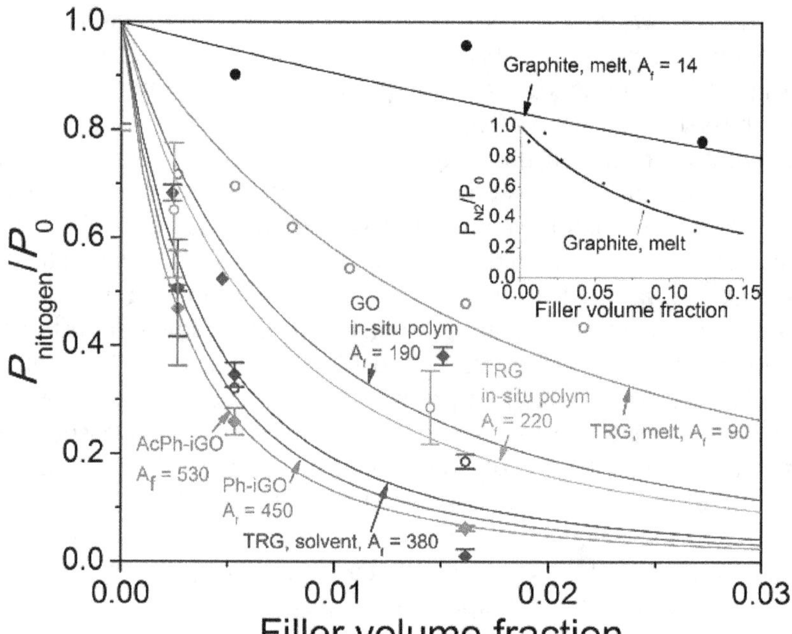

Figure 5.4 N_2 permeability of TPU/graphitic platelets nanocomposite films at 35 °C temperature. The solid curves are forecast based on Lape *et al.* [48]. Reproduced from Reference 47 with permission from American Chemical Society.

5.2.2 Carbon Nanotubes (CNTs)/Polymer Nanocomposites

CNTs, microscopic carbon cylinders, have generated enormous activity in several territories of engineering and science because of their remarkable physicochemical characteristics. The exclusive properties of CNTs and technical advantages that CNT based PNCs provide, including improved thermo-mechanical characteristics, keep on attracting the universal research interest. The diameters of CNTs are in the nanometer scale and these materials are either multi-walled (MWCNT) or single-walled (SWCNT). Due to higher stiffness, aspect ratio and strength at low thickness, the CNTs have captivated the attention of researchers as reinforcing fillers for the preparation of advanced low-weight high-performance polymer based nanomaterials [49]. During the past decade, numerous polymers were utilized as matrices for the generation of CNT based PNCs [50]. The techniques for the dispersion and distribution of the CNTs inside the polymer frameworks perform an essential role in determining the properties of the ultimate product. For the synthesis of CNT based PNCs, different techniques such as solution casting, melt mixing and *in-situ* polymerization have been utilized.

Currently, the petroleum based synthetic polymers are utilized as the packaging materials because of their excellent thermo-mechanical as well as barrier properties and, furthermore, cost-effectiveness. However, these materials are non-biodegradable in nature. Hence, several efforts are in progress for developing the alternate packaging materials, which are ecological friendly. The conventional materials under consideration are poly(glycolic acid), poly(lactic acid) (PLA), polycaprolactone (PCL) and different biopolymer based materials. The biodegradable films generated from the aforementioned materials do not cause any risk to the environment. As a result, several investigations have been undertaken to generate the biodegradable films with physicochemical features corresponding to those of the petroleum-based polymers [51].

PCL is a thermoplastic semi-crystalline biodegradable polyester, and possesses a glass transition temperature of approx. 60 °C. The primary commercial utilization of the PCL polymer is in the production of biodegradable films and bottles. PCL is regularly utilized as an additive for resins in order to enhance their end use properties and processing features. The polymer is employed essentially in adhesives, thermoplastic PUs, resins for the surface coatings and completely biodegradable compostable bags [52]. Khan *et al.* [53] prepared CNT reinforced PCL nanocomposites by compression molding technique. WVP of pristine PCL film was observed to be 1.51 g.mm.m^{-2}.day^{-1}, however the PCL film with 0.2% CNT demonstrated a permeability of 1.08 g.mm.m^{-2}.day^{-1}. Further, the OTR of PCL film was noted to diminish with CNT incorporation. Nevertheless, it was observed that the PCL film's CO_2 transmission rate (CO_2TR) enhanced with CNT addition. The impact of gamma radiation on the pristine PCL films as well as CNT containing PCL nanocomposites was additionally examined and the barrier features of irradiated (10 kGy) and non-irradiated pristine PCL films and 0.2% CNT reinforced PCL nanocomposites were analyzed. PNCs and PCL films demonstrated lesser values of WVP consequent to irradiation and displayed improved water vapor barrier performance. The CO_2TR and OTR of the irradiated polymer films and the PNCs were diminished, as compared to their analogues.

In a related study, Prusty and Swain [54] generated PAN-co-starch nanocomposites with functionalized multi-walled carbon nanotubes (MWCNTs) using *in-situ* polymerization technique. The O_2 permeability analysis of MWCNT/PAN-co-starch nanocomposite was performed to analyze the impact of the concentration of MWCNT. It was observed that at consistent pressure of 1.03 x 10^4 Pa, the flow rate diminished with increment in MWCNT content [55]. The significant decrease in O_2 permeability could be attributed to the MWCNT dispersion inside the PAN-co-starch framework. Further, it was also noted that the thermal stability of the nanocomposites enhanced with diminishing O_2 permeability by increasing content of MWCNT.

The impact of acid treatment of MWCNTs on the permeability of isotactic PP (iPP) was examined by Bikiaris *et al.* [56]. It was observed that the permeability of the O_2 and N_2 gases marginally decreased by the incorporation of the MWCNTs. The pristine iPP exhibited O_2 permeability of almost 7.35 ml.cm.day^{-1}.m^{-2}.atm^{-1}, whereas it was decreased to 5.87 ml.cm.day^{-1}.m^{-2}.atm^{-1} subsequent to the incorporation of 2.5 wt% of MWCNTs. An analogous behavior was also noted in the case of N_2 permeability. Hence, it was confirmed that the incorporation of nanoparticles into the polymer improved its barrier features by compelling the gaseous molecules to pass through a complicated pathway. Similar findings have also been reported for numerous layered silicate/polymer nanocomposites as well as for iPP utilizing fumed silica nanoparticles [57]. In the case of the treated nanotubes, it might be anticipated that the O_2 permeability will be greater because of the poor adhesion between the iPP and MWCNTs because of the fact that the treatment enabled the nanotubes hydrophilic by the generation of polar surface groups. Nevertheless, such phenomenon was not observed by the authors and only for the MWCNTs treated for about 60 minutes, the rate of permeability was higher as compared to pristine iPP. The study additionally confirmed that the improved thermal and oxidative stability of the generated nanocomposites at elevated temperatures are apparently the after effect of the lesser permeation of O_2 in the bulk framework. Therefore, O_2 could react quickly at the surface of iPP in the primary oxidation stages, however, it was blocked from entering the matrix bulk because of the presence of MWCNTs.

5.2.3 Clay/Polymer Nanocomposites

Clay is a widely available and comparatively cheap resource; and has been utilized progressively as a nano-additive in several PNC systems. It is a natural nanomaterial and recognized to be less toxic or completely non-toxic to human health. The enhancement in the polymer properties includes improved thermal stability, barrier properties, environmental degradation stability, flammability resistance and mechanical properties. The investigation on the gas barrier properties of the clay based PNCs relates to mainly O_2, CO_2 and N_2 for packing, storing as well as the preservation of various products [58]. As mentioned earlier, the efficiency of the PNCs with diffusion barrier properties depends on three main factors, i.e., properties of filler (volume fraction and aspect ratio), dispersion quality and inherent barrier property of the polymer matrix. It is also considered that the synergy between the specific polymer matrices and nanoplatelets could influence the molecular orientation, crystallinity and molecular packing in the vicinity of nanoplatelets [59]. The change in the packing of molecules around the nanoplatelets might additionally improve the barrier features of the PNCs. It is also notable that the morphology, in addition to the distribution of clay nanoplatelets in polymer lattices, is another major factor influencing the gas barrier performance [60].

The organically modified layered silicates have been extensively examined for several years to improve the barrier features of polymeric materials [15]. The most widely utilized clay filler for the improvement in the gas barrier performance of PNCs is montmorillonite (MMT), due to its extensive cation exchange ability [61]. MMT is a hydrated form of alumina-silicate layered clay and its framework comprises of an octahedral sheet of aluminum oxide (AlO_6), interposing in the middle of two tetrahedral silicon oxide (SiO_4) layers to generate a platelet framework [61]. MMT is a natural clay where certain aluminum ions, situated in the middle of the plate-shaped layers, exchange with different cations (eg. Mg, Li or Fe) for generating a negative charge at the exterior of the platelet. Organo-modified MMT (O-MMT) is the MMT treated using an organic modifier and possesses good compatibility with organic materials. There are three attainable morphologies in clay based PNC frameworks: exfoliated, phase-separated and intercalated [62]. It is considered that the *in-situ* polymerization is a successful approach for attaining exfoliated clay nanoplatelets in polymer framework, relative to solution and melt intercalation. Dai *et al.* [63] generated O-MMT/EP nanocomposites by *in-situ* ring-opening polymerization (ROP). Quaternary alkyl-ammonium and alkyl-phosphonium salts were utilized for the generation of organophilic clay. Quaternary alkyl-ammonium salt-modified MMT and quaternary alkyl-phosphonium salt-modified MMT at 7 wt% clay loading demonstrated 66% and 86.1% decrease, respectively, in O_2 permeability, as compared to pristine EP. In the study by Nazarenko *et al.* [64], clay/PS nanocomposites were prepared by *in-situ* polymerization of styrene in the presence of MMT modified with zwitterionic octadecyl-dimethyl betaine (C_{18}DMB), vinyl-benzyl-dimethyl-dodecylammonium chloride (VDAC) and unmodified sodium MMT clay. The C_{18}DMB–MMT/PS and VDAC–MMT/PS nanocomposite frameworks exhibited lower O_2 permeability, as compared to the Na^+-MMT/PS nanocomposite with same clay content.

The improved gas and water vapor barrier performance of PNCs containing clay nanoparticles emphasizes major benefit for packaging applications. Undoubtedly, in numerous commercial uses, it is confirmed that the nanoclay particles can reduce the permeability by almost 75%. The PE films are normally utilized (approximately 33% of the world's manufactured plastic) for packaging, as these provide a powerful humidity barrier; nevertheless, these suffer from high O_2 permeability. Osman and Atallah [65] generated high-density polyethylene (HDPE) nano- as well as micro-composites containing platy and spherical fillers by melt mixing method. The authors observed that the plate-like fillers unequivocally diminished the permeability coefficient of polymer, whereas the spherical-shaped fillers had no impact on the polymer. At 3 vol% of an appropriate O-MMT, the HDPE permeability coefficient was decreased by 35%, without the presence of a compatibilizer. Later, Osman *et al.* [66] incorporated O-MMT with various surface coverages as well as packing fractions of alkyl chains and generated HDPE/O-MMTs nanocomposites. O_2 penetrability of 2.8 vol% O-MMT nanocomposites diminished to approximately half, as compared to HDPE.

De Abreu *et al.* [67] generated low density polyethylene (LDPE) and PP based nanocomposites containing modified organic clay, Cloisite 15 A, for preparing films to be utilized in packaging [67]. The filler improved the film's barrier properties against O_2, CO_2 and water vapor, while retaining the transparency and enhancing the hot packaging operation. As an illustration, in PE films with clay filler, the permeability of O_2 was 12.5% lesser as compared to the pristine PE matrix. Identical to other nanocomposites, the reduction in the O_2 permeability took place due to the fact that nanoparticles perform as a physical barrier hindering the mobility of O_2 through the film, diminishing its movement rate and increasing the distance that should be covered for crossing the film. In another study, Xie *et al.* [68] revealed that the O_2 and CO_2 barrier performance of LDPE enhanced by 4 times and 7 times, respectively, consequent to the addition of 0.5 wt% octadecyl-trimethyl ammonium (OTA)-MMT. Further, it was noted that at 2 wt% loading of OTA-MMT, the LDPE WVP was reduced by a factor of 2.5.

For acquiring a properly exfoliated clay based PNC framework for non-polar polymers, the interfacial agents or compatibilizers containing polar groups must be added to the nanocomposite, thereby, enhancing the interfacial synergy between non-polar polymer and clay layers [69]. For instance, Arunvisut *et al.* [70] generated clay/LDPE nanocomposites by melt mixing LDPE with organoclay and compatibilizing agent polyethylene grafted maleic anhydride (PE-g-MA). The authors observed that the O_2 permeability of clay/LDPE nanocomposites diminished by 24% as the content of clay enhanced to 7 wt%. In another study by Lotti *et al.* [71], HDPE was blended in a twin-screw extruder with a compatibilizer (HDPE-g-MA) and organophilic clay. The generated nanocomposites were noted to have 60% decrease in the O_2 permeability by the addition of 5 wt% O-MMT, as compared to pristine HDPE. Durmus *et al.* [72] utilized low molecular weight oxidized PEs (OxPEs) with various acid numbers as well as molecular weights as the compatibilizing agents in linear LDPE (LLDPE)/organoclay nanocomposite. The authors noted that the incorporation of oxidized PE compatibilizer to

LLDPE decreased the O_2 permeability by 44%, as compared to pure LLDPE. O_2 permeability of the nanocomposites generated using Ox-PE was also observed to be lesser as compared to the composites generated using typical compatibilizer like PE-g-MA. In a study by Carvalho *et al.* [73], the impact of ammonium quaternary based compounds as coupling agent in polyethylene nanocomposites was analyzed. Out of different coupling agents utilized, cetyl-propyl-dimethyl-ammonium chloride generated the most effective exfoliation of vermiculite (VER) filler. At 3 wt% filler loading, the O_2 permeability of VER/LLDPE-g-MA based nanocomposites decreased by 18%.

The surfactants can also be utilized to enhance the clay dispersibility and permeability behavior. The amphiphilic block as well as random PE copolymers were utilized as surfactants for improving the O-MMT dispersion in PE framework [74]. The surfactants enhanced the separation between the clay layers to various degrees. The increment in the degree of exfoliation remarkably improved the gas barrier performance and enhanced the strength as well as stiffness of the material [74].

In the packaging industry, PET is one of the most widely utilized polymers. It is, thus, greatly advantageous to enhance its barrier features. In a study by Sanchez-Garcia *et al.* [75], O-MMT/PET nanocomposites were prepared by melt mixing process. At 0% RH, the O_2 permeation of the PNC decreased by 55% with 5% O-MMT, as compared to pristine PET. Hamzehlou and Katbab [76] prepared PET composites incorporating various concentrations of O-MMT nanofiller using melt mixing technique. The O_2 permeability of 5 wt% O-MMT filled PET was observed to be 7.890 $cm^3.m^{-2}.day^{-1}.bar^{-1}$, whereas the value was 25.33 $cm^3.m^{-2}.day^{-1}.bar^{-1}$ for pristine PET. Kim and Kim [77] prepared clay/PET nanocomposites using N-methyl-diethanol amine based organoclays by *in-situ* polymerization process. The nanocomposite exhibited a delaminated and intercalated framework and was noted to display a two-fold decrease in O_2 permeation with just 1 wt% clay, relative to pristine PET. In a study by Choi *et al.* [78], clay/PET nanocomposites were generated via *in-situ* polymerization process using catalyst supported clay. The nanocomposites demonstrated a 11.3-15.6 fold decrease in O_2 permeation with 1-5 wt% clay, relative to pure polymer.

Due to the distinction in the polarity between clay (polar) and PP (non-polar), diverse strategies have been developed to enhance the miscibility, such as, modification of the fillers using organic compounds [79] and use of functional compounds employed as compatibilizing agents [80]. Villaluenga *et al.* [81] prepared PP membranes with natural as well as organically modified MMT clay. The solubility, diffusivity and permeability of N_2, O_2 and helium were examined for the filled as well as unfilled membranes over the temperature range from 25 °C to 65 °C. Permeability coefficient of O-MMT filled PP membrane for O_2 and N_2 exhibited a reduction of 77% and 89%, respectively, while the maximal decrease in the penetrability of helium was 39%. In case of helium, the decreased diffusivity was accountable for the decrease in permeation. For O_2 and N_2, both solubility and diffusivity were decreased by the presence of filler particles.

The gaseous molecular transportation through a polymer relies on the free volume existing because of the chain segment displacement and chain packing. Ammala *et al.* [82] utilized positron annihilation lifetime spectroscopy (PALS) strategy for analyzing the variation in free volume when various O-MMTs were integrated into nylon MXD6. The authors noted that the PALS free volume information had good inter-relationship with the determined OTR in all cases. The incorporation of the nanofiller brought about an enhancement in the glass transition temperature and crystallinity, as compared to the pristine polymer. Further, the authors observed that the enhanced barrier properties of the clay based PNCs have been essentially because of an enhancement in the crystallinity degree of the polymer.

Arora *et al.* [83] generated clay/PS nanocomposites by melt mixing technique utilizing O-MMT and MMT. The TEM and XRD analyses demonstrated that the nanofiller was exfoliated or/and intercalated to a greater degree as compared to MMT. In another study by Wilson *et al.* [84], O-MMT/poly (ethylene-co-vinyl acetate) nanocomposites were generated with different clay loadings. The authors observed that the gas barrier performance of PNCs diminished when the clay loadings were above 3 wt%, essentially because of the clay aggregation at greater loadings. Patra and Swain [85] prepared organically modified clay/poly(methylmethacrylate-co-acrylonitrile) copolymer hybrid nanostructured materials by *in-situ* intercalative emulsion polymerization technique with no emulsifier. With a 3 wt% clay content, the O_2 permeability was decreased by 51.5%, as compared to the pristine copolymer.

Polyamide 6-MMT films with a broad range of clay content (0-18 wt%) were analyzed by Picard *et al.* [15]. From the TEM and wide-angle X-ray scattering (WAXS) analyses, it was observed that a satisfactory dispersion of nanofiller was attained for all films. The nanocomposites displayed effective barrier performance for dioxygen, helium, water and dihydrogen, when compared to pristine PA film. Relative permeability was observed to be independent of the diffusing molecule. The permeation in the nanocomposites was related to the clay dispersion as well as content. From the quantitative depiction of the MMT particle distribution, the capacity of various geometrical models to portray the exploratory relative permeability was also noted.

Osman *et al.* [21] generated O-MMT/epoxy nanocomposites and examined their O_2 as well as water vapor penetrability. It was noted that the exfoliated alumino-silicate layers resulted in a barrier for the penetration of gas molecules, whereas the polymer intercalated tactoids did not contribute to the penetration barrier ability. Gas penetration across the composite films was related to the volume fraction of the inorganic fraction of the O-MMT. The addition of a small volume fraction of the nanoparticles in the polymer lattice diminished the permeability coefficient when the interface between the two heterogeneous phases was effectively developed. It was noted that the long alkyl chains in the filler surface modification improved the intercalation of polymer, yet enhanced the permeability coefficient possibly because of the phase separation at the interface between non-polar surface of the inclusions and polar polymer. The O_2 permeability coefficients of the 3.5 vol% composites were analyzed and the relative permeability was plotted against filler basal spacing (*d*-spacing) in the composites, as presented in Figure 5.5. It was, thus, concluded that the intercalated fraction of the nanofiller did not contribute much to the decrease in permeability and the exfoliated layers solely contributed to the penetration barrier. It was also found that the water vapor transmission rate across the nanocomposite films was more effected by the composite-permeant synergy and, thereby, the hydrophobicity of monolayer around the surface of filler inclusions.

Transparent as well as flexible polymeric super-barrier packaging materials have turned out to be progressively critical in recent past. Layer-by-layer assembly contributes an effortless strategy for the generation of layered, thin clay/polymer super-barrier films. This involves the development of thin multi-functional films through exposure of a substrate to aqueous anionic as well as cationic mixtures. Barrier protection, transparency and transparency make such

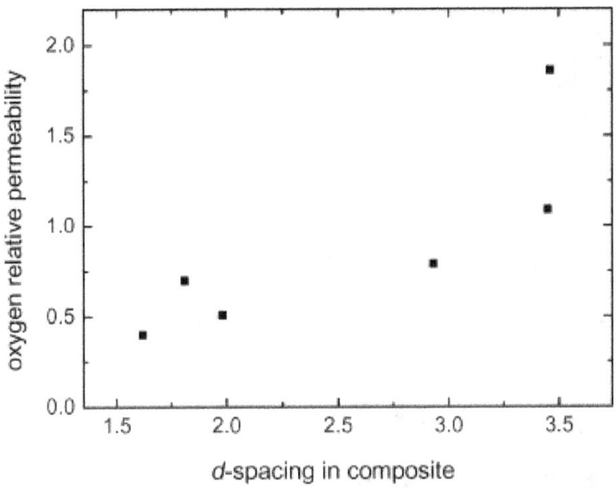

Figure 5.5 O_2 relative penetrability of the composites plotted as a function of the d-spacing. Reproduced from Reference 21 with permission from American Chemical Society.

films promising candidates for different packaging applications. Priolo *et al.* [86] reported that thin nanocomposite films generated using layer-by-layer approach exhibited O_2 permeability an order of magnitude lesser than EVOH (ethylene vinyl alcohol) and SiOx. The clay containing LbL films have been extensively studied because of their O_2 barrier properties, strength, transparency and anti-flammability. It was proposed that clay enabled the penetrating molecules to be captured inside the polymer-filled channels, as demonstrated by Cussler [87]. It was noted that a five QL (one deposition series of polyethylenimine (PEI)-poly(acrylic acid) (PAA)-PEI-MMT is termed as a quadlayer (QL)) film consisted of 26.2 wt% clay. The OTR of the films diminished with the number of QLs deposited on PET film with 179 μm thickness, as presented in Figure 5.6. A four QL film displayed an O_2 transmission rate equivalent to or lower than the limit of detection of commercial instrumentation. The observed barrier efficiency with four layers of clay is exceptional for a PNC, particularly due to a thickness of 51 nm. The capacity to generate low permeability films with use of water based solutions signifies the potential of the layer-by-layer method as a moderately cheap and commercially feasible framework for different packaging applications, where the filler distribution and composite performance can be accurately controlled and tuned.

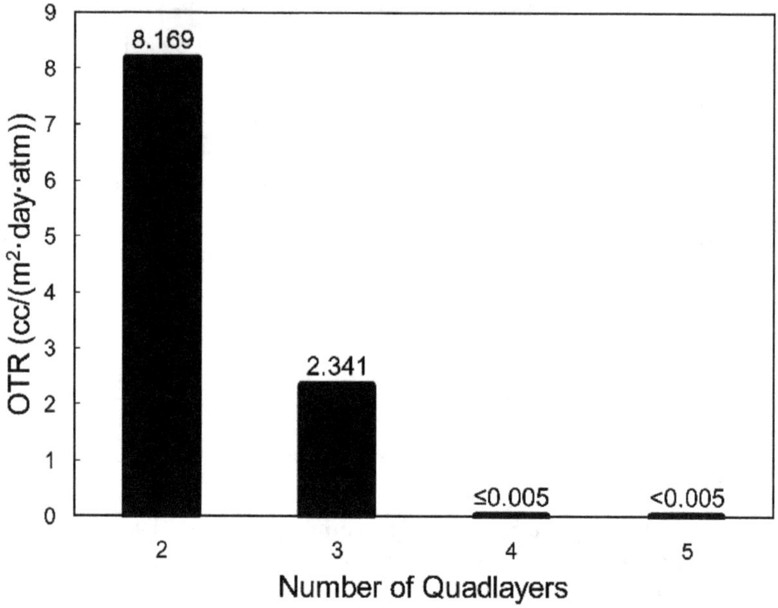

Figure 5.6 OTR as a function of the number of QLs deposited on PET film with 179 μm thickness. Reproduced from Reference 86 with permission from American Chemical Society.

5.2.4 SiO_2/Polymer Nanocomposites

PLA is an aliphatic biodegradable polyester generated from entirely renewable resources. The use of PLA has remarkably intensified due to the simplicity of its processing by traditional techniques (like blow molding, fiber spinning, injection molding, thermoforming, film extrusion and film forming), which allow its use in numerous applications, including food packaging [88]. Nevertheless, due to its poor mechanical and barrier properties, the use of PLA for food packaging application is generally restricted to films, containers and cups for lower shelf-life products [89]. The properties of PLA have also been significantly enhanced by the addition of nanoparticles in the polymeric network. For instance, Sanchez-Garcia *et al.* [90] reported that the incorporation of inorganic nanoparticles to the polymer resulted in the significant enhancement of barrier and mechanical properties. The impact of various types of silica nanoparticles, which are hydrophilic, on the properties of PLA has also been studied [91]. Fumed silica, an amorphous white powder, is generated by the chlorosilane hydrolysis in an O_2–H_2 gas flame. It has non-irritating and non-toxic nature, as compared to the crystalline silica, which is carcinogenic. These particles display hydrophilicity and surface energy because of their greater surface area per unit weight as well as the plentiful silanol groups existing on their surface. It was reported that the mixing of hydrophilic silica nanoparticles in the polymer matrices results in their accumulation. Thus, surface functionalization is performed in order to block the silanol end-groups. The surface modification results in a hydrophobic surface, which is less reactive as compared to the hydrophilic material, with improved distribution in the polymer, along with lower moisture adsorption and capacity to change rheological performance.

Pilic *et al.* [92] analyzed the properties of PLA nanocomposites consisting of silica nanoparticles (SiO_2) as filler. The pristine PLA films and PLA nanocomposites with various proportion of hydrophobic fumed silica were generated by solution casting. Analysis of the O_2, N_2 and CO_2 gas permeability was carried out as per the Lyssy technique. It was noted that the permeability of gases was fundamentally influenced by hindered gas molecule diffusion over the polymer film because of the silica nanoparticle incorporation. For instance, the incorporation of 0.2% SiO_2 decreased the O_2, N_2 and CO_2 permeability by approximately 50%, as compared to pristine PLA. As an example, CO_2 permeability was decreased to 5.45 x 10^{-18} m^2 Pa^{-1} s^{-1} from 8.16 x10^{-18} m^2 Pa^{-1} s^{-1} with 0.2 wt% silica content. Further, for 0.5 wt%, 1.0 wt% and 2.0 wt% of silica, the CO_2 permeability of the PNCs decreased to 4.45 x 10^{-18} m^2 Pa^{-1} s^{-1}, 4.14 x 10^{-18} m^2 Pa^{-1} s^{-1} and 4.78 x 10^{-18} m^2 Pa^{-1} s^{-1}, respectively. Nevertheless, in the case of higher loadings (3 wt% and 5 wt%), the CO_2 permeability was greater (6.10 x 10^{-18} m^2 Pa^{-1} s^{-1} and 6.25 x 10^{-18} m^2 Pa^{-1} s^{-1}, respectively), yet at the same time lesser than pristine PLA. Analogous patterns were noted for all analyzed gases. The barrier performance could be attributed to the efficient filler dispersion in the polymer. Moreover, the crystallinity degree could also influence the decrease in permeability. The gas transportation performance, thus, corresponded to binary stage model- impermeable crystalline state, distributed in the permeable amorphous state [93].

As mentioned earlier, isotactic-PP is an attractive commodity thermoplastic because of its properties and low cost. It is broadly utilized in the generation of bottles, films for food packaging, fibers and tubes. The capacity of SiO_2 to enhance the gas barrier properties of i-PP was assessed by Vladimirov *et al.* [57]. The authors prepared fumed silica/i-PP nanocomposites through melt mixing technique utilizing a twin-screw extruder. For enhancing the degree of nanoparticle dispersion, maleic anhydride grafted polypropylene was utilized as a compatibilizing agent. It was noted that the O_2 and N_2 permeability was diminished with increment in the content of SiO_2. As i-PP consists of amorphous as well as crystalline regions and the crystalline areas are regarded impervious for gases, thus, the diffusion of gas molecules occurs through the amorphous regions. The authors also reported in another study that SiO_2 nanoparticles could perform as nucleating agents for the crystallization of i-PP [94]. For decreasing the impact of crystallization, the generated films were water quenched from a temperature 200 °C. As the nanocomposites are regarded as multi-phase frameworks, the presence of various phases (amorphous, crystalline and inorganic) might bring about complex phenomena for the gas permeation process. The presence of SiO_2 nanoparticles generated a labyrinth, a complex pathway that the gaseous molecules should go through for passing through the entire film width. Thus, the decrease in the permeability was due to the presence of such more convoluted pathways.

5.2.5 ZnO/Polymer Nanocomposites

As mentioned earlier, out of the different PNCs, the composites based on i-PP have found extensive application due to low cost and optimal mechanical properties, along with processability and chemical stability [95]. PP films are employed in different packaging applications due to their excellent barrier characteristics, transparency and strength [96]. The biaxial elongation of the material either in a single step or sequentially is a basic measure for the generation of PP films for packaging. The orientation process includes a morphological conversion of the spherulites into a system of microfibrils, thus, decreasing the PP gas permeation. To enhance the properties further, i-PP has been often incorporated with a variety of fillers, such as clay, zinc oxide (ZnO) [97], calcium carbonate [98] and silica [99], among others [100-102]. Lepot *et al.* [100] generated PP nanocomposites with 2.0, 5.0 and 7.5 wt% of ZnO nanoparticles (elongated and spherical) by melt mixing technique. Subsequent to melt mixing, the PP/ZnO nanocomposites were compression molded as well as biaxially extended. The impact of the incorporation of ZnO on the barrier properties of bi-axially oriented PP (BOPP)/ZnO nanocomposites was analyzed. The addition of ZnO to the oriented polymer matrix was observed to enhance the barrier features. This improvement was due to the consolidation of two phenomena. The first phenomenon is the reduction in the field accessible for diffusion, a consequence of impermeable nanoparticles substituting penetrable polymer and polymer chain biaxial orientation. Second phenomenon is the increment in the distance which a permeant should move for crossing the film as it is compelled to pursue a complicated pathway over the impermeable nanoparticles. For i-PP/ZnO nanocomposites, the O_2 permeability ranged from 51 to 55 $mL.m^{-2}.day^{-1}$; and 30 to 35 $mL.m^{-2}.day^{-1}$ for BOPP/ZnO nanocomposites, with no relationship with the concentration of ZnO. In addition, the spherical and elongated ZnO nanoparticles in BOPP and iPP nanocomposites demonstrated a similar pattern with respect to O_2 permeability.

Elen *et al.* [103] studied the properties of PCL/ZnO nanocomposites with ZnO particles of assorted morphologies, i.e., ZnO nano-disks, nano-rods, and nano-spheres. It was noted that the nanocomposites with ZnO nano-rods demonstrated superior gas barrier properties (O_2 penetrability reduced by almost 65%). The penetrability of the PNCs consisting spherical ZnO was remarkably high. The penetrability of the nanocomposites with ZnO nano-disks was also higher, relative to the ZnO nano-rods. Also, the convoluted pathway model of Nielsen, as observed in clay based PNCs, was less evident. These findings could be described by two justifications. Primarily, in spite of their distinct 2-D morphology, the ZnO nano-disks did not attain a greater aspect ratio as the clay platelets which can be exfoliated into platelets of nanometer size with an aspect ratio of 100–1500 and a thickness of almost 1 nm. Due to the fact that the hexagonal disks were larger than the other ZnO nanoparticles with different morphologies, these overall were less uniformly dispersed within the polymer matrix. Therefore, the possibility that a diffusing gas molecule encounters a ZnO nanoparticle was lesser for the PNCs with hexagonal shaped disks. Secondly, the compression molding technique was not effective in orienting the particles at right angles to the gaseous stream.

5.3 Summary and Outlook

In this chapter, gas barrier and packaging applications of PNCs consisting of different nanoscale fillers, such as graphene, CNT, MMT clay, SiO_2 and ZnO have been reviewed. From the analysis of the literature studies, it was noted that the barrier performance of polymers is remarkably improved by the incorporation of nanofillers as the gaseous molecules are compelled to follow a complicated pathway in the polymer. It was observed that the gas barrier efficiency of the nanocomposites depends mainly on three major factors: intrinsic barrier features of the polymeric matrix, properties of filler (volume fraction, gas diffusion resistance and aspect ratio) and dispersion quality. This chapter also highlighted several processes that have been utilized to generate PNCs, including melt mixing, *in-situ* polymerization, solution intercalation and layer-by-layer assembly.

References

1. Sun, L., Boo, W. J., Clearfield, A., Sue, H. J., and Pham, H. Q. (2008) Barrier properties of model epoxy nanocomposites. *Journal of Membrane Science*, **318**(1), 129-136.
2. Yoo, B. M., Shin, H. J., Yoon, H. W., and Park, H. B. (2014) Graphene and graphene oxide and their uses in barrier polymers. *Journal of Applied Polymer Science*, **131**(1), 39628.
3. Huang, H. D., Ren, P. G., Xu, J. Z., Xu, L., Zhong, G. J., Hsiao, B. S., and Li, Z. M. (2014) Improved barrier properties of poly (lactic acid) with randomly dispersed graphene oxide nanosheets. *Journal of Membrane Science*, **464**, 110-118.
4. Zeng, Q. H., Yu, A. B., Lu, G. Q., and Paul, D. R. (2005) Clay-based polymer nanocomposites: research and commercial development. *Journal of Nanoscience and Nanotechnology*, **5**(10), 1574-1592.
5. Moller, M. W., Kunz, D. A., Lunkenbein, T., Sommer, S., Nennemann, A., and Breu, J. (2012) UV-cured, flexible, and transparent nanocomposite coating with remarkable oxygen barrier. *Advanced Materials*, **24**(16), 2142-2147.
6. Alexandre, M., and Dubois, P. (2000) Polymer-layered silicate nanocomposites: preparation, properties and uses of a new class of materials. *Materials Science and Engineering: R: Reports*, **28**(1), 1-63.
7. Sorrentino, A., Gorrasi, G., and Vittoria, V. (2007) Potential perspectives of bio-nanocomposites for food packaging applications. *Trends in Food Science & Technology*, **18**(2), 84-95.
8. Chaiko, D. J., and Leyva, A. A. (2005) Thermal transitions and barrier properties of olefinic nanocomposites. *Chemistry of Materials*, **17**(1), 13-19.
9. Shah, R. K., Krishnaswamy, R. K., Takahashi, S., and Paul, D. R. (2006) Blown films of nanocomposites prepared from low density polyethylene and a sodium ionomer of poly (ethylene-co-methacr-

ylic acid). *Polymer*, **47**(17), 6187-6201.

10. Choi, W. J., Kim, S. H., Kim, Y. J., and Kim, S. C. (2004) Synthesis of chain-extended organifier and properties of polyurethane/clay nanocomposites. *Polymer*, **45**(17), 6045-6057.
11. Lai, M., and Kim, J. K. (2005) Effects of epoxy treatment of organoclay on structure, thermo-mechanical and transport properties of poly (ethylene terephthalate-co-ethylene naphthalate)/organoclay nanocomposites. *Polymer*, **46**(13), 4722-4734.
12. Kim, J. K., Hu, C., Woo, R. S., and Sham, M. L. (2005) Moisture barrier characteristics of organoclay–epoxy nanocomposites. *Composites Science and Technology*, **65**(5), 805-813.
13. Ogasawara, T., Ishida, Y., Ishikawa, T., Aoki, T., and Ogura, T. (2006) Helium gas permeability of montmorillonite/epoxy nanocomposites. *Composites Part A: Applied Science and Manufacturing*, **37**(12), 2236-2240.
14. Meneghetti, P., and Qutubuddin, S. (2006) Synthesis, thermal properties and applications of polymer-clay nanocomposites. *Thermochimica Acta*, **442**(1), 74-77.
15. Picard, E., Vermogen, A., Gérard, J. F., and Espuche, E. (2007) Barrier properties of nylon 6-montmorillonite nanocomposite membranes prepared by melt blending: influence of the clay content and dispersion state: consequences on modelling. *Journal of Membrane Science*, **292**(1), 133-144.
16. Gain, O., Espuche, E., Pollet, E., Alexandre, M., and Dubois, P. (2005) Gas-barrier properties of poly (ε-caprolactone)/clay nanocomposites: Influence of the morphology and polymer/clay interactions. *Journal of Polymer Science, Part B: Polymer Physics*, **43**(2), 205-214.
17. Lee, J. H., Jung, D., Hong, C. E., Rhee, K. Y., and Advani, S. G. (2005) Properties of polyethylene-layered silicate nanocomposites prepared by melt intercalation with a PP-g-MA compatibilizer. *Composites Science and Technology*, **65**(13), 1996-2002.
18. Ray, S. S., Yamada, K., Okamoto, M., and Ueda, K. (2003) New polylactide-layered silicate nanocomposites. 2. Concurrent improvements of material properties, biodegradability and melt rheology. *Polymer*, **44**(3), 857-866.
19. Osman, M. A., Mittal, V., and Lusti, H. R. (2004) The Aspect Ratio and Gas Permeation in Polymer-Layered Silicate Nanocomposites. *Macromolecular Rapid Communications*, **25**(12), 1145-1149.
20. Osman, M. A., Mittal, V., Morbidelli, M., and Suter, U. W. (2003) Polyurethane adhesive nanocomposites as gas permeation barrier. *Macromolecules*, **36**(26), 9851-9858.
21. Osman, M. A., Mittal, V., Morbidelli, M., and Suter, U. W. (2004) Epoxy-layered silicate nanocomposites and their gas permeation properties. *Macromolecules*, **37**(19), 7250-7257.
22. Triantafyllidis, K. S., LeBaron, P. C., Park, I., and Pinnavaia, T. J.

(2006) Epoxy-clay fabric film composites with unprecedented oxygen-barrier properties. *Chemistry of Materials,* **18**(18), 4393-4398.

23. Wang, Y., Zhang, H., Wu, Y., Yang, J., and Zhang, L. (2005) Preparation and properties of natural rubber/rectorite nanocomposites. *European Polymer Journal,* **41**(11), 2776-2783.
24. Wu, T. M., and Liu, C. Y. (2005) Poly(ethylene 2,6-naphthalate)/layered silicate nanocomposites: fabrication, crystallization behavior and properties. *Polymer,* **46**(15), 5621-5629.
25. Sun, Q., Schork, F. J., and Deng, Y. (2007) Water-based polymer/clay nanocomposite suspension for improving water and moisture barrier in coating. *Composites Science and Technology,* **67**(9), 1823-1829.
26. Jiang, T., Wang, Y. H., Yeh, J. T., and Fan, Z. Q. (2005) Study on solvent permeation resistance properties of nylon6/clay nanocomposite. *European Polymer Journal,* **41**(3), 459-466.
27. Doppers, L. M., Breen, C., and Sammon, C. (2004) Diffusion of water and acetone into poly (vinyl alcohol)–clay nanocomposites using ATR-FTIR. *Vibrational Spectroscopy,* **35**(1), 27-32.
28. Thomassin, J. M., Pagnoulle, C., Bizzari, D., Caldarella, G., Germain, A., and Jerome, R. (2006) Improvement of the barrier properties of Nafion® by fluoro-modified montmorillonite. *Solid State Ionics,* **177**(13), 1137-1144.
29. Karthikeyan, C. S., Nunes, S. P., and Schulte, K. (2006) Barrier properties of poly(benzimidazole)-layered silicates nanocomposite materials. *Advanced Engineering Materials,* **8**(10), 1010-1015.
30. Takahashi, S., Goldberg, H. A., Feeney, C. A., Karim, D. P., Farrell, M., O'leary, K., and Paul, D. R. (2006) Gas-barrier properties of butyl rubber/vermiculite nanocomposite coatings. *Polymer,* **47**(9), 3083-3093.
31. Nielsen, L. E. (1967) Models for the permeability of filled polymer systems. *Journal of Macromolecular Science - Chemistry,* **1**(5), 929-942.
32. Geim, A. K. (2009) Graphene: status and prospects. *Science,* **324**(5934), 1530-1534.
33. Paul, D. R., and Robeson, L. M. (2008) Polymer nanotechnology: Nanocomposites. *Polymer,* **49**(15), 3187-3204.
34. Kalaitzidou, K., Fukushima, H., and Drzal, L. T. (2007) Multifunctional polypropylene composites produced by incorporation of exfoliated graphite nanoplatelets. *Carbon,* **45**(7), 1446-1452.
35. Kim, H., and Macosko, C. W. (2008) Morphology and properties of polyester/exfoliated graphite nanocomposites. *Macromolecules,* **41**(9), 3317-3327.
36. Prusty, G., and Swain, S. K. (2011) Synthesis and characterization of conducting gas-barrier polyacrylonitrile/graphite nanocompos-

ites. *Polymer Composites*, **32**(9), 1336-1342.

37. Minelli, M., Baschetti, M. G., and Doghieri, F. (2009) Analysis of modeling results for barrier properties in ordered nanocomposite systems. *Journal of Membrane Science*, **327**(1), 208-215.
38. Jiao, W., Shioya, M., Wang, R., Yang, F., Hao, L., Niu, Y., Liu, W., Zheng, L., Yuan, F., Wan, L., and He, X. (2014) Improving the gas-barrier properties of Fe_3O_4/graphite nanoplatelet reinforced nanocomposites by a low magnetic field induced alignment. *Composites Science and Technology*, **99**, 124-130.
39. Lin, J., Shenogin, S., and Nazarenko, S. (2002) Oxygen solubility and specific volume of rigid amorphous fraction in semicrystalline poly (ethylene terephthalate). *Polymer*, **43**(17), 4733-4743.
40. Liu, R. Y. F., Hiltner, A., and Baer, E. (2004) Free volume and oxygen transport in cold-drawn polyesters. *Journal of Polymer Science, Part B: Polymer Physics*, **42**(3), 493-504.
41. Goh, K., Setiawan, L., Wei, L., Si, R., Fane, A. G., Wang, R., and Chen, Y. (2015) Graphene oxide as effective selective barriers on a hollow fiber membrane for water treatment process. *Journal of Membrane Science*, **474**, 244-253.
42. Goh, K., Heising, J. K., Yuan, Y., Karahan, H. E., Wei, L., Zhai, S., Koh, J.X., Htin, N.M., Zhang, F., Wang, R., Fane, A. G., Dekker, M., Dehghani, F., and Chen, Y. (2016) Sandwich-architectured poly(lactic acid)-graphene composite food packaging films. *ACS Applied Materials & Interfaces*, **8**(15), 9994-10004.
43. Yang, Y. H., Bolling, L., Priolo, M. A., and Grunlan, J. C. (2013) Super gas-barrier and selectivity of graphene oxide-polymer multilayer thin films. *Advanced Materials*, **25**(4), 503-508.
44. Morimune, S., Nishino, T., and Goto, T. (2012) Ecological approach to graphene oxide reinforced poly (methyl methacrylate) nanocomposites. *ACS Applied Materials & Interfaces*, **4**(7), 3596-3601.
45. Zhu, J., Lim, J., Lee, C. H., Joh, H. I., Kim, H. C., Park, B., You, N. H., and Lee, S. (2014) Multifunctional polyimide/graphene oxide composites via in situ polymerization. *Journal of Applied Polymer Science*, **131**(9), 40177.
46. Chen, J. T., Fu, Y. J., An, Q. F., Lo, S. C., Zhong, Y. Z., Hu, C. C., Lee, K. R., and Lai, J. Y. (2014) Enhancing polymer/graphene oxide gas-barrier film properties by introducing new crystals. *Carbon*, **75**, 443-451.
47. Kim, H., Miura, Y., and Macosko, C. W. (2010) Graphene / polyurethane nanocomposites for improved gas barrier and electrical conductivity. *Chemistry of Materials*, **22**(11), 3441-3450.
48. Lape, N. K., Nuxoll, E. E., and Cussler, E. L. (2004) Polydisperse flakes in barrier films. *Journal of membrane science*, **236**(1), 29-37.
49. Ganesan, Y., Peng, C., Lu, Y., Loya, P. E., Moloney, P., Barrera, E., Yakobson, B.I., Tour, J. M., Ballarini, R., and Lou, J. (2011) Interface

toughness of carbon nanotube reinforced epoxy composites. *ACS applied materials & interfaces*, **3**(2), 129-134.

50. Coleman, J. N., Cadek, M., Blake, R., Nicolosi, V., Ryan, K. P., Belton, C., Fonseca, A., Nagy, J.B., Gun'ko, Y. K., and Blau, W. J. (2004) High performance nanotube-reinforced plastics: Understanding the mechanism of strength increase. *Advanced Functional Materials*, **14**(8), 791-798.
51. Le Tien, C., Letendre, M., Ispas-Szabo, P., Mateescu, M. A., Delmas-Patterson, G., Yu, H. L., and Lacroix, M. (2000) Development of biodegradable films from whey proteins by cross-linking and entrapment in cellulose. *Journal of Agricultural and Food Chemistry*, **48**(11), 5566-5575.
52. Khan, R. A., Salmieri, S., Dussault, D., Sharmin, N., and Lacroix, M. (2012) Mechanical, barrier, and interfacial properties of biodegradable composite films made of methylcellulose and poly (caprolactone). *Journal of Applied Polymer Science*, **123**(3), 1690-1697.
53. Khan, R. A., Dussault, D., Salmieri, S., Safrany, A., and Lacroix, M. (2013) Mechanical and barrier properties of carbon nanotube reinforced PCL-based composite films: Effect of gamma radiation. *Journal of Applied Polymer Science*, **127**(5), 3962-3969.
54. Prusty, G., and Swain, S. K. (2013) Dispersion of multiwalled carbon nanotubes in polyacrylonitrile-co-starch copolymer matrix for enhancement of electrical, thermal, and gas-barrier properties. *Polymer composites*, **34**(3), 330-334.
55. Patra, S. K., Prusty, G., and Swain, S. K. (2012) Ultrasound assisted synthesis of PMMA/clay nanocomposites: Study of oxygen permeation and flame retardant properties. *Bulletin of Materials Science*, **35**(1), 27-32.
56. Bikiaris, D., Vassiliou, A., Chrissafis, K., Paraskevopoulos, K. M., Jannakoudakis, A., and Docoslis, A. (2008) Effect of acid treated multi-walled carbon nanotubes on the mechanical, permeability, thermal properties and thermo-oxidative stability of isotactic polypropylene. *Polymer Degradation and Stability*, **93**(5), 952-967.
57. Vladimirov, V., Betchev, C., Vassiliou, A., Papageorgiou, G., and Bikiaris, D. (2006) Dynamic mechanical and morphological studies of isotactic polypropylene/fumed silica nanocomposites with enhanced gas-barrier properties. *Composites Science and Technology*, **66**(15), 2935-2944.
58. Choudalakis, G. A., Kalo, H., Breu, J., and Gotsis, A. D. (2014) CO_2 gas-barrier properties in polymer nanocomposite coatings containing Li-hectorite clays. *Journal of Applied Polymer Science*, **131**(18), 40805.
59. Incarnato, L., Scarfato, P., Russo, G. M., Di Maio, L., Iannelli, P., and Acierno, D. (2003) Preparation and characterization of new melt

compounded copolyamide nanocomposites. *Polymer,* **44**(16), 4625-4634.

60. Zeppa, C., Gouanve, F., and Espuche, E. (2009) Effect of a plasticizer on the structure of biodegradable starch/clay nanocomposites: Thermal, water-sorption, and oxygen-barrier properties. *Journal of Applied Polymer Science,* **112**(4), 2044-2056.
61. De Azeredo, H. M. (2009) Nanocomposites for food packaging applications. *Food Research International,* **42**(9), 1240-1253.
62. Tang, X. Z., Kumar, P., Alavi, S., and Sandeep, K. P. (2012) Recent advances in biopolymers and biopolymer-based nanocomposites for food packaging materials. *Critical Reviews in Food Science and Nutrition,* **52**(5), 426-442.
63. Dai, C. F., Li, P. R., and Yeh, J. M. (2008) Comparative studies for the effect of intercalating agent on the physical properties of epoxy resin-clay based nanocomposite materials. *European Polymer Journal,* **44**(8), 2439-2447.
64. Nazarenko, S., Meneghetti, P., Julmon, P., Olson, B. G., and Qutubuddin, S. (2007) Gas-barrier of polystyrene montmorillonite clay nanocomposites: effect of mineral layer aggregation. *Journal of Polymer Science, Part B: Polymer Physics,* **45**(13), 1733-1753.
65. Osman, M. A., and Atallah, A. (2004) High-density polyethylene micro-and nanocomposites: Effect of particle shape, size and surface treatment on polymer crystallinity and gas permeability. *Macromolecular Rapid Communications,* **25**(17), 1540-1544.
66. Osman, M. A., Rupp, J. E., and Suter, U. W. (2005) Gas permeation properties of polyethylene-layered silicate nanocomposites. *Journal of Materials Chemistry,* **15**(12), 1298-1304.
67. de Abreu, D. P., Losada, P. P., Angulo, I., and Cruz, J. M. (2007) Development of new polyolefin films with nanoclays for application in food packaging. *European Polymer Journal,* **43**(6), 2229-2243.
68. Xie, L., Lv, X. Y., Han, Z. J., Ci, J. H., Fang, C. Q., and Ren, P. G. (2012) Preparation and performance of high-barrier low density polyethylene/organic montmorillonite nanocomposite. *Polymer-Plastics Technology and Engineering,* **51**(12), 1251-1257.
69. Ishida, H., Campbell, S., and Blackwell, J. (2000) General approach to nanocomposite preparation. *Chemistry of Materials,* **12**(5), 1260-1267.
70. Arunvisut, S., Phummanee, S., and Somwangthanaroj, A. (2007) Effect of clay on mechanical and gas-barrier properties of blown film LDPE/clay nanocomposites. *Journal of Applied Polymer Science,* **106**(4), 2210-2217.
71. Lotti, C., Isaac, C. S., Branciforti, M. C., Alves, R. M., Liberman, S., and Bretas, R. E. (2008) Rheological, mechanical and transport properties of blown films of high density polyethylene nanocomposites. *European Polymer Journal,* **44**(5), 1346-1357.

72. Durmus, A., Woo, M., Kasgoz, A., Macosko, C. W., and Tsapatsis, M. (2007) Intercalated linear low density polyethylene (LLDPE)/clay nanocomposites prepared with oxidized polyethylene as a new type compatibilizer: structural, mechanical and barrier properties. *European Polymer Journal*, **43**(9), 3737-3749.
73. Carvalho, J. W. C., Sarantopoulos, C., and Innocentini-Mei, L. H. (2010) Nanocomposites-based polyolefins as alternative to improve barrier properties. *Journal of Applied Polymer Science*, **118**(6), 3695-3700.
74. Osman, M. A., Rupp, J. E., and Suter, U. W. (2005) Effect of non-ionic surfactants on the exfoliation and properties of polyethylene-layered silicate nanocomposites. *Polymer*, **46**(19), 8202-8209.
75. Sanchez-Garcia, M. D., Gimenez, E., and Lagaron, J. M. (2007) Novel PET nanocomposites of interest in food packaging applications and comparative barrier performance with biopolyester nanocomposites. *Journal of Plastic Film & Sheeting*, **23**(2), 133-148.
76. Hamzehlou, S., and Katbab, A. A. (2007) Bottle-to-bottle recycling of PET via nanostructure formation by melt intercalation in twin screw compounder: Improved thermal, barrier, and microbiological properties. *Journal of Applied Polymer Science*, **106**(2), 1375-1382.
77. Kim, S. H., and Kim, S. C. (2007) Synthesis and properties of poly (ethylene terephthalate)/clay nanocomposites by in situ polymerization. *Journal of Applied Polymer Science*, **103**(2), 1262-1271.
78. Choi, W. J., Kim, H. J., Yoon, K. H., Kwon, O. H., and Hwang, C. I. (2006) Preparation and barrier property of poly (ethylene terephthalate)/clay nanocomposite using clay-supported catalyst. *Journal of Applied Polymer Science*, **100**(6), 4875-4879.
79. Kotek, J., Kelnar, I., Studenovsky, M., and Baldrian, J. (2005) Chlorosulfonated polypropylene: preparation and its application as a coupling agent in polypropylene–clay nanocomposites. *Polymer*, **46**(13), 4876-4881.
80. Modesti, M., Lorenzetti, A., Bon, D., and Besco, S. (2006) Thermal behaviour of compatibilised polypropylene nanocomposite: effect of processing conditions. *Polymer Degradation and Stability*, **91**(4), 672-680.
81. Villaluenga, J. P. G., Khayet, M., Lopez-Manchado, M. A., Valentin, J. L., Seoane, B., and Mengual, J. I. (2007) Gas transport properties of polypropylene/clay composite membranes. *European Polymer Journal*, **43**(4), 1132-1143.
82. Ammala, A., Pas, S. J., Lawrence, K. A., Stark, R., Webb, R. I., and Hill, A. J. (2008) Poly(m-xylene adipamide)-montmorillonite nanocomposites: effect of organo-modifier structure on free volume and oxygen barrier properties. *Journal of Materials Chemistry*, **18**(8), 911-916.

83. Arora, A., Choudhary, V., and Sharma, D. K. (2011) Effect of clay content and clay/surfactant on the mechanical, thermal and barrier properties of polystyrene/organoclay nanocomposites. *Journal of Polymer Research*, **18**(4), 843-857.
84. Wilson, R., Plivelic, T. S., Aprem, A. S., Ranganathaiagh, C., Kumar, S. A., and Thomas, S. (2012) Preparation and characterization of EVA/clay nanocomposites with improved barrier performance. *Journal of Applied Polymer Science*, **123**(6), 3806-3818.
85. Patra, S. K., and Swain, S. K. (2012) Effect of organoclays on the thermal, mechanical, and oxygen barrier properties of poly (methylmethacrylate-co-acrylonitrile)/clay nanocomposites. *Polymer Composites*, **33**(5), 796-802.
86. Priolo, M. A., Gamboa, D., Holder, K. M., and Grunlan, J. C. (2010) Super gas-barrier of transparent polymer-clay multilayer ultrathin films. *Nano Letters*, **10**(12), 4970-4974.
87. Cussler, E. L., Hughes, S. E., Ward, W. J., and Aris, R. (1988) Barrier membranes. *Journal of Membrane Science*, **38**(2), 161-174.
88. Zhang, J., Lou, J., Ilias, S., Krishnamachari, P., and Yan, J. (2008) Thermal properties of poly (lactic acid) fumed silica nanocomposites: experiments and molecular dynamics simulations. *Polymer*, **49**(9), 2381-2386.
89. Silvestre, C., Duraccio, D., and Cimmino, S. (2011) Food packaging based on polymer nanomaterials. *Progress in Polymer Science*, **36**(12), 1766-1782.
90. Sanchez-Garcia, M. D., Lopez-Rubio, A., and Lagaron, J. M. (2010) Natural micro and nanobiocomposites with enhanced barrier properties and novel functionalities for food biopackaging applications. *Trends in Food Science & Technology*, **21**(11), 528-536.
91. Wen, X., Zhang, K., Wang, Y., Han, L., Han, C., Zhang, H., Chen, S., and Dong, L. (2011) Study of the thermal stabilization mechanism of biodegradable poly (L-lactide)/silica nanocomposites. *Polymer International*, **60**(2), 202-210.
92. Pilic, B. M., Radusin, T. I., Ristic, I. S., Silvestre, C., Lazic, V. L., Balos, S. S., and Duraccio, D. (2016) Hydrophobic silica nanoparticles as reinforcing filler for poly (lactic acid) polymer matrix. *Hemijska Industrija*, **70**(1), 73-80.
93. Drieskens, M., Peeters, R., Mullens, J., Franco, D., Lemstra, P. J., and Hristova-Bogaerds, D. G. (2009) Structure versus properties relationship of poly (lactic acid). I. Effect of crystallinity on barrier properties. *Journal of Polymer Science, Part B: Polymer Physics*, **47**(22), 2247-2258.
94. Papageorgiou, G. Z., Achilias, D. S., Bikiaris, D. N., and Karayannidis, G. P. (2005) Crystallization kinetics and nucleation activity of filler in polypropylene/surface-treated SiO_2 nanocomposites. *Thermochimica Acta*, **427**(1), 117-128.

95. Vassiliou, A., Bikiaris, D., Chrissafis, K., Paraskevopoulos, K. M., Stavrev, S. Y., and Docoslis, A. (2008) Nanocomposites of isotactic polypropylene with carbon nanoparticles exhibiting enhanced stiffness, thermal stability and gas-barrier properties. *Composites Science and Technology*, **68**(3), 933-943.
96. Lee, J. W., Son, S. M., and Hong, S. I. (2008) Characterization of protein-coated polypropylene films as a novel composite structure for active food packaging application. *Journal of Food Engineering*, **86**(4), 484-493.
97. Tang, J., Wang, Y., Liu, H., and Belfiore, L. A. (2004) Effects of organic nucleating agents and zinc oxide nanoparticles on isotactic polypropylene crystallization. *Polymer*, **45**(7), 2081-2091.
98. Chan, C. M., Wu, J., Li, J. X., and Cheung, Y. K. (2002) Polypropylene/calcium carbonate nanocomposites. *Polymer*, **43**(10), 2981-2992.
99. Garcia, M., van Vliet, G., Jain, S., Schrauwen, B., Sarkissov, A., van Zyl, W. E., and Boukamp, B. (2004) Polypropylene/SiO_2 nanocomposites with improved mechanical properties. *Reviews on Advanced Materials Science*, **6**(2), 169-175.
100. Lepot, N., Van Bael, M. K., Van den Rul, H., D'Haen, J., Peeters, R., Franco, D., and Mullens, J. (2011) Influence of incorporation of ZnO nanoparticles and biaxial orientation on mechanical and oxygen barrier properties of polypropylene films for food packaging applications. *Journal of Applied Polymer science*, **120**(3), 1616-1623.
101. Choi, W. Y., Lee, C. M., and Park, H. J. (2006) Development of biodegradable hot-melt adhesive based on poly-ε-caprolactone and soy protein isolate for food packaging system. *LWT-Food Science and Technology*, **39**(6), 591-597.
102. Choudalakis, G., and Gotsis, A. D. (2009) Permeability of polymer/clay nanocomposites: a review. *European Polymer Journal*, **45**(4), 967-984.
103. Elen, K., Murariu, M., Peeters, R., Dubois, P., Mullens, J., Hardy, A., and Van Bael, M. K. (2012) Towards high-performance biopackaging: barrier and mechanical properties of dual-action polycaprolactone/zinc oxide nanocomposites. *Polymers for Advanced Technologies*, **23**(10), 1422-1428.

6

Deep Eutectic Solvents for Gas Separation, EOR and Other Applications

6.1 Introduction

In chemical processes, the selection of appropriate solvent is very important as it may correspond to almost 80% of the entire volume of chemicals employed in the process [1]. The application of solvents presents various health, safety as well as environmental challenges inclusive of process safety hazards, eco-toxicity, human impact and waste management issues [2]. Following the fundamentals of green chemistry, majority of the organic solvents do not achieve the pre-requisite for their application in the green technology because of the fact that these solvents have greater volatility and inherent toxicity [1]. Thus, the usage of solvents is required to be optimized as well as minimized for generating the processes with minimal operational and environmental concerns.

Over the past several decades, many efforts have been made for replacing the organic solvents. The approaches include employing no solvent at all, usage of easily recyclable systems like supercritical carbon dioxide ($scCO_2$) or fluorous solvents, utilization of involatile systems like ionic liquids (ILs), low melting mixtures (LMMs) and deep eutectic solvents (DESs) [3]. The ILs are salts that are mostly liquid at temperature lesser than 100 °C. The first IL, ethylammonium nitrate ($[EtN^{+}H_3][NO_3^{-}]$), was reported in 1914 [4]. Ever since, exponential growth in the studies concerning ILs has been observed [5]. The greatest benefit of using ILs is that these can be tuned by mixing various anions (Cl^{-}, BF_4^{-}, PF_6^{-}, NTf_2^{-}, etc.) and cations (generally imidazolium based cations) [6]. Nevertheless, their disadvantage is the questionable environmental acceptability with regards to their synthesis and application [5,7]. Many research reports have revealed the poor biodegradability and toxicity of most ILs [8]. Further, their preparation demands enormous quantity of solvents and salts for completing the anion exchange. These disadvantages along with their higher price restrict the industrial evolution of ILs.

Haleema Saleem and Vikas Mittal, Khalifa University of Science and Technology, Abu Dhabi, UAE

In order to overcome these drawbacks, a new class of IL counterpart, i.e., the DESs were discovered and developed recently.

DESs contain a combination of two or more components, which have the ability to correlate with each other [9]. Despite the fact that these low transition temperature mixtures (LTTMs) share several properties and characteristics with ILs, they portray a different class of solvents [10]. Contrary to ILs, which consist of one kind of distinct cation and anion, the DESs are generated from a eutectic mixture of Bronsted and Lewis acids as well as bases. DESs consist of different cationic and/or anionic species and possess a great depression in the melting point, when compared to pure materials [10]. The reduction in DESs' freezing point is linked to the co-operation between mixture components, mainly of hydrogen bonding character [11]. Among the various DES components, choline chloride (ChCl) is regarded as the most common component (Figure 6.1). ChCl is a non-toxic and biodegradable salt and is very similar to the B vitamins. The DESs have numerous benefits when compared to the traditional ILs, since they are: (1) less expensive, (2) easy to synthesize as

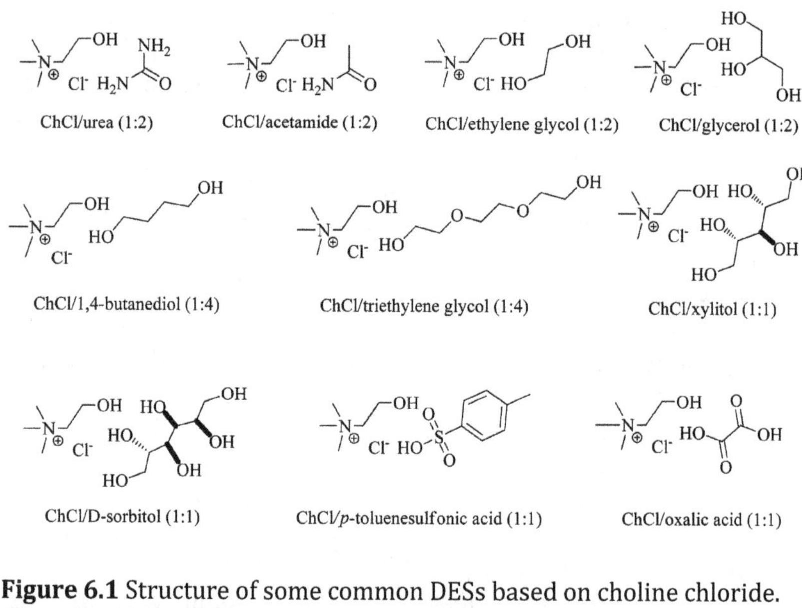

Figure 6.1 Structure of some common DESs based on choline chloride. Reproduced from Reference 13 with permission from American Chemical Society.

materials purification is not required, (3) non- volatile and (4) mostly non-toxic and biodegradable [12]. The term DES indicates the liquids which are close to the mixtures' eutectic content, i.e. molar ratio of ingredients that contributes the least melting point. The DESs are formed as a result of a complexation between a hydrogen bond acceptor (HBA) or a halide salt and a hydrogen bond donor (HBD). Figure 6.1 also depicts the structures of common DESs based on choline chloride [13]. DESs can be also be generated from natural sources (known as natural DESs), specifically through primary metabolites like sugars, amino acids and organic acids. The diversification of attainable consolidations of the starting materials contributes an effective tool for governing the physical characteristics of DESs, which also makes the DESs more suitable than conventional ionic liquids in terms of performance and material properties.

The idea of DES was initially proposed by Abbott *et al.* [14] as a combination of two or more constituents, which generates a eutectic mixture. As mentioned above, the "deep eutectic solvent" term indicates a combination of HBD and a halide salt to generate the liquid [15]. The aforementioned liquid was designated as deep eutectic solvent for discriminating the DES from IL that consists of particularly distinct anion. The eutectic occurrence was initially established by means of a combination of ChCl and urea (molar ratio 1:2) with melting points of 302 °C and 133 °C, respectively. As a result, a eutectic mix with a melting point of 12 °C was generated. Lithium chloride and copper (II) oxide were favorably dissolved in the DES. The utilization of DESs attracted other researchers as their physicochemical properties were same as ILs [16]. As a result, a range of liquids generated from eutectic mix of HBD and salts have now been developed. The Figure 6.2 also illustrates the schematic of schematic diagram of the solid–liquid boundaries of a mixture of two solids forming the DES depending on the composition of the mixture [17]. DESs are expressed by the generic formula $Cat^{+}X^{-}zY$, where Cat^{+} denotes any sulfonium, phosphonium or ammonium cation, X^{-} represents a Lewis base (mostly halide anion), Y is a Bronsted or Lewis acid, and z denotes the number of Y molecules. Complex anion species is generated between X^{-} and either a Bronsted or Lewis acid Y [10].

During the past decade, a quick advancement in the DESs as designer solvents for different applications has been observed [18]. As mentioned earlier, the DESs, as advanced kind of green solvents, have several distinguished characteristics like excellent thermal stability, high viscosity and less vapor pressure [19]. Due to these significant benefits of DESs, there were used in applications in various fields such as gas absorption, enhanced oil recovery (EOR), preparation of materials, solvent development/reaction medium, electro-chemistry, catalytic processes, hydro-metallurgy, etc. [20]. In this review, modern developments made in the field of DESs have been focused along with the different applications of DESs. In addition to this, an overview of the properties as well as the different types of DESs employed has also ben presented. Based on the available literature, it can be suggested that the DESs are potential candidates for gas separation, EOR, generation of new materials as well as novel structures and also in the field of metal extraction and metal oxides processing.

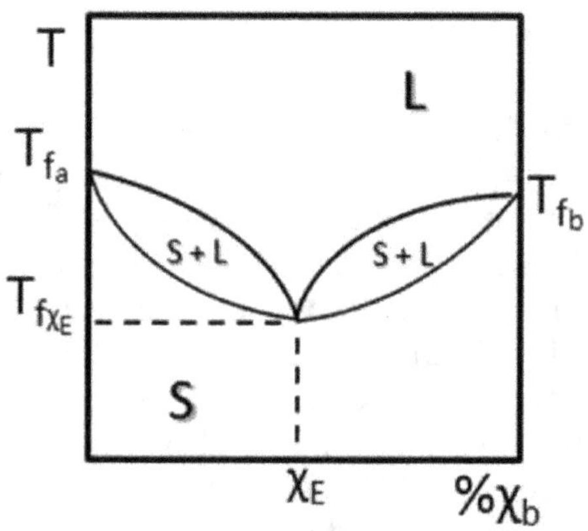

Figure 6.2 Schematic diagram of the solid-liquid phase of a mixture of two solids determined by the mixture composition. Reproduced from Reference 17 with permission from American Chemical Society.

6.2 Properties

DESs are considered to be similar to ILs in reference to the ease of being adoptable in different operations [21], which can be accomplished by the appropriate combination of HBDs and salts. Subsequently, a work specific DES can be designed, which is able to meet the physico-chemical properties necessary for specific application areas [6]. Due to the fact that the DESs have the ability to dissolve certain metal oxides (MOs), these solvents present modern appealing avenues in chemistry exclusively [22]. When compared to other solvents, the DESs possess high surface tensions, viscosities, excellent conductivities, and lower vapor pressures [23]. Due to such advantageous properties of DESs, these solvents have found numerous applications in the oil and gas industries. The particular physicochemical properties of the DESs, which can be tailored according to the requirement are freezing point, viscosity, surface tension, density and conductivity. These properties are reviewed in the below sections.

6.2.1 Freezing Point

DES's freezing point (T_f) is controlled by the proper selection of HBD, the composition as well as the organic salt. Overall, the strength of the interaction between the anion and HBD plays a vital role. As an illustration, together with urea, the T_f of a choline salt based DES reduced in the order $F^- > NO3^- > Cl^- > BF4^-$, revealing an interrelationship with hydrogen bond strength. The NMR spectroscopy confirmed the presence of hydrogen bond network inside the eutectic mixtures. In addition, a deep cross-interrelationship between the fluoride anion (F^-) of choline fluoride and the NH_2 groups existing in urea molecule in a DES was observed by two-dimensional hetero-nuclear NOE (HOESY) measurements [14]. Abbott *et al.* [24] suggested that the T_f of HBD-salt eutectic mixtures is determined by (1) the mechanism how the couple HBD-anion interacts, (2) lattice energies of DESs, and (3) variation in entropy originating from the liquid phase generation. Kareem *et al.* [25] prepared novel eutectic mixtures (Figure 6.3) based on benzyl-triphenyl-phosphonium chloride and methyl-triphenyl-phosphonium bromide as salts and 2,2,2-triflouracetamide, ethylene glycol and glycerin as HBDs for different molar ratios of salt/HBD. As seen in Figure 6.3, it was observed that the salt/HBD molar ratio that contributed the eutectic mixture with least melting temperatures depended on both the HBD and the salt.

6.2.2 Viscosity

For the practical applications, the viscosity of a mixture or a substance is of great significance. Majority of the DESs display comparatively greater viscosity at room temperature (greater than 100 cP), when compared to the molecular solvents [26]. As the viscosity is linked to the free volume as well as the probability of discovering holes of appropriate dimensions for the solvent molecules or ions to progress into [26], hence, it can be confirmed that the ionic size also influences the viscosity. Low viscosity liquids are generated employing quaternary ammonium cations like fluorinated HBD and ethylammonium. Further, the presence of a comprehensive hydrogen bond network causes lesser mobility of free species inside the mixture, which provides higher viscosity to the DESs. In addition to this, other forces like van der Waals or electrostatic interactions may generate higher viscosities [6]. The water content, composition and temperature also have a substantial significance [15]. As the viscosity of DESs reduce with increasing temperature, it was confirmed that these follow an Arrhenius-like behavior [27]. Compatible with the recognized density characteristics of various ChCl-glycerol mixtures, the viscosity reduced with rising ChCl quantity,

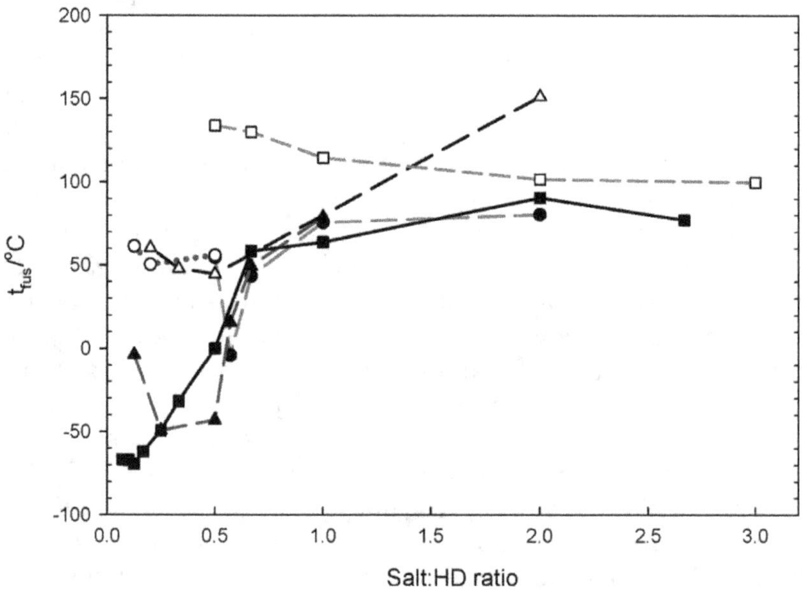

Figure 6.3 Freezing temperatures of different phosphonium-based eutectic mixtures as a function of salt/HBD molar ratio. ●, methyl salt/glycerine; ▲, methyl salt/ethylene glycol; ■, methyl salt/2,2,2-triflouracetamide; ○, benzyl salt/glycerine; △, benzyl salt/ethylene glycol; and □, benzyl salt/2,2,2-triflouracetamide. Reproduced from Reference 25 with permission from American Chemical Society.

though the mixtures with ethylene glycol exhibit the opposite effect [28]. From these examples, it was noted that the viscosity is dependent on the composition as well as resulting interactions. Considering the fact that water has lesser viscosity than the DESs, the viscosity value reduces with rising the content of water [29]. Yadav and Pandey [29] also reported the dynamic viscosities of ChCl/urea [molar ratio 1:2, known as reline] and its aqueous mixtures in the temperature range 293.15-363.15 K, as shown is Figure 6.4. The authors observed that the variation of dynamic viscosity of aqueous mixtures of reline with temperature was better described by a Vogel–Fulcher–Tamman (VFT) model, as opposed to an Arrhenius-like behavior.

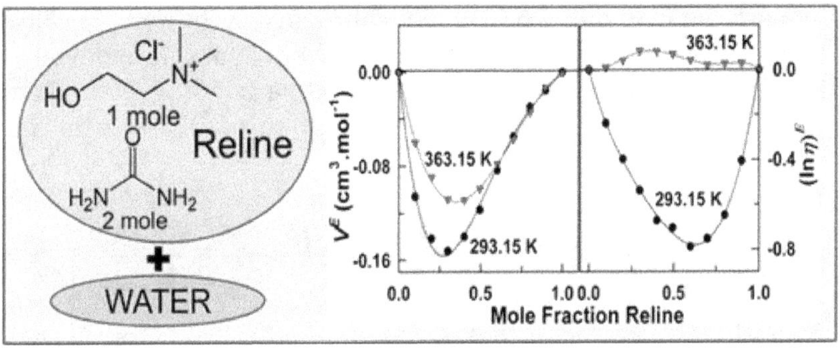

Figure 6.4 Dynamic viscosities of ChCl/urea [molar ratio 1:2, known as reline] and its aqueous mixtures, at a range of temperature 293.15-363.15 K. Reproduced from Reference 29 with permission from American Chemical Society.

6.2.3 Density

It is obvious that the density has a significant importance for the process design applications [30]. Similarly, the pressure and temperature impact on density are necessary for the advancement of appropriate equations of state which contribute immensely in the determination of thermo-dynamic characteristics needed for the industrial processes [31]. Mostly, the DESs display greater density values, when compared to water. The density of the DES is determined by the molecular organization as well as packing of DES [20]. Same as ILs, it was observed that the DESs consists of holes as well as empty vacancies that control the density. Generally, the density values of DESs reduce with temperature increase [32]. Moreover, the water content also affects the density, where the density reduces with increasing the water percentage [33]. In addition to this, the ratio of organic salt and HBD also influences the density. As an illustration, the incorporation of ChCl to glycerol causes a reduction in density, which can also be illustrated in terms of hole theory and free volume [32,15]. Figure 6.5 illustrates the change in density with the amount of HBD for ChCl:LA (lactic acid) based and ChCl:G (glycerol) based DESs. It was noted that for both kinds of DESs, absolute deviation enhanced with increasing the HBD mole fraction in the DES [36]. For analyzing the effect of generation of ester on the properties of DESs, the density of DES ChCl:glutaric acid, prepared by the heating and grinding methods was analyzed, and the results are demonstrated in Figure 6.6 [37]. The difference in the density values obtained using two methods was small (average absolute deviation of 0.15%). The authors also concluded that the grinding method was the more suitable method for the generation of DESs as the formation of ester was not favored.

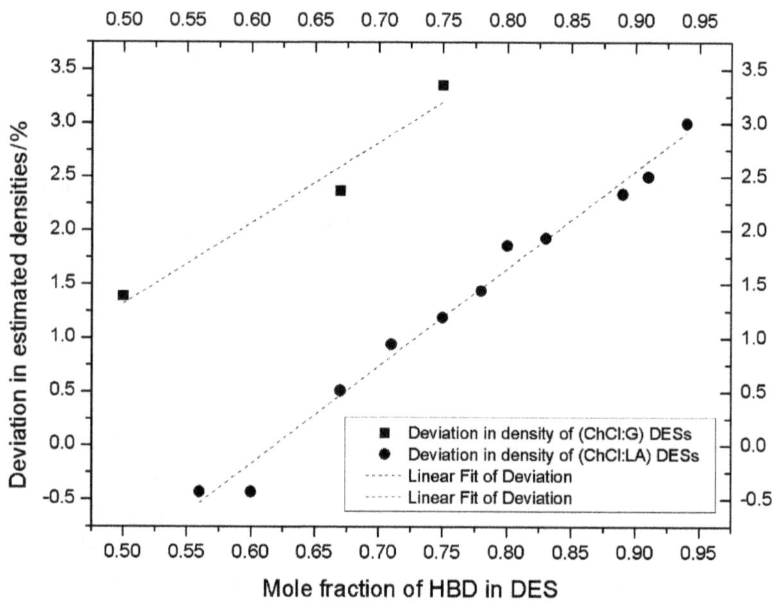

Figure 6.5 Change in density with the amount of HBD for ChCl:LA (lactic acid) based and ChCl:G (glycerol) based DESs. Reproduced from Reference 36 with permission from American Chemical Society.

6.2.4 Electrical Conductivity

Electrical conduction ability of DESs is associated with the viscosity; as the viscosity increases; the conductivity reduces. Hence, majority of the usable DESs possess low conductivities (at ambient temperature, <1 $mS.cm^{-1}$) [6], which might cause challenges for the specific electro-chemical applications. The low viscosity DESs like those accommodating ethylene glycol or imidazole with ChCl display higher conductivities. Hou *et al.* [38] developed a new series of the binary eutectic mixtures with organic salts (tetrabutylammonium bromide (Bu_4NBr), choline chloride (ChCl) and 1-ethyl-3-butylbenzotriazolium hexafluorophosphate (C_2BtPF_6) and organic

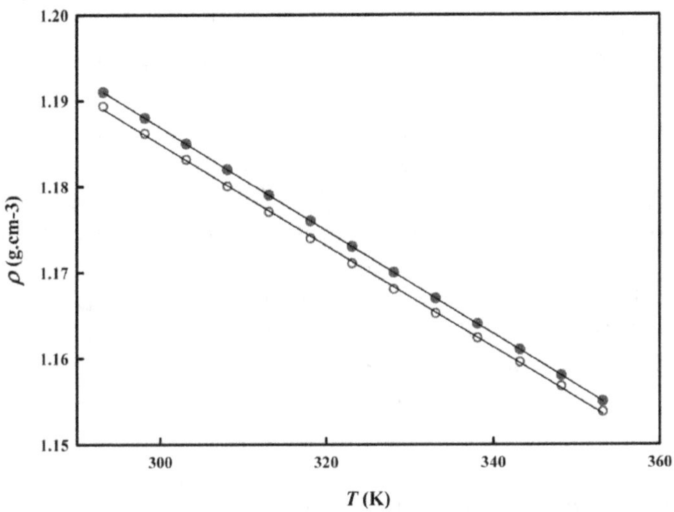

Figure 6.6 Experimental densities (ρ) vs. temperature for the DESs [Ch]Cl:Glu, generated using two distinct techniques: (•) grinding and (○) heating. Reproduced from Reference 37 with permission from American Chemical Society.

neutral molecule (imidazole (IM) solid) (denoted as Bu4NBr–IM, ChCl–IM and C2BtPF6–IM). The conductivities of the binary eutectic mixtures decreased in the order ChCl–IM > C_2BtPF_6–IM > Bu_4NBr–IM below 80 °C. In addition, the conductivity of the eutectic mixtures was also observed to increase with temperature. Abbott *et al.* [28] applied the hole theory for analyzing the conductivity of DES, and despite the fact that significant results were attained in a few cases (e.g. ethylene glycol + ChCl 2:1 DES), feeble predictions were recorded for others (e.g. glycerol + ChCl 2:1 DES). This proved that the ion mobility and conductivity are controlled by appropriate holes' availability as well as the strength and type of HBD-ion interactions. Further, it was observed that the conductivity of DES enhanced with increase in the salt concentration [6]. However, it is not true for all DESs, and the change in the conductivity with the concentration of salt is dependent on the type of both HBD and salt.

6.2.5 Surface Tension

Until now, the studies pertaining to the surface tension of DESs have been very limited. Abbott *et al.* [24] reported the surface tension of $ZnCl_2$ as well as ChCl derived DESs. The surface tension values of the DESs phenylacetic acid/ChCl [2:1 mole ratio] and malonic acid/ChCl [1:1 mole ratio] were almost 41.86 and 65.68 $mN.m^{-1}$, respectively. In addition, the surface tension values of 1,4-butanediol/ChCl [3:1 mole ratio], glycerol/ChCl [3:1 mole ratio] and ethylene glycol/ChCl [3:1 mole ratio] eutectic mixtures were observed to be 47.6, 50.8 and 45.4, and $mN.m^{-1}$ at a temperature 20°C, respectively [28]. The surface tension is anticipated to act in an identical mode as viscosity as it is determined by the stability of inter-molecular synergy which controls the DES generation. In another study, the surface tension values of different glycerol/ChCl DESs were also observed to display a direct inter-relationship with the temperature [15].

Thus, a number of properties of DESs define their tenability and performance under various conditions and a control on these properties is necessary in order to generate DESs of required characteristics suitable for different applications.

6.3 Types of DESs

Depending on the types of complexing agent employed, the DESs can be classified into four types, namely type I, type II, type III and type IV [10]. Among the four types of DESs, type III is the most reported in the literature studies due to the simple preparation, biodegradability, cheaper raw materials, etc. [44]. Also, as mentioned above, many of the developed DESs are based on choline chloride due to non-toxicity as well as low cost. The structural details and other basic characteristics of these types of DESs are explained in the sections below.

6.3.1 Type I DESs

The type I DESs are formed from quaternary ammonium salts and MCl_x. This type could be regarded as a resembling kind to the adequately-researched imidazolium/metal halide salt systems. The different examples of type I DESs comprise the imidazolium/chloroaluminate salt melts, along with solvents generated with imidazolium salts and different metal halides inclusive of $FeCl_2$, and the consecutive halides of metals such as LiCl, CuCl, AgCl, $CuCl_2$, $ZnCl_2$, $SnCl_2$, $CdCl_2$, YCl_3, $LaCl_3$ and $SnCl_4$. The type I DESs did not gain as much attention as the type III DESs (generated from HBD and ChCl) due to the inherent hygroscopic nature of the metal salts (e.g., $AlCl_3$) utilized to generate them [44].

6.3.2 Type II DESs

The range of non-hydrated metal halides that have an appropriate melting temperature to generate the type I DES is restricted, nevertheless, the capacity of DESs can be enhanced by employing ChCl and hydrated metal halides. Such DESs are termed as type II DESs. The second type has found superior applications in different large scale industrial operations because of inherent moisture/air insensitivity along with relatively lower cost of several hydrated metal salts [45]. For instance, the type II DESs have been utilized for depositing non-micro-cracked chromium (Cr), as the aqueous electrodeposition of the same is restricted by the legislation [45]. From a 2: 1 molar ratio mixture of $CrCl_3.6H_2O$ and ChCl, a liquid was generated. On the other hand, the formation of liquid from mixture of 6 molar equivalents of water and anhydrous $CrCl_3$ was impossible, indicating that the Cr coordination sphere is vital in the liquid formation. This also implies that the indicated system is IL like in nature and is not commonly a concentrated aqueous solution. More studies revealed the significance of cationic species as grain modifiers. By the incorporation of LiCl, the nano-crystalline black Cr was formed [46].

6.3.3 Type III DESs

The type III DESs, which are generated from HBD and ChCl, are of great interest because of their capability to solvate an extensive spectrum of transition metal species, including oxides and chlorides [14]. A wide range of HBDs have been analyzed till date, with the DESs generated utilizing alcohols, carboxylic acids and amides. These type of DESs are easy to synthesize, relatively cheap and comparatively non-reactive with water, along with majority of them being highly biodegradable. An example for type III DESs is [ChCl]-[amide/carboxylic acids] engineered by Abbott *et al.* [24]. The broad range of accessible HBDs indicates that this class of DESs is specifically adaptable. Physical characteristics of the liquid are greatly dependent on the HBD and it can be readily tailored for particular applications. In spite of the fact that the capabilities of these DESs are remarkably lesser than some imidazolium salt-distinct anionic ILs, these are still adequately broad to permit the metal deposition with greater effectiveness. The type III DESs have been observed to be specifically adaptable, with a broad variety of attainable utilizations inclusive of the metal oxides processing [22], separation of glycerol from biodiesel [47] and the preparation of cellulose derivatives [48]. In addition, as mentioned above, these solvate transition metal species extensively.

6.3.4 Type IV DESs

Most of the ILs that are in a flowing state at room temperature are generated employing an organic cation, mainly using sulfonium, ammonium and phosphonium moieties. Due to the higher charge density of inorganic cations, these are not able to generate low melting point eutectics. Nevertheless, the earlier research studies have presented that the mixtures of urea and metal halides can establish eutectics having melting points <150 °C [49,50]. Abbott *et al.* [20] generated type IV DESs by the incorporation of a variety of transition metals into the room temperature eutectics. It was anticipated that the aforementioned metal salts may not ionize in non-aqueous medium, however, $ZnCl_2$ was observed to generate eutectics with acetamide, 1,6- hexanediol, urea and ethylene glycol [20]. Many studies have described the physical characteristics of the eutectics and it was observed that the properties of these eutectics were similar to other classes of ILs and could be efficiently modelled using the hole theory.

6.4 Applications of DESs

6.4.1 Gas Separation

In this section, the application of DESs for the gas separation has been reviewed. In specific, the removal of CO_2 and SO_2 has been examined in detail. The prevailing knowledge on the gas separation employing the DESs, correlation of the capturing capacities as well as properties of DESs with those of ILs, identification of the defects as well as strengths of DESs, and suggestions related to prospective research directions on this topic have been explored.

CO_2 Capture

Greenhouse gas emissions, specifically the emission of CO_2, result in the serious problem of global warming. Reducing the CO_2 emissions by carbon capture and storage (CCS) has been a focus of intense research in the recent years. The separation of CO_2 is a critical step in CCS and is considered to be an energy intensive process. In addition, efforts have been concentrated on establishing energy sources which do not discharge CO_2, nevertheless, these are not yet at the stage where these could be enforced on a large scale. Also, it is very important to design a feasible synthetic process to generate recyclable solid sorbent which has an enhanced sorption capacity. ILs as the green solvents have been considered for application as liquid absorbents for the separation of CO_2. On the other hand, the combustible character [52], toxicity [53], poor biodegradability [54] and high cost limit the usage of ILs. In recent past, DESs based on ChCl were recommended as a new type of ILs, however, with added benefits in preparation, environmental impact as well as cost.

Gutierrez *et al.* [55,56] utilized resorcinol:ChCl:3-hydroxypyridine accommodating DES as the structure directing agent, liquid medium as well as the source of nitrogen and carbon for generating nitrogen doped hierarchical carbons. Zhu *et al.* [57] stated the greater catalytic effectiveness of a DES urea:ChCl supported on the molecular sieves in order to chemically fix CO_2 to the cyclic carbonates. The green as well as biodegradable catalyst was observed to be active as well as selective, and had also the capability for preparing cyclic carbonates from epoxides and CO_2. Following the chemical process, the products as well as solid catalyst could be effortlessly isolated due to the fact that DES was not soluble in products. The porous carbon nano-sheets prepared in DESs with manageable thicknesses, which could detach the CO_2 from N_2 have also reported [58].

DESs can also be employed themselves as the solvents for capturing CO_2 [59-63]. Leron and Li [60] examined the solubility of CO_2 in a DES consisting of 1 mole ChCl and 2 mole ethylene glycol at pressure till 6 MPa and the temperature from 303.15 to 343.15 K. CO_2

solubility in the DES was determined utilizing a thermo-gravimetric pressure balance that was integrated with a control system, which enabled real-time information monitoring as well as recording. The CO_2 solubility in the solvent was observed to enhance with pressure and reduce with rising temperature. Further, the values were not beyond the range of the CO_2 solubility in 1-butyl-3-methylimidazolium $[Bmim]^-$ as well as 1-ethyl-3-methylimidazolium $[Emim]^-$ based RTILs. The solubility information was favorably depicted by an extended Henry's law equation as a function of pressure and temperature with an average absolute deviation of 1.6%.

One of the most famous DESs is the mixture generated from 1 mole ChCl and 2 moles urea ($CO(NH_2)_2$; commercial name: Reline). Its melting temperature is 12 °C, which is significantly lesser than the melting temperature of its distinctive components [14]. The DES was observed to be very beneficial as a solvent in a variety of applications like catalytic reactions and preparation of micro-porous crystalline zeolites, metal-organic frameworks and coordination polymers [64,65]. Identical to several room temperature ILs (RTILs), the capability of this green solvent for the absorption of CO_2 has also been examined. Li *et al.* [66] stated the CO_2 solubilities in a range of ChCl–urea mixtures, and analyzed the influence of composition on the gas solubility in the DES. It was confirmed that at the same pressure and temperature, solubilities of CO_2 were very identical in mixtures with 1:2.5 and 1:1.5 mole ratios but remarkably greater in the combination with 1:2 mole ratio. In a study by Leron *et al.* [62], experimental analysis on the CO_2 solubility in ChCl–urea (1:2 mole ratio) DES at pressures till 6.0 MPa and temperature range 303.15 - 343.15 K was carried out. It was noted that the CO_2 solubility in the DES reduced with enhancing the temperature. Further, it was also observed that the CO_2 solubility enhanced with increase in pressure at constant temperature. The impact of buoyancy (Δm) on the CO_2 solubility in the DES was remarkable at greater pressures, however, essentially insignificant at lesser pressures. It was confirmed that the effect of buoyancy reduced with increase in the temperature. By utilizing an extended Henry's law equation with an average absolute deviation of 1.0%, the CO_2 solubility information was precisely characterized as a function of pressure and temperature.

Ali *et al.* [63] conducted research on the ammonium- and phosphonium- based DESs having various HBDs. The solubility of CO_2 in the generated DESs at a fixed temperature and pressure was experimentally analyzed. In addition, a mathematical model based on the Peng Robinson equation of state (PR EoS) was generated for correlating the solubility of CO_2 in these categories of DESs. This model was justified with the attained experimental information and then analyzed with additional precise DESs stated in the literature over a broad range of pressure and temperature values. The experimental data as well as the calculated data using PR EoS had a very good agreement. Hence, the generated model could be employed for studying the efficiency in utilizing DES in CO_2 removal processes or any other separation processes.

Sze *et al.* [67] examined the CO_2 capturing as well as releasing ability of ternary DES systems consisting of glycerol, ChCl and one of the three various superbases. Out of the systems examined, the superior performance system was observed to remove CO_2 at a capacity of approximately 10% by weight. Analysis was performed on ternary mixtures ChCl:glycerol:1,5-diazabicyclo[4.3.0]-non-5-ene (DBN) with molar ratios 1:2:x, where x = 3, 6, 7 or 8, respectively. As illustrated is Figure 6.7, the DES system with 1:2:7 molar ratio offered the highest CO_2 capturing ability (105 mg/g of DES). In a study by Francisco *et al.* [68], a new liquid solvent was synthesized by combining two naturally as well as effortlessly accessible solid starting materials (ChCl and lactic acid). ChCl is a HBA, whereas lactic acid is a strong HBD. The combination generated a stable liquid in a broad spectrum of compositions, displaying extremely low transition temperatures. The lactic acid:ChCl combinations were analyzed as the solvents for gas separations. Analysis of the physical characteristics confirmed the tunability of physical characteristics as well as phase behavior by varying the composition. Glass transition temperature, surface tension, density and viscosity were examined as a function of lactic acid:ChCl ratio. The physical characteristics of the new DESs were very identical to that of ILs, however, these did not possess the main drawbacks of typical ILs, for instance, the DESs were renewable, cheap and environmental friendly solvents. Due to the potential characteristics such as tunability of physical as well as thermodynamic properties by varying the composition, excellent solvation properties generated from hydrogen bonding interactions, broad liquid range, recoverability and biodegradability, the lactic acid:ChCl DESs are considered to be potential candidates as solvents for CO_2 separation.

Considering the literature findings, it can be confirmed that the choline-based DESs are promising for the application as liquid absorbents for CO_2 separation. For instance, the DES generated from ChCl and urea (molar ratio 1:2) displayed a better performance with a calculated CO_2 absorption of 3.559 mmol/g at 60 bar pressure and 303.15 K temperature [62]. The above discussion clearly revealed that the DESs have the ability to absorb more CO_2 than the toxic ILs because of their tunable chemical as well as physical properties. However, uncertainties and obstacles still prevail, and further examination on the micro-structure as well as properties has to be performed using experimental measurements and model developments. Further, detailed analysis is needed for determining the CO_2 absorption mechanism.

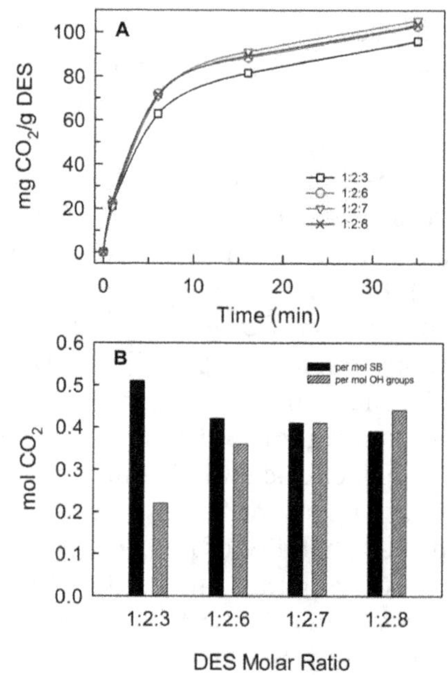

Figure 6.7 (A) Quantity of CO_2 separated per gram of DES as a function of time; (B) total CO_2 captured per mole of DBN and per mole of alcohol (OH) group for the same DES mixtures with increasing relative DBN amounts. Reproduced from Reference 67 with permission from American Chemical Society.

Sulfur Dioxide (SO_2) Capture

The generation of SO_2, principally from the fossil fuels' burning, leads to severe environmental issues like smog and acid rain. Till date, the technologies of flue gas desulfurization (FGD) perform a vital role in the separation of SO_2 [69]. The advancement of efficient and renewable absorbents for the removal as well as recovery of SO_2 is very critical to from health, safety and environment perspectives. Similar to the CO_2 removal studies, majority of the prevailing SO_2 removal technologies that employ DES systems are concentrated on the application of ChCl as the IL, due to the fact that the material is non-toxic, cheap, biodegradable and can be either derived from biomass or easily accessible as bulk commodity material.

Yang *et al.* [70] analyzed the SO_2 absorption using ChCl-glycerol DESs at different temperatures and SO_2 partial pressures. Molar ratios of ChCl and the glycerol varied from 1:4 to 1:1. The solubility of SO_2 was observed to enhance with increase in the ChCl concentration. At 1 atm pressure and 20 °C temperature, the SO_2 absorption ability of the DESs with a glycerol/ChCl mole ratio of 1:1 was as large as 0.678 g SO_2 per g DES. In addition, the absorbed SO_2 could be effortlessly discharged, and the superior characteristics of greater absorption ability as well as accelerated absorption/desorption rates continued for five consecutive absorption/desorption cycles. However, ChCl-glycerol DES did not reach greater absorption abilities of pure SO_2, the absorption process was physical as well as the solubility data adapted to Henry's law, which overall led ChCl-glycerol DES inadequate for absorbing low-partial pressure SO_2 competently. Later, Sun *et al.* [71] generated four types of ChCl-based DESs using four organic compounds namely malonic acid (MA), ethylene glycol (EG), thiourea and urea as the HBDs. All DESs exhibited excellent thermal stability and were observed to be stable at 363 K, which is very advantageous for the usage in FGD. The SO_2 absorption abilities of the DESs were analyzed at varying temperatures as well as SO_2 partial pressures. Results obtained from the absorption analysis confirmed that ChCl-thiourea (1:1) and ChCl-EG (1:2) DESs exhibited superior absorption efficiency, and the absorption abilities were 2.96 and 2.88 mol SO_2 per mol DES at 1 atm and 293 K, respectively, as demonstrated in Figure 6.8. Further, the SO_2 regeneration experiments were performed and it was noted that all the solvents were regenerated at 343 K with N_2 bubbling. Also, the absorption abilities of DESs persisted without any remarkable loss after six absorption/desorption cycles.

In another study by Liu *et al.* [72], DESs based on caprolactam (CPL) and low molecular weight organic compounds, namely benzoic acid, imidazole, acetamide, o-toluic acid and furoic acid were synthesized and designated as CPL-benzoic acid (molar ratio 1:1), CPL-imidazole (molar ratio 1:1), CPL-acetamide (molar ratio 1:1), CPL-o-toluic acid (molar ratio 2:1) and CPL-furoic acid (molar ratio 1:1), respectively. The DESs were illustrated as promising absorbents for the capture of SO_2 in the temperature range 30-70 °C and 1 atmospheric pressure of pure gas. The characteristics of the DESs were identical to the traditional ILs and entirely distinct from the molecular solvents.

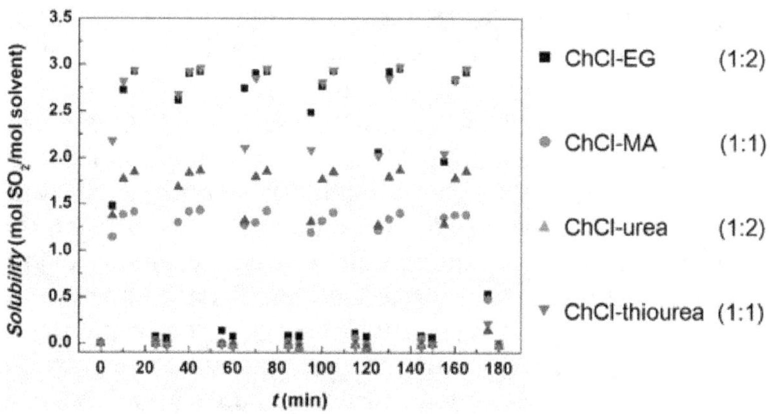

Figure 6.8 SO_2 absorption abilities of different DES systems. Reproduced from Reference 71 with permission from American Chemical Society.

From the examined temperature range, it was noted that the density enhanced linearly with increasing temperature. Further, Vogel–Tammann–Fulcher (VTF) equation was employed for describing the temperature dependence of conductivity or viscosity. At 30 °C, the CPL-imidazole (1:1) DES exhibited highest conductivity (32.3 μs/cm) and lowest viscosity (48.6 mPa·s). It was noticed that SO_2 absorption abilities of CPL-organic amine based DESs were greater, when compared to the CPL-organic acid based DESs. The SO_2 solubility in CPL-acetamide (1:1) DES was greater than the BMImBF4 IL and lower than dimethylformamide (DMF) solvent. The SO_2 solubility in CPL-acetamide (1:1) solvent was 0.497 g/g of mass fraction at temperature 30 °C and the absorption was essentially reversible. Similar to the ChCl based DESs, the SO_2 absorption of CPL-acetamide DES was observed to be a physical process, which induced CPL-acetamide to not absorb low-partial pressure SO_2 conveniently. The SO_2 absorption/desorption in CPL-acetamide

(1:1) was carried out for 5 times and proportionate absorption capacity was observed.

As noted in the earlier literature studies, the DESs for the absorption of SO_2 mainly displayed a physical process and the solubility data in accordance with Henry's law make it challenging for the application of DESs to separate low partial pressure SO_2 in the flue gas. Hence, it is essential to design environmentally benign functional DESs, which have the ability to absorb lesser partial pressure SO_2 in the flue gas. In a study by Zhang *et al.* [73], DESs were designed for absorbing lesser partial pressure SO_2 from the simulated flue gas. Two types of biodegradable functional DESs based on ethylene glycol (EG) as HBD and l-carnitine (L-car) as well as betaine (Bet) as HBA were synthesized with mole ratios of HBA to HBD from 1:3 to 1:5. The aforementioned DESs were analyzed for absorbing SO_2 with varying partial pressures at distinct temperatures. The obtained results confirmed that the two DESs could be used to absorb the low-partial pressure SO_2 effectively. At SO_2 partial pressure of 0.02 atm and 40 °C temperature, the SO_2 absorption abilities of the DESs with HBA/HBD molar ratio of 1:3 were found to be 0.820 mol SO_2/mol HBA for the DES [L-car+EG] and 0.332 mol SO_2/mol HBA for DES [Bet+EG]. In addition, there was no apparent loss of SO_2 absorption abilities in the course of the regeneration cycles. The SO_2 absorption of two DESs at a temperature of 40 °C and partial pressure of 0.0037 atm was performed and it was observed that the two DESs had very quick absorption rates for SO_2 with the low-partial pressure because of the practically linear absorption curve at the introductory absorption period. Further, the absorption mechanism was suggested demonstrating that there were acid–base reactions between the acidic SO_2 and the Lewis base COO− present in HBA. DESs based on L-car and betaine as HBA are potential absorbents for SO_2 removal in the FGD application.

Guo *et al.* [74] synthesized economically as well as environmental friendly DES systems having distinct molar ratios of tetrabutylammonium bromide (TBAB) and caprolactam (CPL) and examined the system for the separation of SO_2 at atmospheric pressure and different temperatures (298.2 to 403.2 K). From Figure 6.9, it can be observed that the equilibrium solubility of SO_2 in distinct mixtures exhibited a firm dependence on the temperature as well as the mole ratio of parent chemicals at the atmospheric pressure. Mole fraction solubility of SO_2 reduced remarkably both with enhancing temperature (at specified mole ratio) and rising mole fraction of CPL in the DES [CPL][TBAB] at a particular temperature. Utmost solubility noticed for the DES with a molar mixing ratio of 1:1 was 0.680 g/g at temperature 298.2 K, however, it decreased to 0.351 g/g at temperature 373.2 K and constant atmospheric pressure. Duan *et al.* [75] made efforts to improve the viscosity of [CPL][TBAB] by the mixing water to enhance the affinity as well as solubility of SO_2 in the binary mixture. Nevertheless, the addition of water reduced the viscosity, and also minimized the SO_2 removing ability of [CPL][TBAB]:H_2O. By heating the densest phase, the removal of SO_2 could be performed effortlessly, and the materials could be reused for absorption.

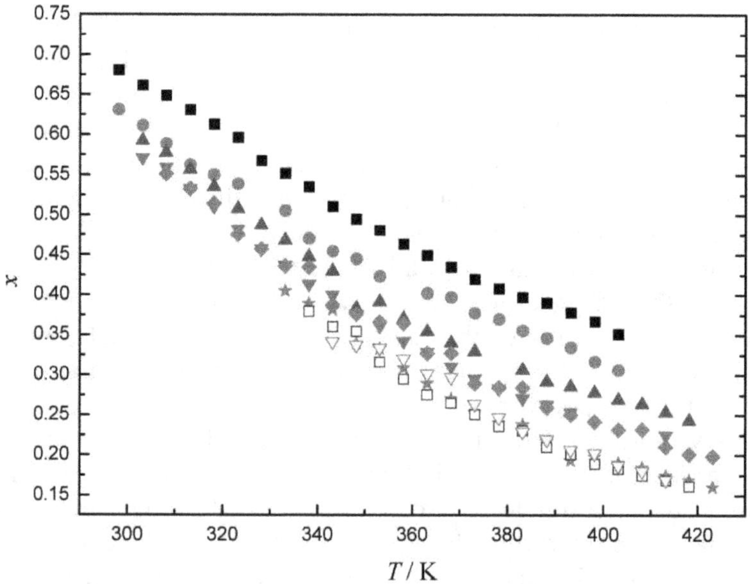

Figure 6.9 Solubility of SO_2 as a function of temperature in [CPL][TBAB] DES system with varying molar mixing ratios of CPL and TBAB. Reproduced from Reference 74 with permission from American Chemical Society.

6.4.2 Enhanced Oil Recovery

The demand for global energy has been developing firmly because of the fast growth in world's population. The growing energy requirement develops into the major justification for applying various technologies for improving the effectiveness of petroleum production. Field as well as laboratory studies reported that just 30-50% of oil is generally recovered succeeding the water injection from original oil in place (OOP) [76]. For extracting the remaining oil, it is essential to employ effective methods and the EOR or tertiary oil recovery is the technique achieving greater importance in this respect [77]. Currently, the three main methods employed for EOR are (1) thermal recovery, (2) gas and (3) chemical injections. Out of the aforementioned techniques, chemical flooding is considered as the superior EOR technique [78]. It can be classified into three major processes namely (i) alkaline flooding, (ii) surfactant flooding and (iii) polymer flooding [76]. Various studies have been carried out on laboratory scale and as field applications on various thermal methods like steam injection and *in-situ* combustion, analyzing their efficiency in increasing the heavy oil recovery. Thermal methods have the benefit of transferring the thermal energy to the reservoir as well as heating the heavy oil. Thus, the viscosity of heavy oil reduces remarkably, thereby increasing the mobility ratio [79]. In order to design advanced steam flooding with efficient heavy oil production, different methodologies have been further developed. One of the methodologies is to co-inject the steam using additives like solvents, caustic steam, hydrocarbon and non-hydrocarbon gases, foam and surfactant [80]. A substantial concern in the usage of chemical additives in steam flooding is the thermal decomposition of chemicals occurring at higher temperature which remarkably decreases the efficiency of these chemicals. The chemicals employed in conventional EOR have been observed to have lesser thermal stability. In contemplation of addressing the challenges related to polymers as well as the surfactants in the flooding operations, the exploration for fresh thermally-stable chemicals or advancement in the current ones has become critical. In the recent past, some research studies have employed ILs in upgrading as well as enhancing the heavy crude oil recovery. Results attained from the upgrading experiments revealed that ILs have a better viscosity lowering property and, thus, can cause higher oil recovery [81]. The capability of ILs was also examined as alternate surfactants in chemical flooding. As mentioned earlier, the analogs of ILs, i.e., DESs have many benefits like low price, biodegradability, non-toxicity, chemical compatibility with water, biocompatibility as green solvents and easy preparation. Besides, the aforestated characteristics, existing polar components, greater viscosity, as well as surface active agents in DESs provide them capacity for the usage in heavy oil recovery.

In recent years, the DESs from group IV have been analyzed as new EOR agents for the heavy oil recovery improvement from sandstone cores at the reservoir environment. These DESs include ChCl:glycerol, ChCl:urea [82,83,84], ChCl:malonic acid [85] and ChCl:ethylene glycol [86]. In all the aforementioned events, the DES flooding exhibited assuring outcomes in terms of recovery improvement through enhancement in viscosity ratio as well as benign wettability amendment of rock surface as the major mechanisms. In a study by Mohsenzadeh *et al.* [82], ChCl-based DESs were analyzed experimentally for the first time for studying the impact on various oil recovery mechanisms in heavy oil/formation brine/Berea sandstone rock system. The authors compared the efficiency of two distinct types of DESs, namely ChCl:glycerol (DES1 1:2 molar ratio) and ChCl:urea (DES2 1:2 molar ratio), in improving the heavy oil recovery. The impact of the two DESs diluted with formation brine on the surface and interfacial tensions, altering wettability, impulsive imbibitions, emulsification, as well as tertiary residual heavy oil saturation minimization were analyzed experimentally at varying temperatures. The heavy oil having 16 °API and formation brine from an Omani heavy oil field were employed. Core flood analysis was carried out at the reservoir conditions utilizing the Berea sandstone core plugs. The results confirmed that in the course of brine flooding, the enhancement in the heavy oil recovery occurred by raising the temperature. Cumulative oil recovery by brine injection was 42%, 40% and 36.7% at temperatures 80 °C, 60 °C and 45 °C, respectively. The viscosity ratios of the brine to heavy oil enhanced with increase in temperature. Further, the results from the Amott tests, contact angle analysis and oil-brine interfacial tension (IFT) examinations confirmed that as the temperature enhanced, the IFT reduced, and the wettability alteration increased. It was concluded that the DES2 consistently outperformed the DES1 in increasing the oil recovery. This was due to the fact that DES1 was less alkaline, when compared to the DES2.

In another study by Mohsenzadeh *et al.* [87], sequential DES and steam flooding was suggested and experimentally analyzed for the enhancement in the heavy oil recovery and achieving capability for *in-situ* heavy oil upgrading. ChCl:glycerol (DES1) and ChCl :urea (DES2) DESs were employed in the study. By utilizing the 16.5 °API heavy oil as well as Berea sandstone core plugs, the primary DES as well as secondary DES flooding at varying concentration followed by high-temperature steam flooding were performed. From the TGA analysis of the undiluted DESs, it was noted that the two DESs initiated to decompose at approximately 210 °C. For DES1 and DES2, the ultimate decomposition temperatures were recorded at 320°C and 370 °C, respectively. Thus, the thermal stability analysis confirmed that the selected DESs exhibited greater thermal stability when compared to other EOR chemicals, thus, confirming high potentials to be utilized as the additives in the thermal EOR processes. The steam flooding occurring after the undiluted DES injection as well as 2-fold diluted DES injection led to an extra heavy oil recovery of 12%, when compared to the primary or secondary steam flooding alone. Nevertheless, by employing greater dilution of DES, i.e., about 20 fold diluted solution, same or even lesser total recovery factors in certain cases were obtained, when compared to the secondary steam flooding. In terms of the recovery increase in the period of subsequent steam flooding, DES1 was found to be less effective when compared to DES2 at the same concentrations. The examination of physico-chemical properties of produced oil for various cases indicated the supportive role of DES in improving the *in-situ* heavy oil. Relatively, the DES1 exhibited superior comprehensive performance than the DES2, in terms of the *in-situ* heavy oil upgrading.

Studies were also performed to investigate the critical parameters influencing the heavy oil recovery performance by DESs injection [83]. Mohsenzadeh *et al.* [83] carried out a study for examining the impacts of DESs concentration, brine salinity, secondary as well as tertiary DES injections on the recovery performance. These were investigated through core flooding experiments, contact angle measurements and IFT measurements. The same DESs namely ChCl:glycerol (DES1 with molar ratio 1:2) and ChCl:urea (DES2 with molar ratio 1:2) were used for the analyses. The IFT between the heavy oil and various solutions were determined by the spinning drop tensiometer at 45 °C. It was observed that both DES1 and DES2 solutions enhanced the IFT of the formation brine due to the powerful synergy between the functional groups present in DES2 [(–NH_2 and C=O) and DES1 (OH–)] and the functional groups existing in the oil because of the van der Waals forces. Nevertheless, greater IFT was seen during the usage of DES1 mainly due to the existence of stronger force (HO– vs –NH_2) as well as larger population (3HO– vs 2 –NH_2) in DES1, when compared to DES2 [82]. In addition, the results also confirmed that the IFT enhanced with increasing DES concentration in brine for both DES1 and DES2 solutions. The contact angle measurements were performed between aged rocks and oil droplet in distinct DES solutions and expressed as a qualitative index of rock wettability. DES2 was observed to have more impact to enhance the contact angles, when compared to DES1, at the identical concentration of DES in brine. In addition, the contact angle values revealed that reducing the concentration of both DESs from 50 vol% to 5 vol% led to minor variations in wettability alterations. The authors concluded that the two DES solutions had good potential for changing the wettability from the oil-wet state towards neutral-wet condition. It was confirmed that the wettability alteration in the period of DES injection had remarkable effect on the oil recovery and it could be the governing as well as superior mechanism for the heavy oil recovery increase. For both DESs, the total recovery at the tertiary mode was almost 3-6% greater at 5 vol% and 16% greater at 50 vol% concentration, when compared to secondary mode. Further, it was seen that the DESs could allow broad range of reservoir salinities and no remarkable variation in oil recovery was observed at distinct brine salinity in the period of injection of tertiary DESs.

Despite the fact that the chemical injection into oil reservoir is regarded as an efficient EOR method, there are certain negative effects on reservoir like formation damage because of fluids incompatibility or the chemical interactions occurring with reservoir rocks as well as fluids. In the oil and gas reservoirs, the formation damage is considered to be one of the most critical justifications for dropping well injectivity and productivity. This damage might occur by drilling fluid invasion close to the wellbore or during the EOR tasks in the field like chemical flooding, gas injection, water flooding as well as thermal methods. Mohsenzadeh *et al.* [84] also analyzed the possible impact of the DESs on sandstone formation damage. The authors employed the same DESs i.e., DES1 (ChCl:glycerol with molar ratio 1:2) and DES2 (ChCl:urea with molar ratio 1:2) for the analysis. The core flooding analysis was carried out at varying temperatures and reservoir pressure to determine possible permeability variation by DESs injection. In order to gain insights about the formation damage mechanism, various analyses such as critical injection rate after which fines transfer occurs, scanning electron microscopy (SEM), water shock impact on the formation damage, X-ray diffraction (XRD), dry core weight analysis after and before each core flooding examination and computed tomography (CT) scanning investigation of fresh as well as treated core samples were carried out. The permeability analysis by the core flooding experiments confirmed that water shock event led to uncompromising damage (65%

decline in absolute permeability) when the distilled water was infused succeeding brine flooding. The DES1 and DES2 displayed positive effect on lowering the water shock loss at distinct temperatures. DES1 was observed to exhibit superior performance when compared to DES2. The core weight analysis proved that when the permeability losses were greater, the extent of core weight enhancement was also observed to be higher. In addition, results of quantitative XRD and SEM confirmed that the precipitation as well as deposition within the pores were the major sources of formation damage. Moreover, the CT-scanning images exhibited better comparisons between damaged and fresh cores. It was also noted that in spite of the DES solutions' function in prohibiting serious water shock deterioration as well as clay stabilization, there was still certain formation damage brought about by the re-crystallization and precipitation processes lowering the core samples' permeability.

For enhancing the heavy oil recovery using ChCl-malonic acid DES, several studies have also been performed. Al-Weheibi *et al.* [85] examined the capability of DES for recovering the residual heavy oil left after the waterflooding. For this study, two DESs with ChCl-malonic acid of molar ratios 1:1 and 1:0.5 were examined. The two solvents were preliminarily analyzed by calculating the conductivity, viscosity, density and pH at distinct temperatures (20-80 °C). The core flood test was performed at reservoir condition (temperature = 45-80 °C, pressure = 1200 psi,) utilizing Berea sandstone core samples as well as fluids from the field of interest (formation brine and crude oil). The results obtained from the core flood test revealed that the solvents generated 7-14% of the residual heavy oil after the brine flooding as tertiary recovery stage. Further, it was also confirmed that both the DESs exhibited superior performance in increasing the oil recovery at elevated temperatures. The analysis of absolute permeabilities after and before the injection of DESs/brine solutions displayed no deterioration to the formation. For the tertiary oil recovery enhancement, the wettability alteration was the dominant mechanism.

In addition to the above-mentioned DESs for EOR, the DES based on ChCl-ethylene glycol has also be utilized for this purpose. Shuwa *et al.* [86] analyzed the ability of the ethylene glycol-ChCl DES solution on increasing the heavy oil recovery by conducting core flood tests. The test was carried out at reservoir conditions (temperature = 45-80 °C and pressure = 1200 psi) with Berea sandstone core specimens as well as fluids from the field (crude oil and formation brine). Enhancement in IFT between the oil and DES/brine solution was observed. Further, from the spontaneous water imbibition analysis, it was confirmed that a reduction in oil wetness brought about an enhancement in the oil production. Succeeding the water flooding process, almost 52% of residual oil was recovered employing the DES in the core flooding process. The obtained results confirmed an advancement in oil recovery with an increase in the temperature of the reservoir. Formation damage did not occur due to the interaction between the DES and core materials, confirmed from the permeability analysis of the DES/brine solution after and before the injection process. The dominant mechanisms for the tertiary oil recovery advancement were observed to be wettability alteration and the viscous forces. Overall, on the basis of the analyses performed and, in comparison with the conventional methods of EOR, the DESs have been proved to be effective and promising.

6.4.3 Preparation of Materials

The traditional synthesis of inorganic materials is generally performed in organic solvents or polar solvents like water. For instance, solvo-thermal synthesis is the method of developing single crystals by thermal treatment under pressure, from a non-aqueous solution [6]. In recent past, the iono-thermal preparation has been developed, where ILs or DESs are employed as both solvent as well as template. Iono-thermal preparation can be easily performed in low pressure conditions because of the lesser volatility of the DESs or ILs. Mostly, the capability for tuning the ionic characteristics of DESs offers modulating reaction surroundings under which the novel materials having beneficial frameworks might be generated. A recently developed method in this field is the substitution of ILs by safe as well as cheap DESs. As an illustration, in the iono-thermal strategy, the DESs based on renewable carboxylic acids/ChCl or urea/ChCl were employed as an alternate medium to the ILs. As stated in the previous sections, the DESs have numerous marked benefits, when compared to ILs. In addition, the existence of neutral components in DESs having greater boiling points such as urea, carboxylic acids or carbohydrates contribute new as well as complementary features, when compared to the ILs which are entirely ionic. The DESs also have the capability to perform multiple functions during materials preparation, inclusive of the classic role of solvent, reactant, templating agent, etc. Due to the aforementioned characteristics, the

DESs are currently attaining remarkable recognition for the iono-thermal preparation of materials as well as the design of novel structures. Till date, several inorganic materials with substantial structures and properties, inclusive of metal oxides, metal-organic frameworks, nanomaterials, carbon and zeolite-type materials have been generated utilizing iono-thermal process. In the following sections, the materials generated in DESs for different usages like gas separation, catalysis and hydrogen storage have been illustrated.

Metal-organic Frameworks

The preparation of metal-organic frameworks (MOFs) acquired importance because of their promising utilization in fields like gas separation, hydrogen storage, etc. [88,89]. Taking benefits of the ecological, physico-chemical and economic benefits of DESs in iono-thermal preparation of materials, the eutectic mixtures are currently considered as potential interesting media for the huge-scale preparation of advanced MOF frameworks. Unlike the ILs (the binary system containing cations and anions), which were utilized for the preparation of MOF, the DESs consist of neutral ligands like urea which can bring about structure-directing impact, thereby, tuning the DESs to be the appropriate solvents for the MOF synthesis. Till date, several efforts have been made for the iono-thermal preparation of advanced porous MOFs in DESs, because of the diversified roles which DESs can perform in such preparation. Zhang *et al.* [90] studied the multifaceted function of the DESs in the creation of open metal sites as well as porosity in the MOF structures for the gas storage application. The authors initially described the preparation of a range of MOFs in the DES, consisting of 1,4-benzenedicarboxylate (bdc), trivalent metal (Sm, Ln, Nd, Y) and a single or two elements from the DES. From the analysis, it was noted that about six various topologies were attained, and the DES performed as structure-directing agent as well as solvent. With reference to the Ln-accommodating structure, the extra-framework choline ion was maintained with Cl^- confined to the layers of polymers. Urea molecules were chemically linked to the structure, along with external choline ion. Further, choline could additionally be held by a chemical bond to MOF framework by way of its OH groups. The urea molecules, which are neutral, could perform as an external structure-directing media. In addition, the authors reported the role of DES in the straight chemical linkage of ethylene urea hemihydrate (e-urea) and m-urea to the polymer structure. Entire novel structures have revealed that the DES can perform as a solvent, and can also act as a ligand supplier and/or template. In another study by Himeur *et al.* [91], the iono-thermal preparation of three advanced lanthanide (Ln)/organic based MOFs present in a DES dimethyl-urea/ChCl were reported. Usage of hydrated mineral Ln precursors inevitably induced the contamination of DES with water. Durable attachment of water molecules, which were isolated, in DES by way of the hydrogen bonding system of connections extremely restricted its reactivity.

Recently, Zhan *et al.* [92] synthesized eight exclusive three dimensional (3-D) Ln-thiophene-2,5-di-carboxylate materials in the DES e-urea/ChCl. In the group of the prepared materials, six compounds were observed to be iso-structural, crystallizing in the $P2_1/c$ space group of monoclinic conformity [Ln(TDC)$_2$](choline) where Ln = Nd, Gd, Er, Eu, Dy, Tb]. In the mentioned frameworks, the choline cation compensated the anionic frameworks' charge. An identical material was achieved employing the IL 1-methyl-3-ethyl-imidazolium bromide, confirming that the guest cation did not perform the template's role. The structure of [Nd$_2$(TDC)$_3$(ethylene-urea)$_4$]·3(ethylene-urea) exhibited three types of guest ethylene-urea molecules as indicated in Figure 6.10. Ethylene-urea A generated durable hydrogen bonds with host structure, ethylene-urea B displayed hydrogen bonding with host structure, however, ethylene-urea C had only poor interactions with host structure. Chen *et al.* [93] stated the preparation of Zn(II)-B(III)-imidazole structure with exceptional pentagonal channels by preparation in DES m-

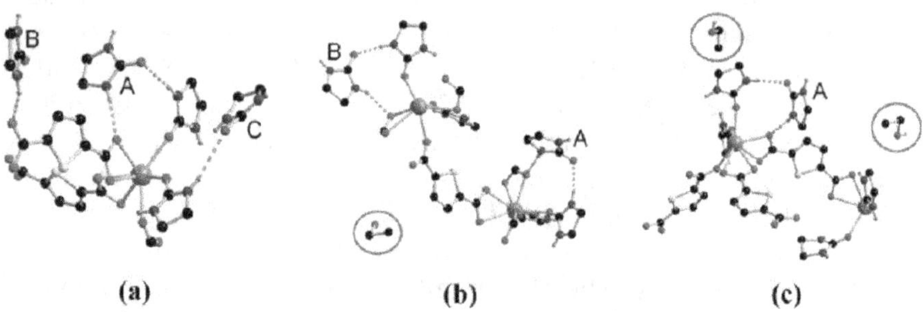

Figure 6.10 (a) Three types of ethylene-ureas (A, B and C) in [Nd$_2$(TDC)$_3$(ethylene-urea)$_4$]·3(ethylene-urea); (b) e-urea C exchanged by ethanol and (c) e-urea B and C exchanged by ethanol. Reproduced from Reference 92 with permission from American Chemical Society.

urea/ChCl. In $Zn_2(im)Cl_2[B(im)_4]$, the neutral ligand and the cation did not incorporate the material framework, however , the DES assisted for stabilizing the halogen-metal bonds. Later, the DES solvent permitted the preparation of material which integrated two different imidazolate structures, with Zn–im connection as well as $B(im)_4^-$ complex.

Metal Phosphates as well as Phosphites with Open-framework

Zeolites or zeolite-like materials are of considerable significance because of their utilization in oil and gas industries. Currently, most of the solid acid catalysts employed in refinery as well as petrochemical industries are based on zeolites. The zeolite-analogous structure materials are generally prepared in organic solvents or water by a conventional solvo-thermal method by employing the structure directing agents. Cooper *et al.* [64] firstly introduced the DES-mediated iono-thermal method for the generation of microporous crystalline zeolite analogues. For the preparation of zeotypes, the DESs based on IL ([EMIm]Br) or urea/ChCl were employed as alternate solvents. Four various alumino-phosphate zeotypes, namely SIZ-5, SIZ-4, SIZ-3, and SIZ-1, were generated in the presence of [EMIm]Br, under various experimental surroundings, where the imidazolium-based IL acted both as template as well as solvent. Unlike IL, the application of DES ChCl/urea brought about a different kind of framework, i.e. SIZ-2. Al–O–P interchange was retained and the structural composition $(NH_4)_3[Al_2(PO_4)_3]$ confirmed that the SIZ-2 too exhibited a disrupted framework. Comparison with the preparation in the IL, it was observed that the incorporation of fluoride to the DES urea/ChCl mix induced the generation of a non-zeolitic alumino-phosphate termed as AlPO-CJ2. Overall, the application of DESs or ILs could adjust the way how the material precursors are put together. As stated by Copper *et al.* [64], this is a critical particularity of these preparations. Depending on the favorable synthesis of SIZ-n type materials, Parnham *et al.* [94] stated the iono-thermal preparation of materials employing different DESs as the template delivery agents. 2-imidazolindone(e-urea), 1,3-Dimethyl urea (DMU) as well as tetrahydro-2-pyrimidione derivatives were examined as the eutectic combinations in consolidation with tetra-ethylammonium bromide or ChCl. The synthesis of materials based on aluminum-phosphate with advanced structures has been achieved utilizing this approach.

Even though the iono-thermal preparation was developed in a comparatively 'dry' and especially 'ionic' surroundings, existence of water may not be eliminated. If existing in very small quantity, water did not exhibit a disadvantageous influence on the DES-mediated iono-thermal preparation of certain materials. Lesser reactivity of water (in very small quantity) was due to the powerful interaction with the anion present in DES that remarkably restricted its reactivity [95]. Particularly, it signifies that advanced inorganic compounds which are reactive to water could be hypothetically synthesized by the iono-thermal method in DESs. In this respect, three distinct DESs based on ChCl/carboxylic acid were employed for the generation of different alumino-phosphate layered materials having chemical configurations which were improbable to be available employing conventional hydro-thermal methods. It was confirmed that the structure of material attained by iono-thermal preparation was affected by the DES chemical composition (e.g. charge balance, template species, mineralizers etc.).

Zinc phosphates are also a critical subset of the large phosphate family. Influenced by the methods stated for the synthesis of alumino-phosphate-based materials, Liao *et al.* [65] illustrated the iono-thermal generation of a $[Zn(O_3PCH_2CO_2)].NH_4$ framework utilizing a urea/ChCl DES as the medium of reaction. Thorough analysis confirmed that NH^{4+}, produced *in situ* by the urea partial decomposition at greater temperature, performed an important task in the generation of the targeted material.

As already stated, the application of DESs with distinct ionic compositions contribute noticeable benefits as green solvents, and also provides rational design of fresh materials with open frameworks. Similar to the zinc-phosphate and alumino-phosphate, zirconium-phosphates having one-dimensional (1 D), two-dimensional (2 D) and three-dimensional (3 D) connectivities are generally synthesized utilizing a solvo-thermal method in organic solvents or water. Liu *et al.* [96] reported the preparation of three zirconium phosphate structures employing H_3PO_4, HF and zirconium (IV) oxychloride in DESs consisting of tetra-methylammonium chloride (TMA)/oxalic acid dehydrate as well as urea/TMA. Recently, Liu *et al.* [97] also prepared a unique open-structure Zr phosphate using small 7- as well as 8-ring channels (ZrPOF-EA, $[(C_2H_7NH)_8(H_2O)_8]$-$[Zr_{32}P_{48}O_{176}F_8(OH)_{16}]$) by the iono-thermal preparation in ethylammonium chloride-oxalic acid DES. In recent studies, the ChCl-1,3-dimethyl urea DES has also been stated as an effective solvent for the synthesis of cobalt phosphate-borate which includes the $[B_5O_6(OH)_4]^-$ anion among the layers of cobalt phosphate [98].

Carbon Materials

The carbon materials are widely employed for different applications such as gas storage and separation, electrochemical devices, catalyst supports, etc. The iono-thermal preparation process is considered as an effective method for the synthesis of crystalline porous materials. In this respect, the ILs have been established to be flexible precursors for the synthesis of porous carbon materials [99]. As reported by Gutierrez *et al.* [100], the generation of carbon materials can also be attained in an identical way in the DESs. As an illustration, the polycondensation of resorcinol-formaldehyde (RF) in the DESs for the synthesis of carbon as well as carbon–carbon nanotube (CNT) composites was performed [100]. During the polycondensation of formaldehyde and resorcinol, the hydroxyl-methyl compounds were generated. The hydroxyl-methyl species reacted to generate methylene as well as methylene-ether joined compounds. The authors systematically examined the water impact, catalyst impact and the function of the DES. During the preparation of the material, the RF polycondensation initially took place in a DES ethylene glycol/ChCl. Later, the resultant RF gels underwent carbonization for the generation of monolithic carbons. The RF gels generated in the DESs exhibited identical surface characteristics, which is considered to be an appealing characteristic of the DESs. The carbon content as well as the conversion rate in the final material was identical to those achieved utilizing the conventional method. Additional appealing characteristic was the capability of DES for dispersing the carbon nanotubes, causing the generation of homogeneous carbon-CNT RF gels. Similar high composite homogeneity is not possible to be achieved in water because of the lower dispersibility of nanotubes in the medium.

Subsequently, Gutierrez *et al.* [56] reported the application of ternary DESs consisting of resorcinol, ChCl and 3-hydroxypyridine as both structure directing agents and precursors for the preparation of nitrogen-doped hierarchic carbon monoliths. The structure of the generated carbons invariably consisted of a bi-continuous porous system of connections composed of immensely cross-linked groups combined into inter-related framework. The DES was illustrated to perform the three distinct roles as (1) a solvent that provides homogeneity of reactants, (2) a source of nitrogen as well as carbon and (3) a structure-directing agent that controls the hierarchical pore framework generation.

In another study, Gutierrez *et al.* [101] illustrated the efficient preparation of hierarchical porous multi-walled CNT (MWCNT) composites in the DESs. The preparation was carried out by means of a furfuryl alcohol (FA) condensation employing the DES ChCl/*p*-toluene sulfonic acid. Further, for attaining a homogeneous dispersion of the CNTs, the DES acted as an excellent medium. The examination of the porosity of the generated carbon-MWCNT composites confirmed that it possessed higher surface areas (400-550 $m^2\ g^{-1}$). In addition, the authors also stated the significant green nature of the recommended procedure as the entire DESs could be used again, because of its complete recovery succeeding the condensation reaction. Thus, it underlined the functional nature of the DES for the generation of advanced materials.

6.4.4 Metal Extraction and Metal Oxides Processing

The metals as well as metal oxides dissolution is essential for a series of extensive procedures like catalyst preparation, remediation of corrosion and metal winning [102]. Processing as well as reprocessing act as the source of large quantities of aqueous waste, with treatment of basic as well as acidic byproducts both chemical in nature and energy intensive. In majority of molecular solvents, the metal oxides are insoluble and are usually soluble only in aqueous alkali or acid. The extraction method is dependent on the value of end product, the original matrix and the demanded purity. In accordance with the current standards for environmental conservation, the entire solutes should be separated before the solvent is released to the environment. The four major techniques employed for recovering the metallic solutes from the solution are ion exchange, precipitation, cementation and electro-deposition. For recovering chromium and aluminum, these should be electro-chemically or chemically reduced in higher temperature operations requiring large energy input.

Currently, research is focused towards employing ILs as well as DESs for the solvato-metallurgical processing. Considerable studies have been carried out on the metal oxides' solubility in a range of DESs. It was confirmed that the type III DESs have the ability to solvate a range of metal oxides [103]; ligands like oxalate, urea and thiourea are famous complexants for a series of metals and could be used with DESs. Abbott *et al.* [103] examined the solubility of a series of metal oxides in a DES of ChCl/urea, and noted that using electrodeposition, the dissolved metals could be reclaimed from the mixed metal oxide matrix.

Separation of metals from a complex mixture can be performed employing electro-chemistry [104]. Solubility of 17 metal oxides present in the elemental mass series Ti through to Zn were reported in three DESs based on ChCl [22], with considerate selection of HBD inducing the selectivity for the extraction of some metals from complicated matrices. In aqueous solutions, the redox potentiality could be judiciously shifted by selective choice of a suitable ligand, although, in principle, it can be attained utilizing the same ligand in DESs. Relative strength of metal–anion complex was mostly obscure, however, still remained a critical parameter to quantify. The first commensurable electro-chemical series in two DESs was developed recently by Abbott *et al.* [102]. It was proved that the metals could be electro-won into their composite materials by employing the dissimilarity in the metals' redox potentials in a model experiment employing Cu/Zn mixtures. The authors also illustrated the application of iodine as a reversible, stable electro-catalyst used for the dissolution of a range of metals. The redox potential of I_2/I^- in a DES ChCl:ethylene glycol was observed to be more positive, when compared to majority of common metals and was proved to be capable of oxidizing as well as recovering gold.

A hybrid DES of urea, ChCl and ethylene glycol was also reported for use in the large scale extraction of zinc (Zn) and lead (Pb) from the electric arc furnace (EAF) dust [105]. DES exhibited good selectivity in the EAF dust towards lead and zinc oxides only, whereas the aluminum and iron oxides were insoluble. The experiment results indicated that Zn could be latterly electro-deposited with an effectiveness of 75%. Nevertheless, this was not cost-effective for a comparatively cheap metal, hence, the most economically feasible solution was the precipitation of ZnCl using ammonia (NH_3), then filtering off the precipitate, and boiling away the NH_3, thus, leaving behind the DES for reusing. A pilot plant was also developed for testing the productiveness of utilizing the DESs for large scale extraction of metals. For the separation of copper oxides as well as fluorides from post etch residues, the DESs ChCl:urea and ChCl:malonic acid were proved to be effective [51]. Further, for the separation of residues from CF_4/O_2 etching of a Cu coated DUV photoresist, the malonic acid based DESs were observed to be efficient. The DESs have also been reported to be promising candidates for the removal of organic sulfides from fuels [106].

6.5 Summary and Outlook

This review outlines the state of the art on the DES types, properties as well as application in the oil and gas industries ranging from gas absorption and enhanced oil recovery (EOR) to the preparation of materials as well as metal extraction and metal oxides processing. The literature studies revealed majority of the DESs to be based on ChCl, with a few research studies also reporting the generation of DESs from other classes of compounds. The DESs have been revealed to display close physico-chemical characteristics (conductivity, density, viscosity) to those of conventional ILs.

In comparison with the conventional methods of EOR, the DESs are proved to be effective and promising candidates for the EOR process. It was observed that the DESs ChCl:glycerol and ChCl:urea have high potential to be utilized as additives in the thermal EOR processes. The ChCl:glycerol DES exhibited superior comprehensive performance in terms of the *in-situ* heavy oil upgrading. Due to the potential characteristics of DESs as CO_2 and SO_2 absorbents, and primarily as gas separating agents, the information on the physico-chemical properties of DESs with regards to their effect on gas separation mechanisms and gas solubility are also presented in the review. The capturing capabilities and properties of DESs have also been compared with ILs. Overall, DESs have the ability to absorb more CO_2 than the toxic ILs because of their tunable chemical as well as physical properties. Regarding the parameters affecting the solubility of CO_2 and SO_2 in DESs, the literature revealed that at constant pressure, the solubility reduced with enhancing the temperature and increased with increasing the pressure at consistent temperature. The solvent composition as well as the mixture ratio also assists the absorption capacity regardless of the preparation method. The literature studies on gas solubility are restricted to CO_2 and SO_2, but studies on the carbon capture from complex gas mixtures like flue gases is generally non-existent. Further, studies reporting the separation of other gases such as H_2, CO, and methane using DESs are very restricted, and there is a requirement for experimental analyses to generate a database on the nature of DESs in co-existence with these gases.

In the area of materials chemistry, for the iono-thermal preparation of a broad series of materials like MOF, carbon, metal phosphates and phosphites with different frameworks and textures, it is possible that the ILs can be favorably substituted by safe as well as cheap DESs. Despite the fact that an appropriate choice of the DES still continues to be a great challenge, the illustrated studies in the review have clearly revealed that substantial materials, from microporous zeo-types to the carbon materials, could be prepared in the DESs. Current application of DESs for the material preparation illustrates the remarkable capability of these solvents for the preparation of novel structures as well as engineered materials. During the preparation process, the DESs might perform multiple roles like solvent, reactant for structure crystallization, structure-directing agent, water inhibitor, etc. Inorganic materials, which are reactive to water, could be effectively synthesized by the iono-thermal method in DESs. Apart from the superior dissolution properties for CO_2, organic molecules and inorganic salts, the DESs could also judiciously dissolve various metal oxides, which signifies capacity for the pure metals recovery, specifically in electro-chemistry.

In conclusion, it can be confirmed that the DESs are an appropriate alternative platform to ILs for technical applications like CO_2 removal, SO_2 removal, EOR and preparation of various materials like MOF, carbon materials as well as metal phosphates and phosphites with open-framework. This is due to the feasibility of tailoring the properties of DES by the proper selection of HBDs, salts and molar ratios. In all the DES-based processes analyzed in this review, it was observed that the usage of DESs not only permitted the design of secure and green processes but also contributed a genuine access to advanced chemicals as well as materials. Even though DESs are not able to currently replace the ILs in entire fields of chemistry, it is envisaged that the attractive price as well as less ecological footprint of DESs will certainly support the industrial development of this advanced medium in a near future.

References

1. Anastas, P. T., and Kirchhoff, M. M. (2002) Origins, current status, and future challenges of green chemistry. *Accounts of Chemical Research*, **35**(9), 686-694.
2. Gani, R., Jimenez-Gonzalez, C., and Constable, D. J. (2005) Method for selection of solvents for promotion of organic reactions. *Comp-*

uters & Chemical Engineering, **29**(7), 1661-1676.

3. Clark, J. H., and Tavener, S. J. (2007) Alternative solvents: shades of green. *Organic Process Research & Development,* **11**(1), 149-155.
4. Walden, P. (1914) Molecular weights and electrical conductivity of several fused salts. *Bull Acad Imp Sci,* **1800**, 405-422.
5. Petkovic, M., Seddon, K. R., Rebelo, L. P. N., and Pereira, C. S. (2011) Ionic liquids: a pathway to environmental acceptability. *Chemical Society Reviews,* **40**(3), 1383-1403.
6. Zhang, Q., Vigier, K. D. O., Royer, S., and Jerome, F. (2012) Deep eutectic solvents: syntheses, properties and applications. *Chemical Society Reviews,* **41**(21), 7108-7146.
7. Deetlefs, M., and Seddon, K. R. (2010) Assessing the greenness of some typical laboratory ionic liquid preparations. *Green Chemistry,* **12**(1), 17-30.
8. Romero, A., Santos, A., Tojo, J., and Rodriguez, A. (2008) Toxicity and biodegradability of imidazolium ionic liquids. *Journal of Hazardous Materials,* **151**(1), 268-273.
9. Abo-Hamad, A., Hayyan, M., AlSaadi, M. A., and Hashim, M. A. (2015) Potential applications of deep eutectic solvents in nanotechnology. *Chemical Engineering Journal,* **273**, 551-567.
10. Smith, E. L., Abbott, A. P., and Ryder, K. S. (2014) Deep eutectic solvents (DESs) and their applications. *Chemical Reviews,* **114**(21), 11060-11082.
11. Hadj-Kali, M. K., Al-khidir, K. E., Wazeer, I., El-blidi, L., Mulyono, S., and AlNashef, I. M. (2015) Application of deep eutectic solvents and their individual constituents as surfactants for enhanced oil recovery. *Colloids and Surfaces A: Physicochemical and Engineering Aspects,* **487**, 221-231.
12. Dai, Y., van Spronsen, J., Witkamp, G. J., Verpoorte, R., and Choi, Y. H. (2013) Natural deep eutectic solvents as new potential media for green technology. *Analytica Chimica Acta,* **766**, 61-68.
13. Zhao, B.-Y., Xu, P., Yang, F.-X., Wu, H., Zong, M.-H., and Lou, W.-Y. (2015) Biocompatible deep eutectic solvents based on choline chloride: Characterization and application to the extraction of Rutin from *Sophora japonica. ACS Sustainable Chemistry & Engineering,* **3**(11), 2746-2755.
14. Abbott, A. P., Capper, G., Davies, D. L., Rasheed, R. K., and Tambyrajah, V. (2003) Novel solvent properties of choline chloride/urea mixtures. *Chemical Communications,* 70-71.
15. Abbott, A. P., Harris, R. C., Ryder, K. S., D'Agostino, C., Gladden, L. F., and Mantle, M. D. (2011) Glycerol eutectics as sustainable solvent systems. *Green Chemistry,* **13**(1), 82-90.
16. Ru B, C., and Konig, B. (2012) Low melting mixtures in organic synthesis-an alternative to ionic liquids? *Green Chemistry,* **14**(11), 2969-2982.

17. Paiva, A., Craveiro, R., Aroso, I., Martins, M., Reis, R. L., and Duarte, A. R. C. (2014) Natural deep eutectic solvents – Solvents for the 21st century. *ACS Sustainable Chemistry & Engineering*, **2**(5), 1063-1071.
18. Tang, B., and Row, K. H. (2013) Recent developments in deep eutectic solvents in chemical sciences. *Monatshefte für Chemie-Chemical Monthly*, **144**(10), 1427-1454.
19. Maugeri, Z., and de Maria, P. D. (2012) Novel choline-chloride-based deep-eutectic-solvents with renewable hydrogen bond donors: levulinic acid and sugar-based polyols. *RSC Advances*, **2**(2), 421-425.
20. Abbott, A. P., Barron, J. C., Ryder, K. S., and Wilson, D. (2007) Eutectic-based ionic liquids with metal-containing anions and cations. *Chemistry - A European Journal*, **13**(22), 6495-6501.
21. Tang, S., Baker, G. A., and Zhao, H. (2012) Ether-and alcohol-functionalized task-specific ionic liquids: attractive properties and applications. *Chemical Society Reviews*, **41**(10), 4030-4066.
22. Abbott, A. P., Capper, G., Davies, D. L., McKenzie, K. J., and Obi, S. U. (2006) Solubility of metal oxides in deep eutectic solvents based on choline chloride. *Journal of Chemical & Engineering Data*, **51**(4), 1280-1282.
23. Chakrabarti, M. H., Mjalli, F. S., AlNashef, I. M., Hashim, M. A., Hussain, M. A., Bahadori, L., and Low, C. T. J. (2014) Prospects of applying ionic liquids and deep eutectic solvents for renewable energy storage by means of redox flow batteries. *Renewable and Sustainable Energy Reviews*, **30**, 254-270.
24. Abbott, A. P., Boothby, D., Capper, G., Davies, D. L., and Rasheed, R. K. (2004) Deep eutectic solvents formed between choline chloride and carboxylic acids: versatile alternatives to ionic liquids. *Journal of the American Chemical Society*, **126**(29), 9142-9147.
25. Kareem, M. A., Mjalli, F. S., Hashim, M. A., and AlNashef, I. M. (2010) Phosphonium-based ionic liquids analogues and their physical properties. *Journal of Chemical & Engineering Data*, **55**(11), 4632-4637.
26. Abbott, A. P., Capper, G., and Gray, S. (2006) Design of improved deep eutectic solvents using hole theory. *ChemPhysChem*, **7**(4), 803-806.
27. D'Agostino, C., Harris, R. C., Abbott, A. P., Gladden, L. F., and Mantle, M. D. (2011) Molecular motion and ion diffusion in choline chloride based deep eutectic solvents studied by 1 H pulsed field gradient NMR spectroscopy. *Physical Chemistry Chemical Physics*, **13**(48), 21383-21391.
28. Abbott, A. P., Harris, R. C., and Ryder, K. S. (2007) Application of hole theory to define ionic liquids by their transport properties. *The Journal of Physical Chemistry B*, **111**(18), 4910-4913.

29. Yadav, A., and Pandey, S. (2014) Densities and viscosities of (choline chloride+ urea) deep eutectic solvent and its aqueous mixtures in the temperature range 293.15 K to 363.15 K. *Journal of Chemical & Engineering Data*, **59**(7), 2221-2229.
30. Franca, J. M., Nieto de Castro, C. A., Lopes, M. M., and Nunes, V. M. (2009) Influence of thermophysical properties of ionic liquids in chemical process design. *Journal of Chemical & Engineering Data*, **54**(9), 2569-2575.
31. Valderrama, J. O. (2003) The state of the cubic equations of state. *Industrial & Engineering Chemistry Research*, **42**(8), 1603-1618.
32. Shahbaz, K., Mjalli, F. S., Hashim, M., and AlNashef, I. M. (2011) Prediction of deep eutectic solvents densities at different temperatures. *Thermochimica Acta*, **515**(1), 67-72.
33. Yadav, A., Trivedi, S., Rai, R., and Pandey, S. (2014) Densities and dynamic viscosities of (choline chloride+ glycerol) deep eutectic solvent and its aqueous mixtures in the temperature range (283.15–363.15) K. *Fluid Phase Equilibria*, **367**, 135-142.
34. Leron, R. B., and Li, M. H. (2012) High-pressure density measurements for choline chloride: Urea deep eutectic solvent and its aqueous mixtures at T=(298.15 to 323.15) K and up to 50MPa. *The Journal of Chemical Thermodynamics*, **54**, 293-301.
35. Abbott, A. P., Ahmed, E. I., Harris, R. C., and Ryder, K. S. (2014) Evaluating water miscible deep eutectic solvents (DESs) and ionic liquids as potential lubricants. *Green Chemistry*, **16**(9), 4156-4161.
36. Mirza, N. R., Nicholas, N. J., Wu, Y., Kentish, S., and Stevens, G. W. (2015) Estimation of normal boiling temperatures, critical properties, and acentric factors of deep eutectic solvents. *Journal of Chemical & Engineering Data*, **60**(6), 1844-1854.
37. Florindo, C., Oliveira, F. S., Rebelo, L. P. N., Fernandes, A. M., and Marrucho, I. M. (2014) Insights into the synthesis and properties of deep eutectic solvents based on cholinium chloride and carboxylic acids. *ACS Sustainable Chemistry & Engineering*, **2**(10), 2416-2425.
38. Hou, Y., Gu, Y., Zhang, S., Yang, F., Ding, H., and Shan, Y. (2008) Novel binary eutectic mixtures based on imidazole. *Journal of Molecular Liquids*, **143**(2), 154-159.
39. Hurley, F. H., and Wler, T. P., Jr. (1951) The electrodeposition of aluminum from nonaqueous solutions at room temperature. *Journal of the Electrochemical Society*, **98**(5), 207-212.
40. Xu, W. G., Lu, X. M., Zhang, Q. G., Gui, J. S., and Yang, J. Z. (2006) Studies on the thermodynamic properties of the ionic liquid BMIGaCl4. *Chinese Journal of Chemistry*, **24**(3), 331-335.
41. Yang, J. Z., Tian, P., He, L. L., and Xu, W. G. (2003) Studies on room temperature ionic liquid InCl 3–EMIC. *Fluid Phase Equilibria*, **204**(2), 295-302.

42. Abbott, A. P., Capper, G., Davies, D. L., and Rasheed, R. (2004) Ionic liquids based upon metal halide/substituted quaternary ammonium salt mixtures. *Inorganic Chemistry*, **43**(11), 3447-3452.
43. Abbott, A. P., Capper, G., Davies, D. L., Munro, H. L., Rasheed, R. K., and Tambyrajah, V. (2001) Preparation of novel, moisture-stable, Lewis-acidic ionic liquids containing quaternary ammonium salts with functional side chains. *Chemical Communications*, 2010-2011.
44. Mbous, Y. P., Hayyan, M., Hayyan, A., Wong, W. F., Hashim, M. A., and Looi, C. Y. (2017) Applications of deep eutectic solvents in biotechnology and bioengineering - Promises and challenges. *Biotechnology Advances*, **35**(2), 105-134.
45. Abbott, A. P., Capper, G., Davies, D. L., and Rasheed, R. K. (2004) Ionic liquid analogues formed from hydrated metal salts. *Chemistry - A European Journal*, **10**(15), 3769-3774.
46. Abbott, A. P., Capper, G., Davies, D. L., Rasheed, R. K., Archer, J., and John, C. (2004) Electrodeposition of chromium black from ionic liquids. *Transactions of the Institute of Metal Finishing*, **82**, 14-17.
47. Abbott, A. P., Cullis, P. M., Gibson, M. J., Harris, R. C., and Raven, E. (2007) Extraction of glycerol from biodiesel into a eutectic based ionic liquid. *Green Chemistry*, **9**(8), 868-872.
48. Abbott, A. P., Bell, T. J., Handa, S., and Stoddart, B. (2006) Cationic functionalisation of cellulose using a choline based ionic liquid analogue. *Green Chemistry*, **8**(9), 784-786.
49. Gambino, M., Gaune, P., Nabavian, M., Gaune-Escard, M., and Bros, J. P. (1987) Enthalpie de fusion de l'uree et de quelques melanges eutectiques a base d'uree. *Thermochimica Acta*, **111**, 37-47.
50. Gambino, M., and Bros, J. P. (1988) Capacite calorifique de l'uree et de quelques melanges eutectiques a base d'uree entre 30 et 140 C. *Thermochimica Acta*, **127**, 223-236.
51. Taubert, J., and Raghavan, S. (2014) Effect of composition of post etch residues (PER) on their removal in choline chloride–malonic acid deep eutectic solvent (DES) system. *Microelectronic Engineering*, **114**, 141-147.
52. Smiglak, M., Reichert, W. M., Holbrey, J. D., Wilkes, J. S., Sun, L., Thrasher, J. S., Kirichenko, K., Singh, S., Katritzky, A. R., and Rogers, R. D. (2006) Combustible ionic liquids by design: is laboratory safety another ionic liquid myth? *Chemical Communications*, 2554-2556.
53. Biczak, R., Pawłowska, B., Bałczewski, P., and Rychter, P. (2014) The role of the anion in the toxicity of imidazolium ionic liquids. *Journal of Hazardous Materials*, **274**, 181-190.
54. Pham, T. P. T., Cho, C. W., and Yun, Y. S. (2010). Environmental fate and toxicity of ionic liquids: a review. *Water Research*, **44**(2), 352-372.
55. Carriazo, D., Gutiérrez, M. C., Ferrer, M. L., and del Monte, F. (2010)

Resorcinol-based deep eutectic solvents as both carbonaceous precursors and templating agents in the synthesis of hierarchical porous carbon monoliths. *Chemistry of Materials*, **22**(22), 6146-6152.

56. Gutierrez, M. C., Carriazo, D., Ania, C. O., Parra, J. B., Ferrer, M. L., and del Monte, F. (2011) Deep eutectic solvents as both precursors and structure directing agents in the synthesis of nitrogen doped hierarchical carbons highly suitable for CO_2 capture. *Energy & Environmental Science*, **4**(9), 3535-3544.
57. Zhu, A., Jiang, T., Han, B., Zhang, J., Xie, Y., and Ma, X. (2007) Supported choline chloride/urea as a heterogeneous catalyst for chemical fixation of carbon dioxide to cyclic carbonates. *Green Chemistry*, **9**(2), 169-172.
58. Hao, G. P., Jin, Z. Y., Sun, Q., Zhang, X. Q., Zhang, J. T., and Lu, A. H. (2013) Porous carbon nanosheets with precisely tunable thickness and selective CO_2 adsorption properties. *Energy & Environmental Science*, **6**(12), 3740-3747.
59. Leron, R. B., and Li, M. H. (2013) Solubility of carbon dioxide in a eutectic mixture of choline chloride and glycerol at moderate pressures. *The Journal of Chemical Thermodynamics*, **57**, 131-136.
60. Leron, R. B., and Li, M. H. (2013) Solubility of carbon dioxide in a choline chloride–ethylene glycol based deep eutectic solvent. *Thermochimica Acta*, **551**, 14-19.
61. Lin, C. M., Leron, R. B., Caparanga, A. R., and Li, M. H. (2014) Henry's constant of carbon dioxide-aqueous deep eutectic solvent (choline chloride/ethylene glycol, choline chloride/glycerol, choline chloride/malonic acid) systems. *The Journal of Chemical Thermodynamics*, **68**, 216-220.
62. Leron, R. B., Caparanga, A., and Li, M. H. (2013). Carbon dioxide solubility in a deep eutectic solvent based on choline chloride and urea at T= 303.15–343.15 K and moderate pressures. *Journal of the Taiwan Institute of Chemical Engineers*, **44**(6), 879-885.
63. Ali, E., Hadj-Kali, M. K., Mulyono, S., Alnashef, I., Fakeeha, A., Mjalli, F., and Hayyan, A. (2014) Solubility of CO 2 in deep eutectic solvents: experiments and modelling using the Peng–Robinson equation of state. *Chemical Engineering Research and Design*, **92**(10), 1898-1906.
64. Cooper, E. R., Andrews, C. D., Wheatley, P. S., Webb, P. B., Wormald, P., and Morris, R. E. (2004) Ionic liquids and eutectic mixtures as solvent and template in synthesis of zeolite analogues. *Nature*, **430**(7003), 1012-1016.
65. Liao, J. H., Wu, P. C., and Bai, Y. H. (2005) Eutectic mixture of choline chloride/urea as a green solvent in synthesis of a coordination polymer: [Zn $(O_3PCH_2CO_2)$]·NH_4. *Inorganic Chemistry Communications*, **8**(4), 390-392.

66. Li, X., Hou, M., Han, B., Wang, X., and Zou, L. (2008) Solubility of CO_2 in a choline chloride+ urea eutectic mixture. *Journal of Chemical & Engineering Data*, **53**(2), 548-550.
67. Sze, L. L., Pandey, S., Ravula, S., Pandey, S., Zhao, H., Baker, G. A., and Baker, S. N. (2014) Ternary deep eutectic solvents tasked for carbon dioxide capture. *ACS Sustainable Chemistry & Engineering*, **2**(9), 2117-2123.
68. Francisco, M., van den Bruinhorst, A., Zubeir, L. F., Peters, C. J., and Kroon, M. C. (2013) A new low transition temperature mixture (LTTM) formed by choline chloride+ lactic acid: characterization as solvent for CO_2 capture. *Fluid Phase Equilibria*, **340**, 77-84.
69. Gutierrez Ortiz, F. J., Vidal, F., Ollero, P., Salvador, L., Cortes, V., and Gimenez, A. (2006) Pilot-plant technical assessment of wet flue gas desulfurization using limestone. *Industrial & Engineering Chemistry Research*, **45**(4), 1466-1477.
70. Yang, D., Hou, M., Ning, H., Zhang, J., Ma, J., Yang, G., and Han, B. (2013) Efficient SO_2 absorption by renewable choline chloride–glycerol deep eutectic solvents. *Green Chemistry*, **15**(8), 2261-2265.
71. Sun, S., Niu, Y., Xu, Q., Sun, Z., and Wei, X. (2015) Efficient SO_2 absorptions by four kinds of deep eutectic solvents based on choline chloride. *Industrial & Engineering Chemistry Research*, **54**(33), 8019-8024.
72. Liu, B., Zhao, J., and Wei, F. (2013) Characterization of caprolactam based eutectic ionic liquids and their application in SO_2 absorption. *Journal of Molecular Liquids*, **180**, 19-25.
73. Zhang, K., Ren, S., Hou, Y., and Wu, W. (2017) Efficient absorption of SO_2 with low-partial pressures by environmentally benign functional deep eutectic solvents. *Journal of Hazardous Materials*, **324**, 457-463.
74. Guo, B., Duan, E., Ren, A., Wang, Y., and Liu, H. (2009) Solubility of SO_2 in caprolactam tetrabutyl ammonium bromide ionic liquids. *Journal of Chemical & Engineering Data*, **55**(3), 1398-1401.
75. Duan, E., Guo, B., Zhang, M., Guan, Y., Sun, H., and Han, J. (2011) Efficient capture of SO_2 by a binary mixture of caprolactam tetrabutyl ammonium bromide ionic liquid and water. *Journal of Hazardous Materials*, **194**, 48-52.
76. Benzagouta, M. S., AlNashef, I. M., Karnanda, W., and Al-Khidir, K. (2013) Ionic liquids as novel surfactants for potential use in enhanced oil recovery. *Korean Journal of Chemical Engineering*, **30**(11), 2108-2117.
77. Hezave, A. Z., Dorostkar, S., Ayatollahi, S., Nabipour, M., and Hemmateenejad, B. (2013) Dynamic interfacial tension behavior between heavy crude oil and ionic liquid solution (1-dodecyl-3-methylimidazolium chloride ([C 12 mim][Cl]+ distilled or saline

water/heavy crude oil]) as a new surfactant. *Journal of Molecular Liquids*, **187**, 83-89.

78. Xie, X., Weiss, W. W., Tong, Z. J., and Morrow, N. R. (2005) Improved oil recovery from carbonate reservoirs by chemical stimulation. *SPE Journal*, **10**(3), 276-285.
79. Mohsenzadeh, A., Nabipour, M., Asadizadeh, S., Nekouie, M., Ameri, A., and Ayatollahi, S. (2011) Experimental investigation of different steam injection scenarios during SAGD process. *Special Topics & Reviews in Porous Media: An International Journal*, **2**(4), 283-291.
80. Canbolat, S., Akin, S., and Kovscek, A. R. (2004) Non-condensable gas steam-assisted gravity drainage. *Journal of Petroleum Science and Engineering*, **45**(1), 83-96.
81. Nares, R., Schacht-Hernandez, P., Ramirez-Garnica, M. A., and Cabrera-Reyes, M. D. C. (2007) Upgrading Heavy and Extra Heavy Crude Oil with Ionic Liquid. *International Oil Conference and Exhibition,* Mexico. Online: https://www.onepetro.org/conference-paper/SPE-108676-MS [assessed 29th April 2017].
82. Mohsenzadeh, A., Al-Wahaibi, Y., Jibril, A., Al-Hajri, R., and Shuwa, S. (2015) The novel use of deep eutectic solvents for enhancing heavy oil recovery. *Journal of Petroleum Science and Engineering,* **130**, 6-15.
83. Mohsenzadeh, A., Al-Wahaibi, Y., Al-Hajri, R., and Jibril, B. (2015) Effects of concentration, salinity and injection scenario of ionic liquids analogue in heavy oil recovery enhancement. *Journal of Petroleum Science and Engineering,* **133**, 114-122.
84. Mohsenzadeh, A., Al-Wahaibi, Y., Al-Hajri, R., Jibril, B., Joshi, S., and Pracejus, B. (2015). Investigation of formation damage by Deep Eutectic Solvents as new EOR agents. *Journal of Petroleum Science and Engineering,* **129**, 130-136.
85. Al-Weheibi, I., Al-Hajri, R., Al-Wahaibi, Y., Jibril, B., and Mohsenzadeh, A. (2015). Oil Recovery Enhancement in Middle East Heavy Oil Field using Malonic Acid based Deep Eutectic Solvent. *SPE Middle East Oil & Gas Show and Conference,* Bahrain. Online: https://www.onepetro.org/conference-paper/SPE-172592-MS [assessed 20th April 2017].
86. Shuwa, S., Jibril, B., Al-Wahaibi, Y., and Al-Hajri, R. (2014) Heavy-oil-recovery enhancement with choline chloride/ethylene glycol-based deep eutectic solvent. *SPE Journal*, **20**(01), 79-87.
87. Mohsenzadeh, A., Al-Wahaibi, Y., Al-Hajri, R., Jibril, B., and Mosavat, N. (2017) Sequential deep eutectic solvent and steam injection for enhanced heavy oil recovery and in-situ upgrading. *Fuel*, **187**, 417-428.
88. Ramsahye, N. A., Maurin, G., Bourrelly, S., Llewellyn, P. L., Serre, C., Loiseau, T., Serre, C., and Ferey, G. (2008) Probing the adsorption sites for CO_2 in metal organic frameworks materials MIL-53 (Al,

Cr) and MIL-47 (V) by density functional theory. *The Journal of Physical Chemistry C*, **112**(2), 514-520.

89. Trung, T. K., Trens, P., Tanchoux, N., Bourrelly, S., Llewellyn, P. L., Loera-Serna, S., Serre, C., Loiseau, T., Fajula, F., & Ferey, G. (2008) Hydrocarbon adsorption in the flexible metal organic frameworks MIL-53 (Al, Cr). *Journal of the American Chemical Society*, **130**(50), 16926-16932.
90. Zhang, J., Wu, T., Chen, S., Feng, P., and Bu, X. (2009) Versatile structure-directing roles of deep-eutectic solvents and their implication in the generation of porosity and open metal sites for gas storage. *Angewandte Chemie International Edition*, **48**(19), 3486-3490.
91. Himeur, F., Stein, I., Wragg, D. S., Slawin, A. M., Lightfoot, P., and Morris, R. E. (2010) The ionothermal synthesis of metal organic frameworks, Ln $(C_9O_6H_3)((CH_3NH)_2CO)_2$, using deep eutectic solvents. *Solid State Sciences*, **12**(4), 418-421.
92. Zhan, C. H., Wang, F., Kang, Y., and Zhang, J. (2011) Lanthanide-thiophene-2,5-dicarboxylate frameworks: ionothermal synthesis, helical structures, photoluminescent properties, and single-crystal-to-single-crystal guest exchange. *Inorganic Chemistry*, **51**(1), 523-530.
93. Chen, S., Zhang, J., Wu, T., Feng, P., and Bu, X. (2010) Zinc (II)-boron (III)-imidazolate framework (ZBIF) with unusual pentagonal channels prepared from deep eutectic solvent. *Dalton Transactions*, **39**(3), 697-699.
94. Parnham, E. R., Drylie, E. A., Wheatley, P. S., Slawin, A. M., and Morris, R. E. (2006) Ionothermal materials synthesis using unstable deep-eutectic solvents as template-delivery agents. *Angewandte Chemie*, **118**(30), 5084-5088.
95. Cammarata, L., Kazarian, S. G., Salter, P. A., and Welton, T. (2001) Molecular states of water in room temperature ionic liquids. *Physical Chemistry Chemical Physics*, **3**(23), 5192-5200.
96. Liu, L., Li, Y., Wei, H., Dong, M., Wang, J., Slawin, A. M., Li, J., Dong, J., and Morris, R. E. (2009) Ionothermal synthesis of zirconium phosphates and their catalytic behavior in the selective oxidation of cyclohexane. *Angewandte Chemie, International Edition*, **48**(12), 2206-2209.
97. Liu, L., Yang, J., Li, J., Dong, J., Sisak, D., Luzzatto, M., and McCusker, L. (2011) Ionothermal synthesis and structure analysis of an open-framework zirconium phosphate with a high CO_2/CH_4 adsorption ratio. *Angewandte Chemie International Edition*, **50**(35), 8139-8142.
98. Lin, Z., Wragg, D. S., Lightfoot, P., and Morris, R. E. (2009) A novel non-centrosymmetric metallophosphate-borate compound via ionothermal synthesis. *Dalton Transactions*, 5287-5289.

99. Paraknowitsch, J. P., Zhang, J., Su, D., Thomas, A., and Antonietti, M. (2010) Ionic liquids as precursors for nitrogen-doped graphitic carbon. *Advanced Materials*, **22**(1), 87-92.
100. Gutierrez, M. C., Rubio, F., and del Monte, F. (2010) Resorcinol-formaldehyde polycondensation in deep eutectic solvents for the preparation of carbons and carbon–carbon nanotube composites. *Chemistry of Materials*, **22**(9), 2711-2719.
101. Gutierrez, M. C., Carriazo, D., Tamayo, A., Jimenez, R., Pice, F., Rojo, J. M., Ferrer, M. L., and del Monte, F. (2011) Deep-eutectic-solvent-assisted synthesis of hierarchical carbon electrodes exhibiting capacitance retention at high current densities. *Chemistry - A European Journal*, **17**(38), 10533-10537.
102. Abbott, A. P., Frisch, G., Hartley, J., and Ryder, K. S. (2011) Processing of metals and metal oxides using ionic liquids. *Green Chemistry*, **13**(3), 471-481.
103. Abbott, A. P., Capper, G., Davies, D. L., Rasheed, R. K., and Shikotra, P. (2005) Selective extraction of metals from mixed oxide matrixes using choline-based ionic liquids. *Inorganic Chemistry*, **44**(19), 6497-6499.
104. Abbott, A. P., Capper, G., Davies, D. L., and Shikotra, P. (2006) Processing metal oxides using ionic liquids. *Mineral Processing and Extractive Metallurgy*, **115**(1), 15-18.
105. Abbott, A. P., Collins, J., Dalrymple, I., Harris, R. C., Mistry, R., Qiu, F., Scheirer, J., and Wise, W. R. (2009) Processing of electric arc furnace dust using deep eutectic solvents. *Australian Journal of Chemistry*, **62**(4), 341-347.
106. Li, C., Li, D., Zou, S., Li, Z., Yin, J., Wang, A. L., Cui, Y. N., Yao, Z. L., and Zhao, Q. (2013) Extraction desulfurization process of fuels with ammonium-based deep eutectic solvents. *Green Chemistry*, **15**(10), 2793-2799.

7

Polyurethane Membranes for Gas Separation

7.1 Introduction

Gas separation is one of the main processes in gas treatment plants in order to purify the gas streams or to recover useful gases. [1-4]. Sipek *et al.* [5] reviewed polymeric membranes as effective materials for separation of gases and vapors. A variety of glassy and rubbery polymers have been explored as membrane materials for separations of gaseous streams. Recently, Bernardo *et al.* [6] has also explored industrial applications and process intensification options for membrane gas separations. Membrane separation systems, being simpler in operation and maintenance, reliable, compact, efficient, are being widely used as an alternative to other cumbersome techniques such as absorption, adsorption or cryogenic distillation. Polymeric membranes for gas separation are widely used in diverse areas like carbon dioxide recovery, helium gas removal in natural gas purification, hydrogen recovery in ammonia plant purge streams, oxygen and nitrogen separations, CO_2 recovery from biogas, oxygen upgrading from air, etc. [7-9].

Polyurethanes (PU) are a versatile category of polymers possessing good physical and tensile strength, chemical resistance, biocompatibility and mechanical properties. These properties make polyurethanes promising candidates for membrane separation. To underline the importance of polyurethane membranes for separation processes, a large number of reviews have been generated exploring the synthesis [10-12], versatility [13], and applications of polyurethanes as rigid foams [14], in scaffolds [15,16], in pharmacy [17], in composites [18], for separation applications in the form of foam sorbents [19], in medicine for bio-stability and carcinogenicity [20], for drug delivery [21], for cancer therapy [22], in spine surgery [23], for water purification [24], among others [25-27]. Polyurethanes consist of hard (glassy) and soft (elastomeric) segments [28]. The hard segments act as physical crosslinks as well as fillers and

Gigi George, Nidhika Bhoria and Vikas Mittal, Khalifa University of Science and Technology, Abu Dhabi, UAE*
**Current address: CMS College, Mahatma Gandhi University, Kerala, India*

are in an amorphous glassy or crystalline state. On the other hand, the soft segments are rubbery which provide the polyurethanes both flexibility as well as elasticity. Polyurethanes are generally synthesized by the reaction of polyfunctional isocyanates with polyols. By variation in reagents' molecular chain length, chemical nature and functionalities, a wide range of linear or cross-linked polyurethanes with different physiochemical properties can be obtained. Accordingly, polyurethane can vary in terms of chain length, free volume, density of polar groups, etc. Thus, various polyurethane grades though have the same chemical urethane group, however, behave differently depending on the reactants selected.

The soft segments in polyurethanes are generally made of polyether. By chain extending a terminal diisocyanate with a low molecular weight diol or diamine, the hard segment is usually prepared. Occasionally, polyurethanes undergo phase separation because of the hard and soft segment incompatibility. The intermolecular hydrogen bonding interactions between hard segments is the major reason for domain formation. A number of characterization techniques like dynamic mechanical analysis [29-31], thermal analysis [32-37], infrared spectroscopy [35-40] and scattering techniques [41] have been used for the study of polyurethanes morphology and hydrogen bonding. Phase separation into hard and soft segment domains is possible due to their type and process parameters used [31,42,43]. Howarth *et al.* [44] has also reviewed the various synthesis techniques of polyurethanes and suggested some modifications for the future. Overall, with different chemical characteristics and microstructures, polyurethanes are of high potential in gas separation [45]. Polyurethanes can behave as thermoplastic and thermosetting materials depending upon the chemical and morphological fabrication. Recently, George *et al.* [46] reviewed the polymer membranes for their acid gas separation applications including polyurethanes. For instance, Figure 7.1 also summarizes the permeability of various membranes for hydrogen sulfide, H_2S. As can be seen that there is still a large degree of advancement needed to generate high permeability PU membranes to match existing membrane systems.

In this review, various literature studies specifically reporting the recent developments in the synthesis and structure-property correlations of the polyurethane membranes for specific applications as gas separation materials have been reviewed in detail, along with presenting the future prospects.

7.2 Theory of Membrane Gas Separation

In membrane separation performance, the essential characterizing features are the permeability coefficient, P_X, and the selectivity $\alpha_{A/B} = P_A/P_B$, where P_A is the more permeable gas's permeability while P_B

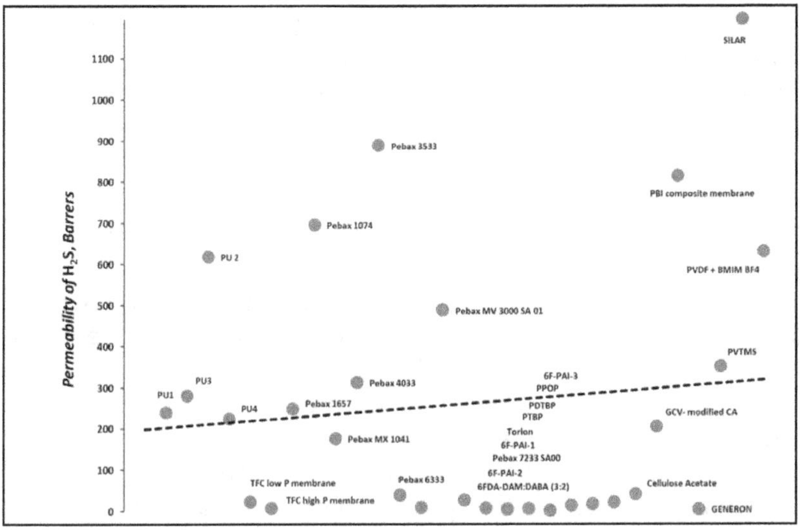

Figure 7.1 H_2S permeability of various membranes. Reproduced from Reference 46 with permission from Elsevier.

is that of the lesser permeable gas [47]. The solution-diffusion phenomenon describes the gas transport behavior of polymeric membranes. P_X is the contributed result of solubility and diffusivity through the membranes. Both solubility and diffusivity are polymer and penetrant dependent. Apart from polymer structure, a number of other parameters are responsible for its gas transport properties. The open volume available between polymer chains through which gas penetrant can pass through is an important parameter. Altering the polymer structure changes the T_g and the free volume which in turn can significantly influence the solubility and diffusivity. As the polymer chain packing efficiency increases, the free volume decreases thereby decreasing the permeability. The increasing penetrant size of gas molecules decreases the diffusion coefficient. Solubility is related to the condensability of the penetrant. The gas and vapor transport are strongly influenced by prevailing pressure and temperature conditions. According to the dual-mode sorption in glassy polymers and Henry's law in rubbery polymers, solubility increases with increase in pressure, while the changes in temperature have a reverse effect [48]. Many research studies have reported the gas permeation behavior through polyurethane membranes and established relationships between diffusion coefficient and the free volume using William-Landel-Ferry theory [49-51]. Schneider *et al.* [52] also observed that the changed glass transition temperature of polyurethanes due to variation in soft segment contents does not impact the gas and vapor permeation behavior [52].

7.3 Ideal Membrane Characteristics

As mentioned earlier, in polymeric membranes, the permeability can be evaluated based on diffusivity and solubility of the permeant into the polymer. Diffusivity is a kinetic term which is a function of the permeant molecules mobility inside the polymer. It depends on the free volume and mobile chain length of the polymer as well as the molecular diameter of the permeant. Solubility shows the affinity between the polymer and the permeant and is a thermodynamic term. Gas transport through membranes is strongly influenced by the polymer's state, i.e., glassy or rubbery. Generally, the gas sorption process in rubbery polymers, being in equilibrium state, ($T>T_g$) is simpler than that of glassy polymers ($T<T_g$) [53-55].

The selectivity, permeability and life-time are the evaluating parameters for polymeric membrane efficiency [56]. The extent of the recovery of the separation process is directly impacted by the selectivity and indirectly by the feed gas flow requirements and the total membrane area. The permeability directly affects the amount of membrane requirement, while the membrane durability affects the maintenance costs. For making membrane-based separations economical than other conservative processes, the polymeric membrane materials should have both high permeability and selectivity [57,58]. The efficiency and performance of the membrane systems also relies on the membrane configurations and modules, irrespective of the inherent properties of the polymeric material. An integration of all these features is important for generating a commercially viable product. High gas fluxes are characteristics of thin layers because of which packing of large membrane areas per unit volume can be achieved. The higher the degree of crosslinking in the membranes, the lower will be the permeability as the gas diffusion coefficients are much lower [59].

7.4 Influence of Polyurethane Structure on Permeability and Selectivity

7.4.1 Influence of Hard and Soft Segments

The majority of the permeability data so far for gas separation via polyurethane membranes are below Robeson's upper bound limit for gas pairs (CO_2/CH_4) and (O_2/N_2) [60]. There exists a tradeoff between the permeability and selectivity and a large number of studies have been carried out to overcome permeability-selectivity tradeoff relationship [61-63]. In polyurethanes, the gas permeation mainly occurs through the soft segments, while the hard segment act as physical crosslinks or impermeable filler in rubbery soft segment phase. An increase in the gas separation performance of polyurethane membranes has been a challenging task. Attempts have been made to enhance the permeability and selectivity of the membranes by optimizing the hard segment content [51]. Few of this include, studying the dependence on hard and soft segments in polyurethane membranes for gas permeability behavior [64], dependence on polymerization methods [65], using metal chelated polyurethane membranes [66-68], etc. Attempts to enhance the performance have also been made through the addition of amine and carboxyl functionalities [69] including the use of PDMS [70-72], epoxidation of hydroxyl terminated polybutadiene [73], polycarbonate–polyurethane membrane [74], etc.

Another attempt to advance the gas permeability and selectivity also included the usage of polymer blends [75-77]. In homogeneous blends, the interaction between the polymers influences the diffusion process, while in the case of heterogeneous blends, degree of heterogeneity influences the permeability significantly [78]. Different studies have reported on structure-morphology-property relationship of the thermoplastic polyurethanes by studying the synthesis, hard segment effect, hydrogen bonding effect, and properties [64,69,79-82]. Mohammadi *et al.* [83] studied the effects of temperature, pressure and stage cut on the gas transport properties through poly(ester urethane urea) for pure gases (CO_2, N_2, CH_4) as well as ternary gas mixtures of CO_2, CH_4, and H_2S. H_2S/CH_4 and CO_2/CH_4 selectivities of 43 and 16 and average permeabilities of 95 and 45 for H_2S and CO_2 were reported.

Wolinska-Grabczyk *et al.* [84] studied poly(acrylonitrile-*co*-butadiene)-based polyurethanes with varied extent of nitrile groups for CO_2 membrane gas separations. It was observed that with higher extent of polar nitrile group polymers, the permeability decreased, while the permselectivity increased. The polymer had less severe trade-off between the gas transport properties and exhibited much higher permeabilites. The polymer structure was analyzed and a relationship was established between the permeability and glass transition temperature.

Talakesh *et al.* [85] studied the polyether based polyurethanes with different hard and soft segments which were prepared by thermal phase inversion method. The soft segments were PEG (2000g/mol), PTMG (2000g/mol) and PTMG/PEG mixture, etc. It was reported that the chain mobility got restricted as the phase separation of hard and soft segments decreased. This led to a hike in the T_g values of the soft segment. By altering the physical conditions, the gas transport properties were studied using constant pressure method. With increase in the ether group content of the polymer structure, gas permeability decreased for pure gases, while CO_2/N_2 ideal selectivity increased. A drop in CO_2 permeability from nearly 130 Barrer in polyurethane containing PTMG in soft segment to 20 barrer in the PU containing PEG in soft segment was observed. An increase from 28 to 90 in CO_2/N_2 selectivity was reported. It was observed that highest selectivity for the polyurethane membranes with 75:25 wt% ratio of PEG:PTMG. Poreba *et al.* [86] also studied nanocomposites based on polycarbonate-based polyurethane with bentonite for thermal, mechanical and gas transport properties. Hexamethylene diisocyanate and butane-1,4-diol were used for hard segment formation. High degree of phase separation was observed for polyurethane based polymer and its nanocomposites. The gas permeation properties exhibited dependence on hard segment content, though not much significant change was observed after varying the polymer structures.

Khosravi *et al.* [87] studied the effect of polyurethane membrane with different polyol, diisocyanate, and chain extender on gas permeability. More phase interaction was observed by the changing of polyol type. The hydrogen bonding, which caused the packing density of the hard segments, was increased by the phase separation of hard and soft segments. This was achieved by changing the diisocyanate groups from cyclic aliphatic to linear aliphatic ones. The phase separation of hard and soft segments was increased on changing the chain extender from a diol one to a diamine. It was observed that more condensable gases permeate more in rubbery polymers and the solubility governs the selectivity of polyurethane. As the microphase separation increased, the permeability and rubbery behavior increased along with their selectivity. Due to the higher rubbery property, polypropylene glycol based polymers exhibited maximum permeability. The high C_3H_8/CH_4 selectivity of 5.47 and permeability of 200 Barrer was reported. Wang *et al.* [88] considered a series of polyurethane films consisting of hydroxyl terminated polybutadiene/acrylonitrile as soft segment, and consisting of TDI and butanediol as hard segments. It was observed that a direct relationship existed between the gas permeability and free-volume. This relation was based on the free-volume parameters and gas diffusivity. The free-volume played a significant part in determining the gas permeability. Marques *et al.* [89] also studied the free volume in polyurethane membranes using positron annihilation spectroscopy. Further, the authors studied the gas permeability and temperature-dependent free volume correlation in polyurethane membranes. Scholten *et al.* [90] also reported electrospun fibers based on polyurethane for removing volatile organic compounds from air. Polyurethanes had 4,4-methylene-bis(phenylisocyanate) (MDI) and aliphatic isophorone diisocyanate as hard segments, whereas butanediol and tetramethylene glycol were used to form the soft segments (Figure 7.2). The sorption performance and capacity of the generated polyurethane fibers was observed to be similar to activated carbon, thus, indicating advanced performance due to balance of hard and soft segments.

Sadeghi *et al.* [91] studied the effect on the hard and soft segment microphase separation with the changes in diisocyanate from aromatic to linear aliphatic, and CO_2/N_2 selectivity of 45 and permeability of 186 Barrer were reported. Ruaan *et al.* [92] studied the microstructure behavior of hydroxyl terminated polybutadiene (HTPB) based polyurethane membranes. With increase in hard segment content, the oxygen and nitrogen permeability was observed to decrease. The authors suggested that the low O_2/N_2 selectivity was because of hydrogen bonding. It was indicated that the key to success for selectivity improvement is to avoid the hydrogen bond formation between the hard segments. Queiroz *et al.* [71] also studied the structural characteristics and gas permeation properties of PDMS/poly (propylene oxide) urethane urea bi-soft segment membranes. The authors reported increased CO_2, O_2, N_2 permeabilities with increase in PDMS membrane content. Lower degree of cross-linking and lower hydrogen bonding contribution between the hard segments exhibited higher permeability. Semsarzadeh *et al.* [93] investigated the effect of hard segments on the gas permeabilities for polyether based urethanes. As the hard segment content was increased, there was a decrease in the permeabilities of the gases. Gas permselectivity of the membranes with same hard segment content increased with increase in soft segment. These studies underlined the importance of optimization of hard and soft segment contents for optimum permeability and selectivity.

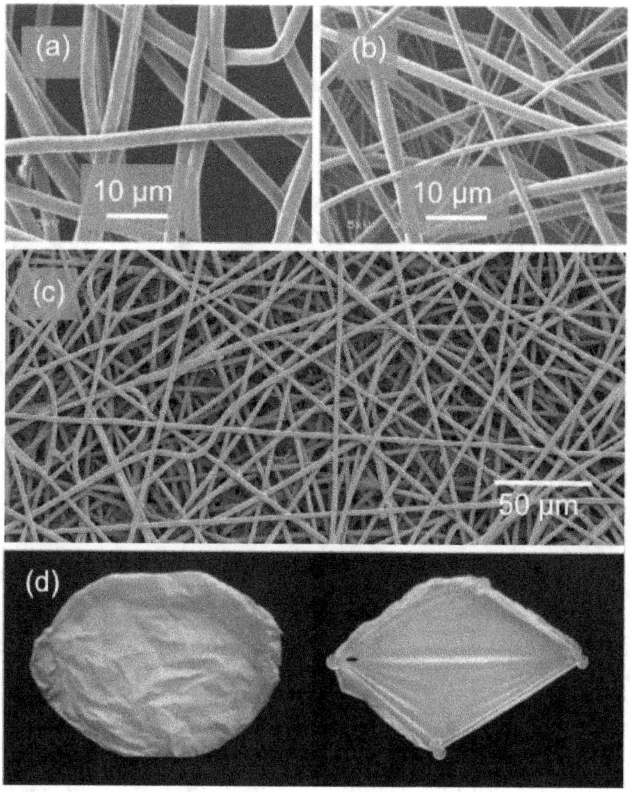

Figure 7.2 SEM images of (a) MDI-based and (b) isophorone-based nonwoven fiber mats; (c) SEM image indicating the uniformity of MDI-based PU fiber diameter and mat density; (d) stretched fiber mat exhibiting lack of tearing or breaking. Reproduced from Reference 90 with permission from American Chemical Society.

Galland *et al.* [94] studied the soft segment molecular weight as a major factor controlling the diffusion. It was reported the gas permeability depended on the chemical composition due to the nature of chain packing and degree of phase segregation. Huang *et al.* [95] recently studied thermoplastic polyurethane (TPU) films synthesized using layer-multiplying co-extrusion. The authors studied the morphological effect of confinement, along with gas barrier and mechanical properties. The soft TPU having the hard segment of 52% exhibited phase separation, whereas the rigid polymer having 100% hard-segment TPU exhibited amorphous structure. A multilayer structure with elasticity ratio of 100 and the viscosity ratio of 10 was generated. A significant reduction in oxygen permeability was observed when stretched at 75% which was due to the micro-confinement occurring during orientation. Park *et al.* [70] studied urethane urea membranes based on polysiloxane/polyether mixed soft segment for gas separation properties. The authors reported that small addition of PDMS into polyurethane matrices based on polyether increased the N_2, O_2, CO_2 permeabilities and N_2 selectivity. Also, the small addition of polyethers like PPO, PEO, PTMO and PEO-PPO-PEO inside the PDMS based polyurethane urea matrices decreased the gas permeabilities, but had no effect on the gas selectivities. In other studies, the authors studied segmented PU and PUU membranes with different soft segments for the separation of toluene and nitrogen. The poly(tetramethylene oxide) (PTMO)/PDMS mixed soft segment based polyurethane membranes exhibited good performance for toluene separation. The authors reported toluene/N_2 selectivity of about 70-140 and permeability as high as 17,500 Barrer. Increased gas permeability was observed with an increase of polysiloxane molecular weight [70,96,97]. Wolinska-Grabczyk *et al.* [98] studied segmented polyurethanes for gas properties with variation in the soft segments. PU membranes exhibited lower permeability when synthesized with more polar or shorter macro-diol segments. The authors observed that the diffusivity and solubility of oxygen and nitrogen were in correlation with glass transition temperature of macro-diol segments. Gomes *et al.* [72] generated poly(ether siloxane urethane urea) with varying content of polysiloxane. Gas permeation properties for O_2, N_2, CO_2, CH_4, n-C_4H_{10} were analyzed. Soft segments of polysiloxane and permeation properties were observed to have strong correlation. Gnanasekaran *et al.* [99] studied the structure-transport and microstructure of mixed soft-segmented poly(urethane-imide) membranes. Polycaprolactone diol, polypropylene glycol, and bis(3-aminopropyl)-terminated polydimethylsiloxane were used as soft segments for membrane synthesis. The membranes exhibited the potential application for n-C_4H_{10}/CH_4 separation because of higher selectivity for a mixture of gases as compared to single gases. Li *et al.* [100] studied a series of polyurethane ureas synthesized using various polyether diols. The polyethers were terathane (R) 2000, terathane (R) 2900, PEG 2000, PPG 2700, and a mixture of PEG 2000 and terathane (R) 2000. The fractional free volume increased with increase in soft segment content and the increase in polyether molecular weight, which increased the gas permeability.

McBride *et al.* [51] studied linear polyurethane membranes and the relation between the aromatic content present in diisocyanates and gas diffusion. The authors observed that the motion of the soft segment chains could be controlled, as hard domains at temperatures below the T_g acted as crosslinks. As the aromatic content increased, the motion of the soft segments was further restricted because of the increased crosslinking effectiveness. On the other hand, on increasing the length of soft segment, it increased the soft segment mobility. With increase in hard segment content, the activation energy increased which further reduced the permeability of the gas. Matsunaga *et al.* [101] studied the influence of chemical structure of thermoplastic polyurethane elastomers on CO_2 and O_2 gas permeation. Both the soft segment as well hard segments' chemical structure affected the gas permeabilities. The authors reported that there was an increase in gas permeabilities with soft segment chain lengths for membranes containing poly(oxytetramethylene)glycol. The diffusion process predominated for dissolution-diffusion gas permeation process.

7.4.2 Effect of Urethane/Urea Content

Teo *et al.* [102] studied the effect of chain extenders and PEG molecular weight on urethane moiety association. Polymeric membranes prepared from PEG with molecular weight of 600 exhibited promising performance for permeation of gas pairs like CO_2/CH_4, He/CH_4, H_2/N_2 and O_2/N_2. Sadeghi *et al.* [103] reported that the content of urea influenced the gas transport characteristics of PU membranes. Polytetramethyleneglycol (PTMG) and isophorone diisocyanate (IPDI) prepolymers were used for the synthesis. Butanediol (BDO) and butanediamine (BDA) were used as chain extenders to synthesize PUU with urethane/urea linkage. With increase in the urea linkages, the hard and soft segment micro-phase separation increased. Increasing urethane content in the polymers decreased the permeability, while with increasing urea content the gas selectivity decreased. The CO_2 permeability was reported to be 128 Barrer and CO_2/N_2 selectivity of 27. Molecular dynamics studies were carried out by Amani *et al.* [104] for understanding the effect of urethane and urea contents on gas separation properties of poly(urethane–urea) membranes. The membranes were synthesized from PTMG, IPDI and designed ratios of 1,4-butanediamine to 1,4-butanediol as chain extenders. For nanostructure characterizations of the membranes, the fractional free volume, X-ray diffraction patterns, glass transition temperature, density, and radial distribution function (RDF) were calculated. The gas permeations studied were done for O_2, N_2, CO_2, CH_4, and H_2S. Phase separation of hard and soft segments increased with increasing urea contents in the membranes and, thus, the *d*-spacing and fractional free volume. The gas permeability of the membranes proportionally increased with increasing urea linkages in the polymer. These studies underlined the importance of attaining balance of urethane and urea contents for effective gas permeation performance.

7.4.3 Effect of Temperature and Pressure

With temperature variations, the permeability and diffusivity change which allows the determination of activation energy. From this, the temperature dependence of the selectivity can be calculated. In case of large difference in permeation activation energies, the selectivity is also higher for those gas pairs [105,106]. For polyurethane membranes, the temperature variation effects on the transport properties have been studied by many researchers [107,108]. For polyurethane and polyurethane blends [77,109], the observed permeability order is $CO_2 > H_2 > O_2 > CH_4 > N_2$. Kinetic diameter [105], critical temperature and the solubility for the gas molecules have been reported to be responsible for this order. Lee *et al.* [75] studied the effect of synthesis temperature and molecular structure variation on the polymer network and gas transport properties. In specific, the gas permeability dependence of synthesis temperature, composition, aromatic content of diisocyanate, molecular weight of the polyol was analyzed. The increase in synthesis temperature and aromatic content decreased the permeability coefficient. There was an increase in tensile strength of the membrane due to the decreasing synthesis temperature and crosslinking density. The authors also studied further the effect of crosslinked state and annealing for gas transport in interpenetrating polymer network membranes.

The reduction in gas permeability due to the increase in pressure has been predicted by dual-sorption model [110], however, this has not been always observed [111,112]. In most of the glassy polymers [105,113-117], there is a decrease in permeability with pressure, when the permeating gas has high critical temperatures, like CO_2. In case of permeant gases like N_2 and O_2 with lower critical temperature, there is no influence of pressure variation on permeabilities, both for rubbery and glassy polymers [105,112,116,118-120]. However, there are some exceptions as well, which reported decrease in permeability of O_2, N_2, CH_4 with increasing pressures like poly(urethane urea) [121], polyimides [115], polyvinyl pyridine ethyl cellulose blend [114]. Madhavan *et al.* [119] studied a series of poly(dimethylsiloxane -urethane) membranes for gas transport properties. Pressure dependence on gas permeation was studied for oxygen, nitrogen and carbon dioxide gases. CO_2 permeability exhibited dependency on pressure, while O_2 and N_2 permeabilities didn't. The reported value of the O_2/N_2 permselectivity was 2.3 and CO_2/N_2 was 8.5. Thus, mixed effects of pressure have been observed on the gas transport properties of various polymers, depending on the type of permeant gas.

7.4.4 Effect of Molecular Chain Extension

Carboxyl or hydroxyl groups as organic functional groups in polyurethane membranes have direct influence on the gas transport properties of the membranes, molecular crystallinity, density and glass transition temperature. The separation coefficient for oxygen-nitrogen separation increases with increased content of functional groups and is strongly subjective to the character of the functional group. The carboxyl group-containing poly(butylene glycol adipate) (PBA)-type polyurethane membrane was observed to exhibit higher density, glass transition temperature, crystallinity, higher gas permeability, separation coefficient [69].

Knight and Lyman [122] studied block copolyether membranes for the effect of chemical structure and fabrication variables on gas permeation. The authors observed that for copolyether-urethane-urea, a linear relation existed between the gas permeability and propylene glycol segment molecular weight. Copolymers of polypropylene segment were more permeable than those with polyethylene segment due to the crystallizing ability of polyethylene glycol. Chain packing nature of different chain extenders affected the permeability, higher the packing, the lower was the gas permeability. The addition of salts like LiBr or urea and variations in casting solvent did not impact gas permeability. Semsarzadeh *et al.* [123] studied polyether-based polyurethanes for the effects of chain extender length on the gas permeabilities. Toluene diisocyanate with 1000 and 2000 g/mol molecular weight and polytetramethylene glycol were used for polyurethane synthesis. Different chain extenders (1,6-hexane diol, 1,4-butane diol, ethylene glycol and 1,10-decane diol) were used. The permeability and diffusivity studies were performed for N_2, O_2, CH_4, CO_2. The glass transition temperature of the polymers decreased on increasing the chain extender's length. The phase separation was more probable with increased chain extender's length. With increasing the length of the chain extenders, the permeability and diffusivity of gases increased. Selectivity of CO_2/N_2 gets changed by chain extender length, while selectivity of CO_2/CH_4 and O_2/N_2 did not show any remarkable change. Damian *et al.* [124] studied the various hybrid membrane networks based on isocyanate chemistry. The permeability coefficients and the morphology depended upon the soft segment's polarity and chain length along with the composition of the networks.

7.4.5 Effect of Polymer Blending

Polymer blending is an attractive approach for enhancing the performance of polymeric membranes as it is both time and cost-effective method for tuning the properties. In recent years, a variety of polymer blends have been explored for gas separation membranes. Mannaan *et al.* [125] recently reviewed polymer blend membranes for permeability, selectivity and phase behavior. For the enhancement of the transport properties, the morphology of the phase separated polymer blends was reported to play a major role. The fundamentals of polymer blends in transport processes (like gas barrier and separation) has also been discussed by Robeson *et al.* [126]. Kim *et al.* [127] studied the blend membranes of polyurethane with polyetherimides and poly (amide-imide) for CO_2 separations. As the blend ratio was increased, there was a decrease in gas permeability. The selectivity of CO_2/N_2 was improved with reduction in polyurethane content. The volume fraction of the dispersion component is an important factor to be taken care of for gas separation properties [127]. Patricio studied the gas transport behavior of gases such as H_2, N_2, O_2, CH_4, and CO_2 using polyurethane and PU/PMMA blend membranes [77]. In correlation with the decrease in average free volume size, the blends with higher wt% PU demonstrated lower permeabilities when compared with the PU membranes. For H_2/N_2 gas pair, the selectivity increased with increase in PMMA content in blends, while for other gas pairs, no marked changes were observed. The H_2 selectivity improvement could be associated with increased rigidity of amorphous phases. Ghalei *et al.* [128] studied PU/PVAc blend membranes for permeation of O_2, N_2, CO_2 and CH_4. The membranes had higher CO_2 permeability as compared to other gases, and higher CO_2/N_2 and CO_2/CH_4 selectivity. De Sales *et al.* [109] studied the polyurethane and PMMA blend membranes for temperature and pressure dependence of gas permeabilities for CO_2, H_2, O_2, CH_4, N_2 gases. With the addition of 30 wt % PMMA, the gas permeabilities decreased approximately 55%. It was observed that CO_2 had lowest permeation activation energy value (28 kJ/mol) with variation in temperature. For low temperatures, the gas pair selectivity increased, and the selectivity was higher for gas pairs having permeation activation energy value difference of about 15 kJ/mol. For pressure variation study, it was concluded that the permeabilities for CO_2 and H_2 gases through PU and the blend membrane increased by 35% at elevated pressure. CO_2 permeability increased by around 35% at higher pressure for the polyurethane and the blend membranes. Also, the O_2/N_2 selectivity increased with pressure, while the permeability to nitrogen decreased in the case of the 30% PMMA blend. Similar to earlier study, Semsarzadeh *et al.* [129] also studied the PU-PVAc blend membranes synthesized in the presence of various pluronic copolymer contents. Blends with 5 wt% PVAc exhibited higher CO_2 permeability in comparison to the PU membrane. Domain size of the dispersed PVAc was controlled by the addition of pluronic and had a positive effect on the permeability. Saedi *et al.* [130] investigated the blend membranes of polyethersulfone/polyurethane for CO_2 and CH_4 separations. The relative affinity of CO_2, CH_4 and H_2O for PES and PU were obtained using density functional theory calculations. Gas sorption by PU, viscosity of solution and PES membrane's mechanical properties were also analyzed. The fractional free volume and *d*-spacing of casting solution decreased the membrane porosity, glass transition temperature, thermal properties, gas sorption and plasticization. This was attributed to decreased Langmuir capacity and increased PES membrane strength at yield and elongation at yield. This resulted in the decrease of CO_2 permeability, a boost in CO_2/CH_4 selectivity and plasticization pressure of the PES membrane due to the presence of polyurethane. Also, the feed temperature affected the membrane behavior against pressure and the mixed gas composition.

7.4.6 Effect of Modifiers and Fillers

Another functional way to improve the gas transport behaviors of membranes can be improved by incorporating fillers. These fillers can be salts, metals, metal oxides, ions, inorganic silica, layered silicate, zeolites, carbon nanotube, graphene, etc. As the polymeric membranes are mixed with inorganic fillers, these membranes are thus termed as mixed matrix membranes. Incorporation of nanoporous filler particles is important breakthrough for boosting the gas separation capabilities of polymeric membranes. Out of a large variety of filler, metal organic frameworks (MOF) have become popular as a new group of nanoporous materials for enhanced membrane characteristics. Remarkable developments in gas permeability and selectivity have been reported for membranes based on MOFs. Erucar *et al.* [131] reviewed the recent developments in membranes incorporated with MOFs. The authors studied the experimental and computational methods to generate the polymer and MOF selection criteria for efficient gas separation membranes. Following sections summarize studies reporting the incorporation of polyurethane with a variety of reinforcements for generating effective gas separation performance of the membranes,

Zeolites

Zeolite incorporated nanocomposite membranes combine the advantages of both the polymer and the zeolite, thus, overcoming the individual shortcomings of the two materials [132]. Tirouni *et al.* [133] investigated polyurethane/zeolite mixed matrix membranes for the separation of C_2H_6 and C_3H_8 from CH_4. The hard and soft segment phase separation increased with the addition of butanediamine chain extender. With increase in urea groups of the polymer structure, both permeability and selectivity were observed to increase. Gas permeation data of polyurethane-zeolite 4Å membranes exhibited an increase in methane permeability and decrease in C_2H_6/CH_4 and C_3H_8/CH_4 selectivity, as the amount of Zeolite 4Å was increased up to 10 wt%. Polyurethane–zeolite (ZSM-5) membranes also exhibited significant improvements in selectivity and permeability of all hydrocarbons. With 20 wt% filled PU–ZSM 5 membranes, the propane permeability increased from 64.8 to 117.2 Barrer and C_3H_8/CH_4 selectivity increased from 2.6 to 3.64. Ciobanu *et al.* [134] studied the use of zeolite SAPO-5 nanocrystals for the synthesis of polyurethane composite membranes. Zeolite content was ranged from 10 to 70%. The zeolite nanocrystals were observed to act as a cross-linker for the polyurethane matrix.

Silica

A number of PU-silica based composite membranes have been reported in the literatiure. Khudyakov *et al.* [135] reviewed the status of UV-curable polyurethane nanocomposites incorporating nanosilica and organically-modified clay. A large number of studies have been reported related to the structure and characteristics of polyurethane nanocomposites formed through dark reactions as well as by UV-curing of urethane acrylate oligomers. Polyurethane nanocomposites with low loadings (less than 5%) of fillers have been reported to have dramatic property improvements [135,136].

In another study, Petrovic *et al.* [137] studied the polymerization of polypropylene glycol/hexamethylenediisocyanate/1,4-butanediol to gain insights about the gas transport properties. Tetraethoxysilane was used to prepare silica nanoparticles through sol-gel method, while the nanocomposites were synthesized by solution mixing technique. Various characterization techniques were used to confirm the desired nano-scale distribution of silica nanoparticles. Gas permeation studies of the membranes revealed the enhancement in CO/N selectivities with increasing amount of silica nanoparticles. It was also observed that the nanocomposites exhibited nearly two fold increment in the selectivity when compared with pristine polyurethanes membranes. However, the CO permeability exhibited a reduction of nearly 35% for the composite membranes while comparing against pure PU ones. Higuchi model may be used to predict the gas transport properties in polyurethane-silica membranes [138]. In this study, properties such as dielectric permeability of nanocomposite membranes were studied. Studies on the ether-based PU and ester-based PU exhibited that the amalgamation of silica nanoparticles helped the improvement of CO_2/CH_4 selectivity. By taking an account of the nanoparticles interfacial layer leading to the formation of void volumes, a model was introduced to help predicting the nanocomposite membrane performance [139].

Polycaprolactone-based polyurethane membranes were studied for the effect of the addition of silica nanoparticles [140]. The membranes were prepared using conventional methods such as solution mixing and casting. Polycaprolactone/hexamethylene diisocyanate/1,4-butanediol based polymer was produced by polymerization involving a two-step methodology. Membrane performance was predicted using modified Higuchi model to obtain a fair agreement in theoretical and experimental values. In another study, hybrid polyurethane/silica based composite membranes were prepared using tetraethoxysilane, cetyltrimethyl ammonium bromide and polyvinyl alcohol [141]. Spectroscopic techniques were used to verify the presence of silica in the polymer network and SEM for the nanoscale distribution of silica particles. It was observed that with increase in the silica content, the diffusivity of gases and CO_2 gas permeability were reduced. With silica content >10 wt%, an enhancement of CO_2/CH_4 selectivity was observed.

In another study, the transport performance of carbon dioxide and methane gases was investigated in polyesterurethane mixed matrix membranes (MMMs), in separate tests, containing different fumed silica nanoparticles [142]. Non-modified and commercially-modified silica (with octylsilane and polydimethylsiloxane) were used as fillers. The structural features were studied by various microscopic, spectroscopic and calorimetric techniques to confirm the interfaces induced in PU microphase when silica was present. Surface treatment of silica filler with long hydrophobic chains condensed the accumulation of nanoparticles and enhanced dispersion in MMMs. Regardless of silica type, both separation factor and CO_2 permeation were increased with the presence of silica nanoparticles, which was attributed to the interrupted chain packing and improved dynamic free volume. The results revealed that among silica nanoparticles, the unmodified particles with OH groups on the surface exhibited better performance for CO_2/CH_4 separation. A new model of gas permeation through PU/silica membranes was proposed considering the presence of filler aggregation in the matrix as well as free volume at the interface layer [142].

Another study revealed the gas transport of two types of polyurethane membranes which are synthesized from PCL225 and PPG polyether [143]. It was observed that by using silica content up to 2.5%, the permeability increased. However, the permeability exhibited a downward effect on increasing the silica content father. It was observed that the selectivity of propane over methane increased with the increase in the amount of silica particles. The studied membranes were the ones with 12.5% silica and the permeation tests were conducted at 2 bar pressure.

With the addition of nanofillers, the flexural properties of the polymer matrix are required to be retained. Ideally, the hybrid polymer materials are expected to exhibit enhancement in properties such as mechanical strength and polymer flexibility. Enhanced material properties are possible with more homogeneous distributions of the inorganic components leading to the formation of high performance functional membranes [144]. In another study, polyurethane (PU) based mixed matrix membranes were generated with polytetramethylene glycol and polyvinyl alcohol along with silica nanoparticles [145]. The dispersion in the nanocomposite membranes was confirmed using microscopic, spectroscopic techniques. Gas permeation studies for pure CH_4, O_2, CO_2, N_2 and He gases were performed through the composites with varying amounts of silica particles. It was observed that upon increasing the silica content, the permeability of the CO_2 gas was enhanced, while the transport properties of other gases decreased. The permeability of CO_2 increased from 68.4 to 96.7 Barrer, where the membrane contained 10 wt% of nanoparticles.

Layered Silicates

Osman *et al.* [146] studied the gas permeation properties of modified montmorillonite based nanocomposites with polyurethane adhesives. The nanocomposites were generated to act as barrier against oxygen and water vapor. Thus, their application in gas industries can be envisaged as membranes which allow permeation of hydrocarbons, but do not allow any other gases like O_2, CO_2 and water vapor to pass through. A correlation between the gas permeation and volume fraction was established. With the incorporation of small volume fractions of nanoparticles, the permeation rate of O_2 and water vapor was decreased (Figure 7.3 for oxygen permeation). A 30% reduction was observed with 3 vol% of filler, when bis(2-hydroxyethyl) hydrogenated tallow ammonium or alkylbenzyldimethylammonium ions were used for clay modification. The clay modified with dimethyl dihydrogenated tallow ammonium ions increased the O_2 permeation rate with increase in inorganic fraction.

The above variation was attributed to the phase separation between the pure hydrocarbon modification and polar polyurethane. This underlined the need of generating compatibility between the filler and polymer phases in order to achieve optimum performance. Increase in oxygen permeation for dimethyl dihydrogenated tallow ammonium ions modified clay composites could still be used for O_2, water vapor separation as the membranes were impermeable to water vapor, whereas highly permeable to O_2 molecules. In a similar study, Mittal [147] also reported the gas permeation performance of polyurethane-clay nanocomposites and underlined the impact of synthesis procedure as well as filler-matrix compatibility. As shown in Figure 7.4, the composite with Tixogel VZ, the clay surface modified with a benzyl(hydrogenated tallow alkyl)dimethyl ammonium ions, exhibited a decrease in oxygen permeation. On the other hand, the composite with Tixogel VP, the clay surface modified with bis(hydrogenated tallow alkyl)dimethyl ammonium ions, exhibited an increase in oxygen permeation as a function of filler fraction. Similar to the previous study, the water vapor permeation in both

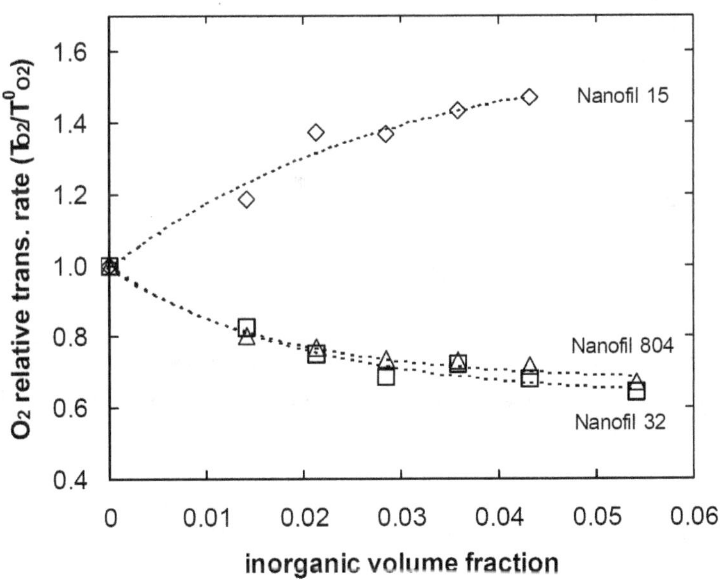

Figure 7.3 Dependence of the oxygen transmission rate through the PU-nanocomposites on the inorganic volume fraction. The dotted lines are guides for the eye. Reproduced from Reference 146 with permission from American Chemical Society.

the composites was reduced as a function of filler fraction. Thus, the chemical nature of the surface coating of the fillers resulted in specific interactions with the polymer matrix, which consequently affected the nanocomposite properties. The authors also compared the oxygen permeation through polyurethane-clay nanocomposites with molecular dynamics predictions in order to gain more insights about the average aspect ratio of the filler platelets in the composites [148]. In polyurethane composites, the clay platelets were observed to be present with an average aspect ratio of 100. In addition, the good agreement with the experimental and predicted values was retained till 3 vol% filler fraction. Afterwards, the experimental values were observed to level off.

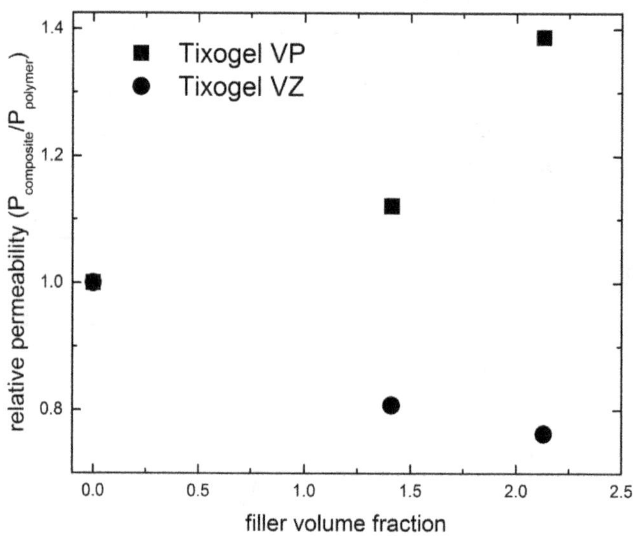

Figure 7.4 Oxygen permeation through the PU and PU nanocomposites as a function of filler volume fraction. Reproduced from reference 147 with permission from Wiley.

Also, as shown in Figure 7.5, the authors generated the relation between the reduction of oxygen permeation through the nanocomposites with filler fraction for both aligned and misaligned filler parties. The aligned filler particles were observed o be significantly effective in tuning the gas transport behavior of the nanocomposites. These studies indicated the successful development of polyurethane-clay nanocomposites with tuned reduction in water vapor permeation as well as increased or decreased O_2 permeation, based on the chosen filler system. Such systems can result in high selectivity membranes for the separation of hydrocarbons from other contaminant gases.

Dense polyurethane-based membranes containing hydrophilic clay were synthesized by Barboza *et al.* [149] and the permeability of carbon dioxide (CO_2) through the membranes was studied. Ethylene diamine was used as the chain extender, which led to the urea linkage network. The nanofiller fraction was optimized as 0.5 and 1% relative to the amount of poly(ethylene glycol). The CO_2 permeability was observed to improve with the amount of PEG, whereas it was noticed to decrease with the amount of clay as the higher clay content contributed to the tortuous pathways for gas diffusion. A novel nanocomposite was reported by Shamini and Yusoh, which consisted of Na^+ montmorillonite which was further modified using transition metal ions such as copper (II) chloride and iron (III) chloride [150]. The presence of metal ions reportedly contributed to better dispersion of nano-filler and also reduced clay agglomeration. The gas transport properties exhibited remarkable decrease. Polyurethane film containing 1% filler modified with iron chloride exhibited the permeability to decrease four fold. Iron and copper exhibited different patterns for the reduction in permeability.

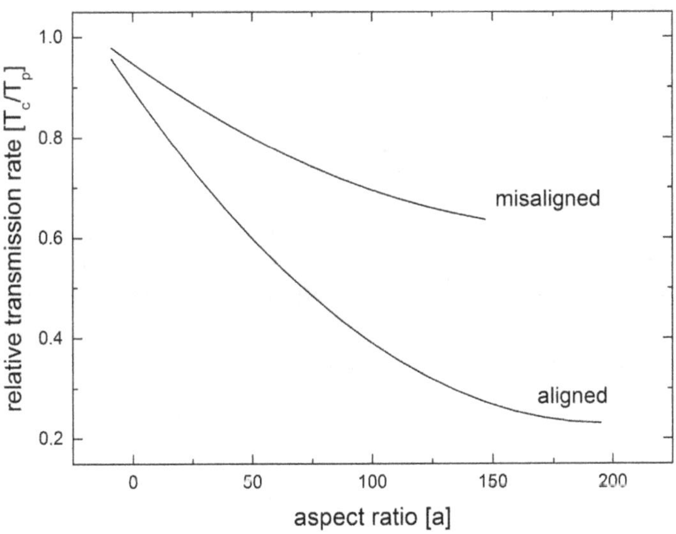

Figure 7.5 Numerical prediction of the effect of filler platelet misalignment on the gas permeation performance at 3 vol% loading. Reproduced from reference 148 with permission from Wiley.

It has to be mentioned that the reduction in permeability of different gases in the composite membranes does not indicate the reduction in the membrane performance, it only signifies the probable enhancement in the selectivity by blocking the passage of certain penetrants and possibly enhancing the permeation of others. In addition, the use of fillers also enhances the mechanical and thermal properties, along with enhancing environmental stability. Most of these developed composite membranes, thus, have strong potential of application in various industries to separate gas mixtures.

Metals, Metal Oxide and Metal Ions

Ameri *et al.* [151] used alumina (Al_2O_3) at different concentrations to prepare polyurethane nanocomposites through a bulk two-step polymerization. Different chain extenders were used to complete the urethane polymer network. Alumina ensured improved O_2/N_2, CO_2/CH_4, and CO_2/N_2 selectivity, but with reduced permeability values. In another study, cobalt chelated plasma treated membranes were generated and the effect of subsequent formamide formation in the membrane matrix on the permeability properties was evaluated [152]. The chelation was obtained by treatment with cobalt(II)/formamide solution. With ethylenediamine plasma treatment, selectivity increased from 2.6 to 3.1 GPU whereas the CoCl •6HO/formamide treatment improved the value to 4.4 GPU. The enhanced O/N selectivity was attributed to the improvement in oxygen affinity achieved through chelation and size sieving effect. Chen *et al.* [153] incorporated TiO_2 nanoparticles at varying concentrations to the polymer to prepare thermo-sensitive polyurethane (TSPU). Membrane formation temperature was optimized in the study to tune the transport behaviors of the nanocomposites. It was observed that the permeability coefficients altered with different membrane formation temperatures. Increasing nano-TiO_2 concentration also favorably helped the transport phenomenon. Conventionally seen tortuous diffusion pathway based low permeation mechanism failed to explain the counter-intuitive phenomenon observed in the study. The authors suggested that the soft segment of the thermo-sensitive polyurethane packed around the TiO_2 nanoparticles, as if these were in the bulk polymer, thus, leading to the observed results.

In another study, PU membranes were synthesized through thermal phase inversion method [154] Polyol:diisocyanate:chain extender was blended at 1:2:1 molar ratio and the interactions at intramolecular level were analyzed. TiO_2 was varied up to 30 wt% and transport behaviors were studied for N_2, O2, CH_4 and CO_2 gases at varying temperatures. It was observed that an increased TiO_2 content led to a proportional rise in selectivity and a proportional fall in permeability values. Membranes of ionic polyurethane were also reported using N-methyldiethanolamine as chain extender, which was later complexed with cupric ions [155]. Different polymerization approaches such as single and double step were employed for the composite preparation. It was observed that the incorporation of CuCl enhanced hard segment aggregation, whereas incorporation of crosslinker hindered the cluster formation by the hard segment. The soft segment aggregation helped to increase the permeability.

Carbon Nanoparticles

The formation of polyurethane nanocomposites have proven the ability to exhibit advanced performance as compared to pure polymer. In one such study, poly(ether urethane) membranes containing multi-walled carbon nanotubes (MWCNTs) were studied for improvements in mechanical strength and permeation capabilities [156]. The MWCNTs were grafted with various functional groups to obtain three types of membranes with fillers such as MWCNT-COOH, MWCNT-OH, MWCNT-IPDI. Fourier-transform infrared spectroscopy was used to confirm the successful grafting of functional groups on to the surface of the MWCNTs. Techniques like SEM, mechanical testing and thermal analysis confirmed the superior properties of MWCNT-IPDI based membranes (Figure 7.6). Despite the development towards the generation of strong, durable, and cost-efficient carbon nanoparticles based polymer membranes, there is a need for the evolution of membrane materials with high gas transport properties [157,158].

Graphene

A large number of studies has reported the generation of polyurethane nanocomposites by incorporating graphene/graphene oxide/modified or functionalized graphene [158-173]. Most of the studies have confirmed these nanocomposites to have significantly superior thermal and mechanical properties. However, only a few studies have reported the gas permeation data for PU-graphene nanocomposites. Kim *et al.* [174] reported 90% decrease in nitrogen

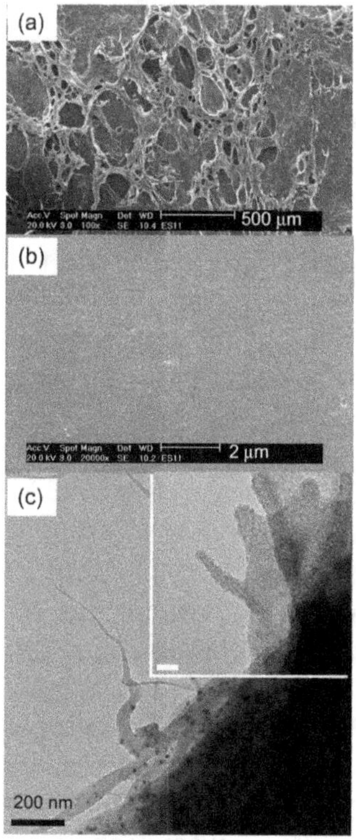

Figure 7.6 SEM and TEM images of PU nanocomposite membrane obtained by the electro spinning technique. Reproduced from Reference 156 with the permission from American Chemical Society.

permeation with 3 wt% isocyanate treated graphene oxide filled thermoplastic polyurethanes. Figure 7.7 shows the mechanism of dispersion of functionalized graphene in polyurethane. In another study, Thermoplastic composite films of polyurethane containing hexadecyl-functionalized low-defect graphene nanoribbons were studied by Xiang *et al.* [175] for improved gas barrier properties. A decrease in nitrogen gas diffusivity by 3 orders of magnitude was observed with only 0.5 wt% of filler. The films were reported to have potential applications in food packaging and light weight gas storage containers. Figure 7.8 also shows the nitrogen pressure drop across the PU membranes and PU composite membranes. The composite with 0.5 wt% filler exhibited strong resistance against nitrogen for a longer period of time, indicating its barrier towards nitrogen permeation. Kaveh *et al.* [176] reported excellent gas barrier properties with an 80% decrease in permeability of Helium for 1 % graphene oxide incorporated thermoplastic polyurethane films. Thus, until now, the graphene nanocomposites have been largely developed to generate potential candidates for gas barrier properties, however, optimization of the graphene based composites for enhanced permeation is still needed. However, as mentioned earlier, decrease permeability of a certain permeant still opens the potential for these membranes to be selective towards the separation of certain gas mixtures.

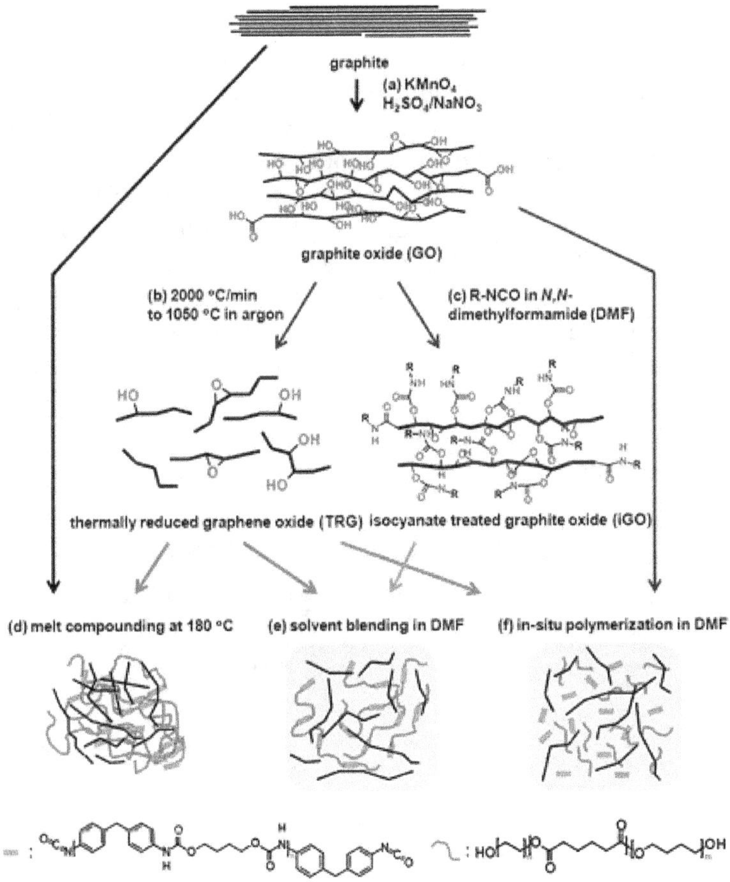

Figure 7.7 Scheme showing the mechanism of functionalized graphene dispersed in polyurethane. Reproduced from Reference 174 with the permission from American Chemical Society.

7.5 Other Natural Gas Specific Applications

Polyether-polyurethane interactions as well as water vapor and methane separation selectivity were studied by Di Landro *et al.* [177]. Water absorption was shown to be higher for urethane rich polymers. Water vapor permeability was 10^3-10^4 times more than methane as well as other permeant gases. This makes these materials a potential membrane material for natural gas dehydration [9,177]. High H_2S/CH_4 selectivities of poly(ether urethane) and poly(ether urethane urea) membranes were reported by Chatterjee *et al.* [178]. The authors reported polymer PU4 as a favorable membrane material for H_2S separation from CO_2 and CH_4 mixtures. H_2S/CH_4 selectivity was reported to be greater than 100 at 20 °C in the pressure range from 4-13.6 atm. Other studies have also reported high permeability, permselectivity, durability for acid gas separtiob from natural gas and carbon dioxide removal from synthesis gas for polyether-urethane or polyether-urea block copolymers [179,180].

Ponangi *et al.* [181] studied for the split-up of volatile organic compounds from nitrogen either in dry or in humidified form using PU membranes. The membranes were examined for the vapor phase separation of a variety of gases such as benzene, hexane, p-xylene, and benzene/toluene/xylene mixtures from nitrogen. The selectivities of organic/dry N_2 ranged from 30-210 and pressure normalized permeabilities of around 1.25x 10^{-3} cm^3 (STP)/(cm^2 s cmHg). The nitrogen permeability was increased with swelling of the polyether membranes, but the organic/N_2 selectivity was lowered. Water

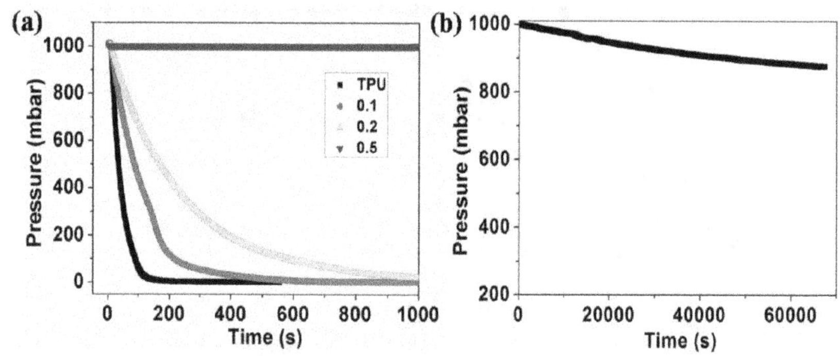

Figure 7.8 (a) Nitrogen pressure drop through PU and PU composite films; (b) nitrogen pressure drop through the composite wit 0.5 wt% filler over a longer period of time. Reproduced from Reference 175 with permission from American Chemical Society.

permeability was observed to be low and did not depend on the organic feed component. The authors also further studied the free volume in polyurethane membranes for organic vapor diffusion [182].

7.6 Conclusion

In this review, various advancements for enhancing the applications of polyurethane membranes for gas separation have been explored. Polyurethanes represent useful class of materials with properties such as mechanical strength, dimensional stability, thermal resistance, resistance towards chemicals and tunable gas transport behavior. Various modifications in the molecular microstructure of polyurethanes have been achieved in order to enhance the permeability as well as selectivity of various gases through these materials. These modifications include optimization of soft and hard segments, tuning of polarity, blending with other polymers, adjusting urea urethane interactions, incorporation of fillers, use of a wide variety of chain extenders and cross-linkers, etc. It is evident that the tunability of gas permeability as well as selectivity in polyurethane membranes would further enhance the industrial application of these materials for gas separation processes.

References

1. Clarizia, G. (2009) Polymer-based membranes applied to gas separation: material and engineering aspects. *Desalination*, **245**(1-3), 763-768.
2. Takht Ravanchi, M., Kaghazchi, T., and Kargari, A. (2009) Application of membrane separation processes in petrochemical industry: a review. *Desalination*, **235**(1-3), 199-244.
3. Yampolskii, Y. (2012) Polymeric gas separation membranes. *Macromolecules*, **45**(8), 3298-3311.
4. Sanders, D. F., Smith, Z. P., Guo, R., Robeson, L. M., McGrath, J. E., Paul, D. R., and Freeman, B. D. (2-13), Energy-efficient polymeric gas separation membranes for a sustainable future: A review. *Polymer*, **54**(18), 4729-4761.
5. Sipek, M., Friess, K., and Hynek, V. (2004) Membrane separation of mixtures of gases and vapors in practice. *Chemicke Listy*, **98**(1), 4-9.
6. Bernardo, P., Drioli, E., and Golemme, G. (2009) Membrane gas separation: a review/state of the art. *Industrial & Engineering Chemistry Research*, **48**(10), 4638-4663.

7. Maier, G. (1998) Gas separation with polymer membranes. *Angewandte Chemie, International Edition*, **37**(21), 2960-2974.
8. Matsuura, T. (1993) *Synthetic Membranes and Membrane Separation Processes*, Taylor & Francis, USA.
9. George, G., Bhoria, N., and Mittal. V. (2015) Improved Polymer Membranes for Sour Gas Filtration and Separation. *Abu Dhabi International Petroleum Exhibition and Conference*, UAE. Online: https://www.onepetro.org/download/conference-paper/SPE-177805-MS?id=conference-paper%2FSPE-177805-MS (assessed 1st March 2017).
10. Krol, P. (2009) Polyurethanes - A review of 60 years of their syntheses and applications. *Polimery*, **54**(7-8), 489-500.
11. Hahn, C., Keul, H., and Moeller, M. (2012) Hydroxyl-functional polyurethanes and polyesters: synthesis, properties and potential biomedical application. *Polymer International*, **61**(7), 1048-1060.
12. Brzeska, J., Dacko, P., Janeczek, H., Janik, H., Sikorska, W., Rutkowska, M., and Kowalczuk, M. (2014) Synthesis, properties and applications of new (bio)degradable polyester urethanes. *Polimery*, **59**(5), 365-371.
13. Engels, H.-W., Pirkl, H.-G., Albers, R., Albach, R. W., Krause, J., Hoffmann, A., Casselmann, H., and Dormish, J. (2013) Polyurethanes: versatile materials and sustainable problem solvers for today's challenges. *Angewandte Chemie, International Edition*, **52**(36), 9422-9441.
14. Yang, C., Fischer, L., Maranda, S., and Worlitschek, J. (2015) Rigid polyurethane foams incorporated with phase change materials: A state-of-the-art review and future research pathways. *Energy and Buildings*, **87**, 25-36.
15. Yao, X., Tuo, X., and Wang, X. (2009) Preparing biodegradable polyurethane porous scaffold for tissue engineering application. *Progress in Chemistry*, **21**(7-8), 1546-1552.
16. Janik, H., and Marzec, M. (2015) A review: Fabrication of porous polyurethane scaffolds. *Materials Science & Engineering C*, **48**, 586-591.
17. Zoltowska, K., Sobczak, M., and Oledzka, E. (2014) Polyurethanes in pharmacy - current state and perspectives of the development. *Polimery*, **59**(10), 689-698.
18. Zia, K. M., Bhatti, H. N., and Bhatti, I. A. (2007) Methods for polyurethane and polyurethane composites, recycling and recovery: A review. *Reactive & Functional Polymers*, **67**(8), 675-692.
19. Braun, T. (1989) Quasi-spherical solid polymer membranes in separation chemistry: polyurethane foams as sorbents. Recent advances. *Fresenius' Zeitschrift fuer Analytische Chemie*, **333**(8), 785-792.
20. Pinchuk, L. (1994) A review of the biostability and carcinogenicity

of polyurethanes in medicine and the new generation of 'biostable' polyurethanes. *Journal of Biomaterials Science, Polymer Edition*, **6**(3), 225-67.

21. Mishra, A., Singh, S. K., Dash, D., Aswal, V. K., Maiti, B., Misra, M., and Maiti, P. (2014) Self-assembled aliphatic chain extended polyurethane nanobiohybrids: Emerging hemocompatible biomaterials for sustained drug delivery. *Acta Biomaterialia*, **10**(5), 2133-2146.
22. Morral-Ruiz, G., Melgar-Lesmes, P., Solans, C., and Garcia-Celma, M. J. (2013) Multifunctional polyurethane-urea nanoparticles to target and arrest inflamed vascular environment: a potential tool for cancer therapy and diagnosis. *Journal of Controlled Release*, **171**(2), 163-171.
23. St. John, K. R. (2014) The use of polyurethane materials in the surgery of the spine: a review. *Spine Journal*, **14**(12), 3038-3047.
24. Teodosiu, C., Wenkert, R., Tofan, L., and Paduraru, C. (2014) Advances in preconcentration/removal of environmentally relevant heavy metal ions from water and wastewater by sorbents based on polyurethane foam. *Reviews in Chemical Engineering*, **30**(4), 403-420.
25. Boretos, J. W., and Pierce, W. S. (1968) Segmented polyurethane: A polyether polymer. An initial evalution for biomedical applications. *Journal of Biomedical Materials Research*, **2**(1), 121-130.
26. Boretos, J. W., Detmer, D. E., and Donachy, J. H. (1971) Segmented polyurethane: A polyether polymer, II. Two years experience. *Journal of Biomedical Materials Research*, **5**(4), 373-387.
27. Hepburn, C. (1982) *Polyurethane* Elastomers, Applied Science Publishers, USA.
28. Chang, Y. J. P., and Wilkes, G. L. (1975) Superstructure in segmented polyether–urethanes. *Journal of Polymer Science: Polymer Physics Edition*, **13**(3), 455-476.
29. Huh, D. S., and Cooper, S. L. (1971) Dynamic mechanical properties of polyurethane block polymers. P*olymer Engineering & Science*, **11**(5), 369-376.
30. Camargo, R., Macosko, C. W., Tirrell, M., and Wellinghoff, S. T. (1985) Phase separation studies in RIM polyurethanes catalyst and hard segment crystallinity effects. *Polymer*, **26**(8), 1145-1154.
31. Cooper, S. L., and Tobolsky, A. V. (1966) Properties of linear elastomeric polyurethanes. *Journal of Applied Polymer Science*, **10**(12), 1837-1844.
32. Clough, S., and Schneider, N. (1968) Structural studies on urethane elastomers. *Journal of Macromolecular Science, Part B: Physics*, **2**(4), 553-566.
33. Seymour, R. W., and Cooper, S. L. (1971) DSC studies of polyurethane block polymers. *Journal of Polymer Science, Part B: Polymer Letters*, **9**(9), 689-694.

34. Seymour, R. W., and Cooper, S. L. (1973) Thermal analysis of polyurethane block polymers. *Macromolecules*, **6**(1), 48-53.
35. Srichatrapimuk, V. W., and Cooper, S. L. (1978) Infrared thermal analysis of polyurethane block polymers. *Journal of Macromolecular Science, Part B*, **15**(2), 267-311.
36. Sung, C. P., and Hu, C. (1981) Orientation studies of segmented polyether poly (urethaneurea) elastomers by infrared dichroism. *Macromolecules*, **14**(1), 212-215.
37. Wang, C. B., and Cooper, S. L. (1983) Morphology and properties of segmented polyether polyurethaneureas. *Macromolecules*, **16**(5), 775-786.
38. Seymour, R., Estes, G., and Cooper, S. (1970) Infrared studies of segmented polyurethan elastomers. I. Hydrogen bonding. *Macromolecules*, **3**(5), 579-583.
39. Ishihara, H., Kimura, I., Saito, K., and Ono, H. (1974) Infrared studies on segmented polyurethane-urea elastomers. *Journal of Macromolecular Science, Part B: Physics*, **10**(4), 591-618.
40. Brunette, C., Hsu, S., and MacKnight, W. (1982) Hydrogen-bonding properties of hard-segment model compounds in polyurethane block copolymers. *Macromolecules*, **15**(1), 71-77.
41. Bonart, R., Morbitzer, L., and Hentze, G. (1969) X-ray investigations concerning the physical structure of cross-linking in urethane elastomers. II. Butanediol as chain extender. *Journal of Macromolecular Science, Part B: Physics*, **3**(2), 337-356.
42. Blackwell, J., and Gardner, K. H. (1979) Structure of the hard segments in polyurethane elastomers. *Polymer*, **20**(1), 13-17.
43. Fridman, I. D., and Thomas, E. L. (1980) Morphology of crystalline polyurethane hard segment domains and spherulites. *Polymer*, **21**(4), 388-392.
44. Howarth, G. A. (2003) Polyurethanes, polyurethane dispersions and polyureas: Past, present and future. *Surface Coatings International, Part B: Coatings Transactions*, **86**(2), 111-118.
45. Huang, S. L., and Lai, J. Y. (1995) Gas permeability of crosslinked HTPB–H12MDI-based polyurethane membrane. *Journal of Applied Polymer Science*, **58**(10), 1913-1923.
46. George, G., Bhoria, N., Alhallaq, S., Abdala, A., and Mittal, V. (2016) Polymer membranes for acid gas removal from natural gas. *Separation and Purification Technology*, **158**, 333-356.
47. Stern, S. A. (1994) Polymers for gas separations - The next decade. *Journal of Membrane Science*, **94**, 1-65.
48. Matteucci, S., Yampolskii, Y., Freeman, B. D., and Pinnau, I. (2006) Transport of gases and vapors in glassy and rubbery polymers. In: *Materials Science of Membranes for Gas and Vapor Separation*, Yampolskii, Y., Pinnau. I., and Freeman, B. (eds.), John Wiley & Sons, UK, doi: 10.1002/047002903X.ch1.

49. Ziegel, K. (1971) Gas transport in segmented block copolymers. *Journal of Macromolecular Science, Part B: Physics*, **5**(1), 11-21.
50. Pegoraro, M., Penati, A., and Zanderighi, L. (1986) Polyurethane membrane for gas fractionation. *Journal of Membrane Science*, **27**(2), 203-214.
51. McBride, J. S., Massaro, T. A., and Cooper, S. L. (1979) Diffusion of gases through polyurethane block polymers. *Journal of Applied Polymer Science*, **23**(1), 201-214.
52. Schneider, N., S., Dusablon, L. V., Snell, E. W., and Prosser, R. A. (1969) Water vapor transport in structurally varied polyurethans. *Journal of Macromolecular Science, Part B: Physics*, **3**(4), 623-644.
53. Barrer, R., Barrie, J., and Slater, J. (1958) Sorption and diffusion in ethyl cellulose. Part III. Comparison between ethyl cellulose and rubber. *Journal of Polymer Science*, **27**(115), 177-197.
54. Vieth, W., Howell, J., and Hsieh, J. (1976) Dual sorption theory. *Journal of Membrane Science*, **1**, 177-220.
55. Chern, R. T., Koros, W. J., Sanders, E. S., and Yui, R. (1983) "Second component" effects in sorption and permeation of gases in glassy polymers. *Journal of Membrane Science*, **15**(2), 157-169.
56. Spillman, R. W. (1989) Economics of gas separation membranes. *Chemical Engineering Progress*, **85**(1), 41-62.
57. Ghosal, K., and Freeman, B. D. (1994) Gas separation using polymer membranes: an overview. *Polymers for Advanced Technologies*, **5**(11), 673-697.
58. Koros, W. J., and Fleming, G. K. (1993) Membrane-based gas separation. *Journal of Membrane Science*, **83**(1), 1-80.
59. Roualdes, S., Sanchez, J., and Durand, J. (2002) Gas diffusion and sorption properties of polysiloxane membranes prepared by PECVD. *Journal of Membrane Science*, **198**(2), 299-310.
60. Robeson, L. M. (1991) Correlation of separation factor versus permeability for polymeric membranes. *Journal of Membrane Science*, **62**(2), 165-185.
61. Mohr, J. M., and Paul, D. R. (1991) Surface fluorination of composite membranes. Part I. Transport properties. *Journal of Membrane Science*, **55**(1), 131-148.
62. Lin, X., Qiu, X., Zheng, G., and Xu, J. (1995) Gas permeabilities of poly (trimethylsilylpropyne) membranes surface modified with CCI4 plasma. *Journal of Applied Polymer Science*, **58**(11), 2137-2139.
63. Robeson, L. M. (2008) The upper bound revisited. *Journal of Membrane Science*, **320**(1-2), 390-400.
64. Hsieh, K. H., Tsai, C. C., and Tseng, S. M. (1990) Vapor and gas permeability of polyurethane membranes. Part I. Structure-property relationship. *Journal of Membrane Science*, **49**(3), 341-350.
65. Huang, S.-L., and Lai, J.-Y. (1995) On the gas permeability of hydro-

xyl terminated polybutadiene based polyurethane membranes. *Journal of Membrane Science*, **105**(1), 137-145.

66. Johnson, B. M., Baker, R. W., Matson, S. L., Smith, K. L., Roman, I. C., Tuttle, M. E., and Lonsdale, H. K. (1987) Liquid membranes for the production of oxygen-enriched air: II. Facilitated-transport membranes. *Journal of Membrane Science*, **31**(1), 31-67.
67. Bicak, N., Koza, G., and Atay, T. (1996) Metal chelating resins by condensation of ethylene diamine with p-dichloromethyl benzene. *Journal of Applied Polymer Science*, **61**(5), 799-804.
68. Park, C., Choi, M., and Lee, Y. M. (1995) Chelate membrane from poly (vinyl alcohol)/poly (N-salicylidene allyl amine) blend. I: Synthesis and characterization of Co (II) chelate membrane. *Polymer*, **36**(7), 1507-1512.
69. Hsieh, K. H., Tsai, C. C., and Chang, D. M. (1991) Vapor and gas permeability of polyurethane membranes. Part II. Effect of functional group. *Journal of Membrane Science*, **56**(3), 279-287.
70. Park, H. B., Kim, C. K., and Lee, Y. M. (2002) Gas separation properties of polysiloxane/polyether mixed soft segment urethane urea membranes. *Journal of Membrane Science*, **204**(1-2), 257-269.
71. Queiroz, D. P., and De Pinho, M. N. (2005) Structural characteristics and gas permeation properties of polydimethylsiloxane/poly(propylene oxide) urethane/urea bi-soft segment membranes. *Polymer*, **46**(7), 2346-2353.
72. Gomes, D., Peinemann, K.-V., Nunes, S. P., Kujawski, W., and Kozakiewicz, J. (2006) Gas transport properties of segmented poly(ether siloxane urethane urea) membranes. *Journal of Membrane Science*, **281**(1-2), 747-753.
73. Yang, J. M., Lai, W. C., and Lin, H. T. (2001) Properties of HTPB based polyurethane membrane prepared by epoxidation method. *Journal of Membrane Science*, **183**(1), 37-47.
74. Cao, N., Pegoraro, M., Bianchi, F., Di Landro, L., and Zanderighi, L. (1993) Gas transport properties of polycarbonate–polyurethane membrane. *Journal of Applied Polymer Science*, **48**(10), 1831-1842.
75. Doo, S. L., Dae, S. J., Tae, H. K., and Sung, C. K. (1991) Gas transport in polyurethane-polystyrene interpenetrating polymer network membranes. I. Effect of synthesis temperature and molecular structure variation. *Journal of Membrane Science*, **60**(2-3), 233-252.
76. Kumar, H. (2005) A study of sorption/desorption and diffusion of substituted aromatic probe molecules into semi interpenetrating polymer network of polyurethane/polymethyl methacrylate. *Polymer*, **46**(18), 7140-7155.
77. Patricio, P. S. O., de Sales, J. A., Silva, G. G., Windmoller, D., and Machado, J. C. (2006) Effect of blend composition on microstructure, morphology, and gas permeability in PU/PMMA blends. *Journal of*

Membrane Science, **271**(1-2), 177-185.

78. George, S. C., and Thomas, S. (2001) Transport phenomena through polymeric systems. *Progress in Polymer Science*, **26**(6), 985-1017.
79. He, Y., Xie, D., and Zhang, X. (2014) The structure, microphase-separated morphology, and property of polyurethanes and polyureas. *Journal of Materials Science*, **49**(21), 7339-7352.
80. Ohst, H., Hildenbrand, K., and Dhein. R. (1991) Polymer Structure/Properties Correlation of Polyurethane PV-membranes for Aromatic Aliphatic Separation. *Proceedings 5th International Conference on Pervaporation Processes in the Chemical Industry*, Germany.
81. Pegoraro, M., Zanderight, L., Penati, A., Severini, F., Bianchi, F., Cao, N, Sisto, R., and Valentini, C. (1991) Polyurethane membranes from polyether and polyester diols for gas fractionation. *Journal of Applied Polymer Science*, **43**(4), 687-697.
82. Xiao, H., Ping, Z. H., Xie, J. W., and Yu, T. Y. (1990) Permeation of CO_2 through polyurethane. *Journal of Applied Polymer Science*, **40**(7-8),1131-1139.
83. Mohammadi, T., Tavakol Moghadam, M., Saeidi, M., and Mahdyarfar, M. (2008) Acid gas permeation behavior through poly (ester urethane urea) membrane. *Industrial & Engineering Chemistry Research*, **47**(19), 7361-7367.
84. Wolinska-Grabczyk, A., and Jankowski, A. (2011) CO_2/N_2 separation ability and structural characteristics of poly(butadiene-co-acrylonitrile)-based polyurethanes and hydrogenated nitrile rubbers. *Journal of Applied Polymer Science*, **122**(4), 2690-2696.
85. Talakesh, M. M., Sadeghi, M., Chenar, M. P., and Khosravi, A. (2012) Gas separation properties of poly(ethylene glycol)/poly(tetramethylene glycol) based polyurethane membranes. *Journal of Membrane Science*, **415-416**, 469-477.
86. Poreba, R., Spirkova, M., Brozova, L., Lazic, N., Pavlicevic, J., and Strachota, A. (2013) Aliphatic polycarbonate-based polyurethane elastomers and nanocomposites. II. Mechanical, thermal, and gas transport properties. *Journal of Applied Polymer Science*, **27**(1), 329-341.
87. Khosravi, A., and Sadeghi, M. (2013) Separation performance of poly(urethane-urea) membranes in the separation of C2 and C3 hydrocarbons from methane. *Journal of Membrane Science*, **434**, 171-183.
88. Wang, Z. F., Wang, B., Yang, Y. R., and Hu, C. P. (2003) Correlations between gas permeation and free-volume hole properties of polyurethane membranes. *European Polymer Journal*, **39**(12), 2345-2349.
89. Ferreira Marques, M. F., Lopes Gil, C., Gardo, P. M., Kajcsos, Z., de

Lima, A. P., Queiroz, D. P., and de Pinho, M. N. (2003) Free-volume studies in polyurethane membranes by positron annihilation spectroscopy. *Radiation Physics and Chemistry*, **68**(3-4), 573-576.

90. Scholten, E., Bromberg, L., Rutledge, G. C., and Alan Hatton, T. (2011) Electrospun polyurethane fibers for absorption of volatile organic compounds from air. *ACS Applied Materials and Interfaces*, **3**(10), 3902-3909.
91. Sadeghi, M., Semsarzadeh, M. A., Barikani, M., and Ghalei, B. (2011) Study on the morphology and gas permeation property of polyurethane membranes. *Journal of Membrane Science*, **385-386**(1), 76-85.
92. Ruaan, R.-C., Ma, W.-C., Chen, S.-H., and Lai, J.-Y. (2001) Microstructure of HTPB-based polyurethane membranes and explanation of their low O_2/N_2 selectivity. *Journal of Applied Polymer Science*, **82**(6), 1307-1314.
93. Semsarzadeh, M. A., Sadeghi, M., Barkani, M., and Moadel, H. (2007) The effect of hard segments on the gas separation properties of polyurethane membranes. *Iranian Polymer Journal*, **16**(12), 819-827.
94. Galland, G., and Lam, T. (1993) Permeability and diffusion of gases in segmented polyurethanes: structure–properties relations. *Journal of Applied Polymer Science*, **50**(6), 1041-1058.
95. Huang, R., Chari, P., Tseng, J.-K., Zhang, G., Cox, M., and Maia, J. M. (2015) Microconfinement effect on gas barrier and mechanical properties of multilayer rigid/soft thermoplastic polyurethane films. *Journal of Applied Polymer Science*, **132**(18), DOI: 10.1002/app.41849.
96. Ho, B. P., Choon, K. K., and Young, M. L. (2002) Gas separation properties of polysiloxane/polyether mixed soft segment urethane urea membranes. *Journal of Membrane Science*, **204**(1-2), 257-269.
97. Park, H. B., and Lee, Y. M. (2002) Separation of toluene/nitrogen through segmented polyurethane and polyurethane urea membranes with different soft segments. *Journal of Membrane Science*, **197**(1-2), 283-296.
98. Wolinska-Grabczyk, A., and Jankowski, A. (2007) Gas transport properties of segmented polyurethanes varying in the kind of soft segments. *Separation and Purification Technology*, **57**(3), 413-417.
99. Gnanasekaran, D., and Reddy, B. S. (2014) A facile synthesis of mixed soft-segmented poly(urethane-imide) - polyhedral oligomeric silsesquioxone hybrid nanocomposites and study of their structure-transport properties. *Polymer International*, **63**(3), 507-513.
100. Li, H., Freeman, B. D., and Ekiner, O. M. (2011) Gas permeation properties of poly(urethane-urea)s containing different polyethers. *Journal of Membrane Science*, **369**(1-2), 49-58.

101. Matsunaga, K., Sato, K., Tajima, M., and Yoshida, Y. (2005) Gas permeability of thermoplastic polyurethane elastomers. *Polymer Journal*, **37**(6), 413-417.
102. Teo, L. S., Kuo, J. F., and Chen, C. Y. (1998) Study on the morphology and permeation property of amine group-contained polyurethanes. *Polymer*, **39**(15), 3355-3364.
103. Sadeghi, M., Semsarzadeh, M. A., Barikani, M., and Ghalei, B. (2010) The effect of urethane and urea content on the gas permeation properties of poly(urethane-urea) membranes. *Journal of Membrane Science*, **354**(1-2), 40-47.
104. Amani, M., Amjad-Iranagh, S., Golzar, K., Sadeghi, G. M. M., and Modarress, H. (2014) Study of nanostructure characterizations and gas separation properties of poly (urethane–urea) s membranes by molecular dynamics simulation. *Journal of Membrane Science*, **462**, 28-41.
105. Lin, W.-H., and Chung, T.-S. (2001) Gas permeability, diffusivity, solubility, and aging characteristics of 6FDA-durene polyimide membranes. *Journal of Membrane Science*, **186**(2), 183-193.
106. Marchese, J., Garis, E., Anson, M., Ochoa, N. A., and Pagliero, C. (2003) Gas sorption, permeation and separation of ABS copolymer membrane. *Journal of Membrane Science*, **221**(1), 185-197.
107. Wang, Z. F., Wang, B., Ding, X. M., Zhang, M., Liu, L. M., Qi, N., and Hu, J. L. (2004) Effect of temperature and structure on the free volume and water vapor permeability in hydrophilic polyurethanes. *Journal of Membrane Science*, **241**(2), 355-361.
108. Yang, S. J., Yang, J. M., and Lin, H. T. (2005) Evaluation of poly(N-isopropylacrylamide) modified hydroxyl-terminated polybutadiene based polyurethane membrane. *Journal of Membrane Science*, **258**(1), 97-105.
109. de Sales, J. A., Patricio, P. S. O., Machado, J. C., Silca, G. G., and Windmoller, D. (2008) Systematic investigation of the effects of temperature and pressure on gas transport through polyurethane/poly(methylmethacrylate) phase-separated blends. *Journal of Membrane Science*, **310**(1-2), 129-140.
110. Kanehashi, S., and Nagai, K. (2005) Analysis of dual-mode model parameters for gas sorption in glassy polymers. *Journal of Membrane Science*, **253**(1), 117-138.
111. Chiou, J., and Paul, D. (1986) Sorption and transport of inert gases in PVF2/PMMA blends. *Journal of Applied Polymer Science*, **32**(5), 4793-4814.
112. Li, X.-G., Kresse, I., Xu, Z.-K., and Springer, J. (2001) Effect of temperature and pressure on gas transport in ethyl cellulose membrane. *Polymer*, **42**(16), 6801-6810.
113. Mokdad, A., and Dubault, A. (2000) Transport properties of carbon dioxide through single-phase polystyrene/poly (vinylmethylether)

blends. *Journal of Membrane Science*, **172**(1), 1-8.

114. Li, X.-G., Kresse, I., Springer, J., Nissen, J., and Yang, Y.-L. (2001) Morphology and gas permselectivity of blend membranes of polyvinylpyridine with ethylcellulose. *Polymer*, **42**(16), 6859-6869.
115. Chung, T. S., Cao, C., and Wang, R. (2004) Pressure and temperature dependence of the gas-transport properties of dense poly [2, 6-toluene-2, 2-bis (3, 4-dicarboxylphenyl) hexafluoropropane diimide] membranes. *Journal of Polymer Science, Part B: Polymer Physics*, **42**(2), 354-364.
116. Lopez-Gonzalez, M. M., Compan, V., Saiz, E., Riande, E., and Guzman, J. (2005) Effect of the upstream pressure on gas transport in poly (ether-imide) films. *Journal of Membrane Science*, **253**(1), 175-181.
117. Hu, C.-C., Lee, K.-R., Ruaan, R.-C., Jean, Y. C., and Lai, J.-Y. (2006) Gas separation properties in cyclic olefin copolymer membrane studied by positron annihilation, sorption, and gas permeation. *Journal of Membrane Science*, **274**(1), 192-199.
118. Lin, H., and Freeman, B. D. (2004) Gas solubility, diffusivity and permeability in poly (ethylene oxide). *Journal of Membrane Science*, **239**(1), 105-117.
119. Madhavan, K., and Reddy, B. S. R. (2006) Poly(dimethylsiloxane-urethane) membranes: Effect of hard segment in urethane on gas transport properties. *Journal of Membrane Science*, **283**(1-2), 357-365.
120. Raharjo, R. D., Lin, H., Sanders, D. F., Freeman, B. D., Kalakkunnath, S., and Kalika, D. S. (2006) Relation between network structure and gas transport in crosslinked poly (propylene glycol diacrylate). *Journal of Membrane Science*, **283**(1), 253-265.
121. Teo, L. S., Chen, C. Y., and Kuo, J. F. (1998) The gas transport properties of amine-containing polyurethane and poly(urethane-urea) membranes. *Journal of Membrane Science*, **141**(1), 91-99.
122. Knight, P., and Lyman, D. (1984) Gas permeability of various block copolyether-urethanes. *Journal of Membrane Science*, **17**(3), 245-254.
123. Semsarzadeh, M. A., Sadeghi, M., and Barikani, M. (2008) Effect of chain extender length on gas permeation properties of polyurethane membranes. *Iranian Polymer Journal*, **17**(6), 431-440.
124. Damian, C., Espuche, E., Escoubes, M., Cuney, S., and Pascault, J. P. (1997) Gas permeability of model polyurethane networks and hybrid organic-inorganic materials: Relations with morphology. *Journal of Applied Polymer Science*, **65**(12), 2579-2587.
125. Mannan, H. A., Mukhtar, H., Murugesan, T., Nasir, R., Mohshim, D. F., and Mushtaq, A. (2013) Recent applications of polymer blends in gas separation membranes. *Chemical engineering & Technology*, **36**(11), 1838-1846.

126. Robeson, L. M. (2010) Polymer blends in membrane transport processes. *Industrial & Engineering Chemistry Research*, **49**(23), 11859-11865.
127. Kim, M. J., Sea, B., Youm, K.-H., and Lee, K.-H. (2006) Morphology and carbon dioxide transport properties of polyurethane blend membranes. *Desalination*, **193**(1-3), 43-50.
128. Ghalei, B., and Semsarzadeh, M.-A. (2007) A novel nano structured blend membrane for gas separation. *Macromolecular Symposia*, **249-250**, 330-335.
129. Semsarzadeh, M. A., and Ghalei, B. (2012) Characterization and gas permeability of polyurethane and polyvinyl acetate blend membranes with polyethylene oxide–polypropylene oxide block copolymer. *Journal of Membrane Science*, **401-402**, 97-108.
130. Saedi, S., Madaeni, S. S., Hassanzadeh, K., Shamsabadi, A. A., and Laki. S. (2014) The effect of polyurethane on the structure and performance of PES membrane for separation of carbon dioxide from methane. *Journal of Industrial and Engineering Chemistry*, **20**(4), 1916-1929.
131. Erucar, I., Yilmaz, G., and Keskin, S. (2013) Recent advances in metal-organic framework-based mixed matrix membranes. *Chemistry-an Asian Journal*, **8**(8), 1692-1704.
132. Lue, S. J., Su, I.-M., Lee, D.-T., Chen, H.-Y., Shih, C.-M., Hu, C.-C., Jean, Y. C., and Lai. J.-Y. (2011) Correlation between free-volume properties and pervaporative flux of polyurethane– zeolite composites on organic solvent mixtures. *The Journal of Physical Chemistry B*, **115**(12), 2947-2958.
133. Tirouni, I., Sadeghi, M., and Pakizeh, M. (2015) Separation of C_3H_8 and C_2H_6 from CH_4 in polyurethane–zeolite 4Å and ZSM-5 mixed matrix membranes. *Separation and Purification Technology*, **141**, 394-402.
134. Ciobanu, G., Carja, G., and Ciobanu, O. (2007) Preparation and characterization of polymer-zeolite nanocomposite membranes. *Materials Science & Engineering C*, **27**(5-8), 1138-1140.
135. Khudyakov, I. V., Zopf, R. D., and Turro, N. J. (2009) Polyurethane nanocomposites. *Designed Monomers & Polymers*, **12**(4), 279-290.
136. George, G., Bhoria, N., and Mittal, V. (2016) Supported UV polymerized ionic liquid membranes with block copolymer. *Journal of Membrane Science and Technology*, **6**(166), DOI: 10.4172/2155-9589.1000166.
137. Petrovic, Z. S., Javni. I., Waddon, A., and Banhegyi, G. (2000) Structure and properties of polyurethane–silica nanocomposites. *Journal of Applied Polymer Science*, **76**(2), 133-151.
138. Sadeghi, M., Semsarzadeh, M. A., Barikani, M., and Chenar, M. P. Gas separation properties of polyether-based polyurethane–silica nanocomposite membranes. *Journal of Membrane Science*, **376**(1),

188-195.

139. Hassanajili, S., Masoudi, E., Karimi, G., and Khademi, M. (2013) Mixed matrix membranes based on polyetherurethane and polyesterurethane containing silica nanoparticles for separation of CO_2/CH_4 gases. *Separation and Purification Technology*, **116**, 1-12.
140. Sadeghi, M., Talakesh, M. M., Ghalei, B., and Shafiei, M. (2013) Preparation, characterization and gas permeation properties of a polycaprolactone based polyurethane-silica nanocomposite membrane. *Journal of Membrane Science*, **427**, 21-29.
141. Semsarzadeh, M. A., and Ghalei, B. (2013) Preparation, characterization and gas permeation properties of polyurethane-silica/polyvinyl alcohol mixed matrix membranes. *Journal of Membrane Science*, **432**, 115-125.
142. Hassanajili, S., Khademi, M., and Keshavarz, P. (2014) Influence of various types of silica nanoparticles on permeation properties of polyurethane/silica mixed matrix membranes. *Journal of Membrane Science*, **453**, 369-383.
143. Khosravi, A., Sadeghi, M., Banadkohi, H. Z., and Talakesh, M. M. (2014) Polyurethane-silica nanocomposite membranes for separation of propane/methane and ethane/methane. *Industrial and Engineering Chemistry Research*, **53**(5), 2011-2021.
144. Ribeiro, T., Baleizao, C., and Farinha, J. P. S. (2014) Functional films from silica/polymer manoparticles. *Materials*, **7**(5), 3881-3900.
145. Semsarzadeh, M. A., Ghalei, B., Fardi, M., Esmaeeli, M., and Vakili, E. (2014) Structural and transport properties of polydimethylsiloxane based polyurethane/silica particles mixed matrix membranes for gas separation. *Korean Journal of Chemical Engineering*, **31**(5), 841-848.
146. Osman, M. A., Mittal, V., Morbidelli, M., and Suter, U. W. (2003) Polyurethane adhesive nanocomposites as gas permeation barrier. *Macromolecules*, **36**(26), 9851-9858.
147. Mittal, V. (2014) Polyurethane-bentonite nanocomposites: morphology and oxygen permeation. *Advances in Polymer Technology*, **33**(3), DOI:10.1002/adv.21416.
148. Osman, M. A., Mittal, V., and Lusti, H. R. (2004) The aspect ratio and gas permeation in polymer-layered silicate nanocomposites. *Macromolecular Rapid Communications*, **25**, 1145-1149.
149. Barboza, E. M., Delpech, M. C., Garcia, M. E. F., and Pimenta, F. D. (2014) Evaluation of carbon dioxide gas barrier properties of membranes obtained from aqueous dispersions based on polyurethane and clay. *Polimeros*, **24**(1), 94-100.
150. Shamini, G., and Yusoh, K. (2014) Gas permeability properties of thermoplastic polyurethane modified clay nanocomposites. *International Journal of Chemical Engineering and Applications*, **5**(1), 64-68.

151. Ameri, E., Sadeghi, M., Zarei, N., and Pournaghshband, A. (2015) Enhancement of the gas separation properties of polyurethane membranes by alumina nanoparticles. *Journal of Membrane Science*, **479**, 11-19.
152. Chen, S.-H., Wu, T.-H., Ruaan, R.-C., and Lai, J.-Y. (1998) Effect of top layer swelling on the oxygen/nitrogen separation by surface modified polyurethane membranes. *Journal of Membrane Science*, **141**(2), 255-264.
153. Chen, Y., Wang, R., Zhou, J., Fan, H., Shi, B. (2011) Membrane formation temperature-dependent gas transport through thermo-sensitive polyurethane containing in situ-generated TiO_2 nanoparticles. *Polymer*, **52**(8), 1856-1867.
154. Sadeghi, M., Afarani, H. T., and Tarashi, Z. (2015) Preparation and investigation of the gas separation properties of polyurethane-TiO_2 nanocomposite membranes. *Korean Journal of Chemical Engineering*, **32**(1), 97-103.
155. Huang, S.-L., Ruaan, R.-C., and Lai, J.-Y. (1997) Gas permeability of cupric ion containing HTPB based polyurethane membranes. *Journal of Membrane Science*, **123**(1), 71-79.
156. Sen, R., Zhao, B., Perea, D., Itkis, M. E., Hu, H., Love. J., Bekyarova, E., and Haddon, R. C. (2004) Preparation of single-walled carbon nanotube reinforced polystyrene and polyurethane nanofibers and membranes by electrospinning. *Nano Letters*, **4**(3), 459-464.
157. Deng, J., Zhang, X., Wang, K., Zou, H., Zhang, Q., and Fu, Q. (2007) Synthesis and properties of poly (ether urethane) membranes filled with isophorone diisocyanate-grafted carbon nanotubes. *Journal of Membrane Science*, **288**(1), 261-267.
158. Yi, D. H., Yoo, H. J., Mahapatra, S. S., Kim, Y. A., and Cho, J. W. (2014) The synergistic effect of the combined thin multi-walled carbon nanotubes and reduced graphene oxides on photothermally actuated shape memory polyurethane composites. *Journal of Colloid and Interface Science*, **432**, 128-134.
159. Kumar, M., Chung, J. S., Kong, B.-S., Kim, E. J., and Hui, S. H. (2013) Synthesis of graphene-polyurethane nanocomposite using highly functionalized graphene oxide as pseudo-crosslinker. *Materials Letters*, **106**, 319-321.
160. Rana, S., Cho, J. W., and Tan, L. P. (2013) Graphene-crosslinked polyurethane block copolymer nanocomposites with enhanced mechanical, electrical, and shape memory properties. *RSC Advances*, **3**(33), 13796-13803.
161. Wang, X., Xing, W., Song, L., Yu, B., Hu, Y., and Yeoh, G. H. (2013) Preparation of UV-curable functionalized graphene/polyurethane acrylate nanocomposite with enhanced thermal and mechanical behaviors. *Reactive & Functional Polymers*, **73**(6), 854-858.
162. Dai, Y. T., Qiu, F. X., Xu., J. C., Yu, Z. P., Yang, P. F., Xu, B. B., Jiang, Y.,

and Yang, D. Y. (2014) Preparation and properties of UV-curable waterborne graphene oxide/polyurethane-acrylate composites. *Plastics Rubber and Composites*, **43**(2), 54-62.

163. Han, S., and Chun, B. C. (2014) Preparation of polyurethane nanocomposites via covalent incorporation of functionalized graphene and its shape memory effect. *Composites, Part A -Applied Science and Manufacturing*, **58**, 65-72.
164. Ma, W.-S., Wu, L., Yang, F., Wang, S.-F. (2014) Non-covalently modified reduced graphene oxide/polyurethane nanocomposites with good mechanical and thermal properties. *Journal of Materials Science*, **49**(2), 562-571.
165. Pokharel, P., and Lee, D. S. (2014) Thermal and mechanical properties of reduced graphene oxide/polyurethane nanocomposite. *Journal of Nanoscience and Nanotechnology*, **14**(8), 5718-5721.
166. Sadasivuni, K. K., Ponnamma, D., Kumar, B., Strankowski, M., Cardinaels, R., Moldenaers, P., Thomas, S., and Grohens, Y. (2014) Dielectric properties of modified graphene oxide filled polyurethane nanocomposites and its correlation with rheology. *Composites Science and Technology*, **104**, 18-25.
167. Khanna, S. K., and Phan, H. T. T. (2015) High strain rate behavior of graphene reinforced polyurethane composites. *Journal of Engineering Materials and Technology*, **137**(2), 021005.
168. Li, X., Deng, H., Li, Z., Xiu, H., Qi, X., Zhang, Q., Wang, K., Chen, F., and Fu, Q. (2015) Graphene/thermoplastic polyurethane nanocomposites: Surface modification of graphene through oxidation, polyvinyl pyrrolidone coating and reduction. *Composites, Part A -Applied Science and Manufacturing*, **68**, 264-275.
169. Luo, X., Zhang, P., Ren, J., Liu, R., Feng, J., and Ge, B. (2015) Preparation and properties of functionalized graphene/waterborne polyurethane composites with highly hydrophobic. *Journal of Applied Polymer Science*, **132**(23), DOI:10.1002/app.42005.
170. Pokharel, P., Choi, S., and Lee, D. S. (2015) The effect of hard segment length on the thermal and mechanical properties of polyurethane/graphene oxide nanocomposites. *Composites, Part A - Applied Science and Manufacturing*, **69**, 168-177.
171. Pokharel, P., Lee, S. H., and Lee, D. S. (2015) Thermal, mechanical, and electrical properties of graphene manoplatelet/graphene oxide/polyurethane hybrid nanocomposite. *Journal of Nanoscience and Nanotechnology*, **15**(1), 211-214.
172. Thakur, S., and Karak, N. (2015) A tough, smart elastomeric bio-based hyperbranched polyurethane nanocomposite. *New Journal of Chemistry*, **39**(3), 2146-2154.
173. Yu, B., Shi, Y., Yuan, B., Liu, L., Yang, H., Tai, Q., Lo, S., Song, L., and Hu, Y. (2015) Click-chemistry approach for graphene modification: effective reinforcement of UV-curable functionalized gra-

phene/polyurethane acrylate nanocomposites. *RSC Advances*, **5**(18), 13502-13506.

174. Kim, H., Miura, Y., and Macosko, C. W. Graphene/polyurethane nanocomposites for improved gas barrier and electrical conductivity. *Chemistry of Materials*, **22**(11), 3441-3450.
175. Xiang, C., Cox, P. J., Kukovecz, A., Genorio, B., Hashim, D. P., Yan, Z., Peng, Z., Hwang, C.-C., Ruan, G., Samuel, E. L. G., Sudeep, P. M., Konya, Z., Vajtai, R., Ajayan, P. M., and Tour, J. M. (2013) Functionalized low defect graphene nanoribbons and polyurethane composite film for improved gas barrier and mechanical performances. *ACS Nano*, **7**(11), 10380-10386.
176. Kaveh, P., Mortezaei, M., Barikani, M., and Khanbabaei, G. (2014) Low-temperature flexible polyurethane/graphene oxide nanocomposites: effect of polyols and graphene oxide on physicomechanical properties and gas permeability. *Polymer-Plastics Technology and Engineering*, **53**(3), 278-289.
177. Di Landro, L., Pegoraro, M., and Bordogna, L. (1991) Interactions of polyether-polyurethanes with water vapour and water-methane separation selectivity. *Journal of Membrane Science*, **64**(3), 229-236.
178. Chatterjee, G., Houde, A. A., and Stern, S. A. (1997) Poly(ether urethane) and poly(ether urethane urea) membranes with high H_2S/CH_4 selectivity. *Journal of Membrane Science*, **135**(1), 99-106.
179. Simmons, J. W. (2003) Novel Block Polyurethane-ether and Polyurea-ether Gas Separation Membranes. US Patent US 6843829 B2.
180. George, G., and Mittal, V. (2015) CO_2 & H_2S removal from natural gas using polymer membranes. Virtual Special Issue, Journal of Membrane Science. Online: https://www.journals.elsevier.com/journal-of-membrane-science/virtual-special-issues/co2-h2s-removal-from-natural-gas-using-polymer-membranes (assessed 27th February 2017).
181. Ponangi, R. P., and Pintauro, P. N. (1996) Separation of volatile organic compounds from dry and humidified nitrogen using polyurethane membranes. *Industrial and Engineering Chemistry Research*, **35**(8), 2756-2765.
182. Ponangi, R., Pintauro, P. N., and De Kee, D. (2000) Free volume analysis of organic vapor diffusion in polyurethane membranes. *Journal of Membrane Science*, **178**(1), 151-164.

8

Schizophyllan for Enhanced Oil Recovery

8.1 Introduction

Schizophyllan (SPG), a non-ionic and high molecular weight exopolysaccharide, is generated by white-rot fungus *Schizophyllum commune* [1]. SPG displays beneficial physical properties such as excellent thermal stability, viscosity and film formation, which drive the promising applications of the biopolymer for enhanced oil recovery (EOR) [2]. This homoglucan consists of a backbone of β-1,3-glucose with β-1,6-linked glucose side chains approximately at each third residue of the main chain, as shown in Figure 8.1 [1,3,4]. The molecular weight of schizophyllan is in the range 6-12 $\times 10^6$ g/mol [5]. The mechanical and thermal stability of the biopolymer are associated with the generation of a rigid triple helix framework which has stability till a temperature of 135 °C. Further, the rigid triple helix framework remarkably enhances the viscosity of aqueous phase. The biopolymer also follows various conformations depending on the temperature of the solution, nature of the solvent and pH. At

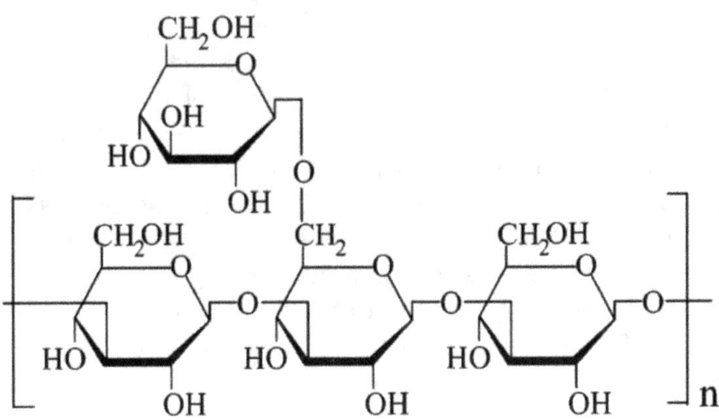

Figure 8.1 Repeating unit of SPG. Reproduced from Reference 4 with permission from American Chemical Society.

Haleema Saleem and Vikas Mittal, Khalifa University of Science and Technology, Abu Dhabi, UAE

25 °C, schizophyllan is distributed as a rigid triple helix in water. However, in dimethyl sulfoxide (DMSO), at pH >12.0 and temperature >135 °C, the triple helical structure of SPG starts melting to generate single as well as randomly coiled chains, which attributes to the decline in the average molecular weight (M_w) by one-third [6]. The aqueous solutions of the biopolymer exhibit viscoelastic, pseudo-plastic as well as thixotropic properties. The native suspensions enhance non-Newtonian behavior due to the filamentous system of connections of the internal woven hyphae. Further, the presence of NaOH at critical concentration 0.2 M generates a conformational change from the triple helix to the coil. These changes are generally associated with the transition in the local framework of the side chain glucose residues and arrangement of water inside the triple helix structure [7,8]. Nominal adsorption of the polymer chains on carbonate minerals and greater salinity tolerance is also predicted due to its non-ionic nature. In recent past, it has been observed that SPG has a novel solution characteristic, i.e., when the renaturing operation is performed in the presence of various polynucleotides, the consequent triple helical structure comprises of one nucleotide chain as well as two single-SPG (s-SPG) chains [9]. These observations indicate that schizophyllan exhibits an exclusive affinity to generate one dimensionally extended stable complexes, which are of commercial importance in many applications.

Several other fungi such as *Sclerotium glucanicum* [10], *Botrytis cinerea* [11] and *Sclerotium rolfsii* [12] have the ability to generate the glucan having same primary molecular structure. Nevertheless, these polysaccharides vary considerably in molecular weights and in their behavior towards micro-gel formation. These features precisely affect their adsorption as well as filtration characteristics, when utilized as additives for the polymer flooding process in the scope of EOR [13-15].

In this chapter, recent studies on the development of biopolymer schizophyllan, especially related to its application (and application potential) in EOR, have been focused. An overview of the structure, synthesis as well as the different physico-chemical properties of schizophyllan has also been presented.

8.2 Synthesis

As mentioned earlier, the *Schizophyllum commune* fungus ATCC 38548 discharges a neutral homoglucan (termed as schizophyllan), which comprises of a base chain of 1,3-β-D-gluco-pyranose units connected with distinct 1,6-bonded β-D-gluco-pyranoses at almost every third glucose molecule present in the back-bone chain. SPG was first discovered by Kikumoto *et al.* [16,17] in late 1960s, generated by the immersed culture of the fungus *Schizophyllum commune*, subsequently precipitated from the fine filtrate of the culture (4-8 days old) and later concentrated to 35-45% by the incorporation of water miscible organic solvents, like methanol. Composition of the culture solution was glucose 3.0%, KH_2PO_4 0.1%, yeast extract 0.3%, $MgSO_4.7H_2O$ 0.05%, $(NH_4)_2HPO_4$ 0.15% and pH 4.8 [17]. Even though glucose is commonly employed for the production of SPG, *S. commune* can actually employ numerous sugars as well as agricultural residues for the production of the polysaccharide [18,19]. The cheap agricultural biomass debris is also suitable for the generation of SPG. Distiller's dried grains with solubles (DDGS), an easily available agricultural biomass, is a co-product of ethanol generation by dry-grind method from corn [20]. Sutivisedsak *et al.* [21] validated the effective application of DDGS for the production of SPG by *S. commune* production strain ATCC 38548. For the generation of SPG, few other strains like CGMCC 5.113 and NRCM have also been studied [22,23]. In another study, Sutivisedsak *et al.* [24] reported the generation of SPG using agricultural debris from ten of eleven diverse strains of *S. commune*. Among these strains, two strains, namely CBS 266.60 and ATCC 20165, generated output proportionate to those obtained from commercial production strain. Physical characteristics of SPG solutions obtained from these strains were also proportionate to that of the commercially produced SPG. Agricultural substrates, as renewable and inexpensive materials, are also of special interest for SPG generation considering that *S. commune* is suited for effective degradation of cellulose, pectin, hemi-cellulose and lignin [25]. Steiner *et al.* [18] reported the production of xylanase-cellulose enzyme complex and extracellular polysaccharide schizophyllan using an isolated strain of *S. commune* [18]. Several other studies have also been performed, where the SPG production from rice hull de-toxificated hydrolysate [26], corn bran [27] and pretreated corn fiber [19] has been examined. However, the studies were reported to have low yields. Sutivisedsak *et al.* [21] compared the soybean hulls, DDGS, pretreated corn fiber and untreated corn fibers as the substrates for the production of SPG in the malt extract basal medium. It was noted that by using 2% w/v malt extract and 1% w/v DDGS, greater production of 15.4 g/L was attained, subsequent to cultivation of 8 days. Thus, these studies confirmed that the cheap agricultural substrates can be used as promising alternatives to glucose. For the advancement of economically viable bioprocesses, the optimization of the fermentation conditions is regarded as a vital step [23]. The response surface methodology (RSM) is an efficient method for the optimization process, examining the influence of different independent variables as well as their interactions, building the empirical models and analyzing the most advantageous conditions [28]. Kumari *et al.* [23] reported the SPG production applying the RSM, where generation of 8.03 g/L was attained at day 8, utilizing sucrose.

8.3 Structure

Polysaccharides are useful macromolecules having functional as well as structural diversity. Chemical structure of this type of carbohydrate is characterized by glycosidic linkage position and configuration, composition and sequence of monosaccharides and number, nature and location of added non-carbohydrate groups [29]. The analysis of monosaccharide composition comprises the breakage of glycosidic linkages by hydrolysis process, fractionation of the resultant monosaccharides and detection as well as quantification of distinctive monosaccharide by gas chromatography-mass spectrometry (GC-MS) [30,31]. The enzymatic degradation research confirmed that the schizophyllan repeating unit consisted of three β-(1→3)-linked D-gluco-pyranose residues, to one of which was attached a single gluco-pyranose residue by means of β-(1→6)-linkage [16]. The structure is similar to scleroglucan (SG), a different β-(1→3)-linked D-glucan, generated by a fungus *Sclerotium rolfsii* [32]. Schizophyllan is more divergent than scleroglucan in its molecular weight and branch distribution [33], however, the biopolymers exhibit indiscernible dissimilarity in the physico-chemical characteristics.

Norisuye *et al.* [6] first recognized the triple helix structure of schizophyllan in water. The performance in water can be defined by the theory of Yamakawa for a rigid rod chain with the largest chain persistence length known for polysaccharides (approximately 200 nm). Slight flexibility originates when the molecular weight of schizophyllan surpasses 500,000 Da. Several X-ray diffraction analyses on the oriented fibers also proved the triple helix structure in the solid state [34]. In the glucose unit, the 6-OH side is a hydrophilic face, whereas the 2-OH side is a hydrophobic face [35]. Due to the fact that the triple stranded helical framework of schizophyllan is sustained by the hydrogen bonding network among the 2-OH groups generated within the helical column [36], it is possible to assume that the inner hollow is significantly hydrophobic, similar to the inner cavity of cyclodextrins (Figure 8.2) [37,38]. Indeed, the driving force for schizophyllan chains to re-organize the triple strand from the individual strand is regarded to be a hydrophobic force along with the hydrogen bonding force [39].

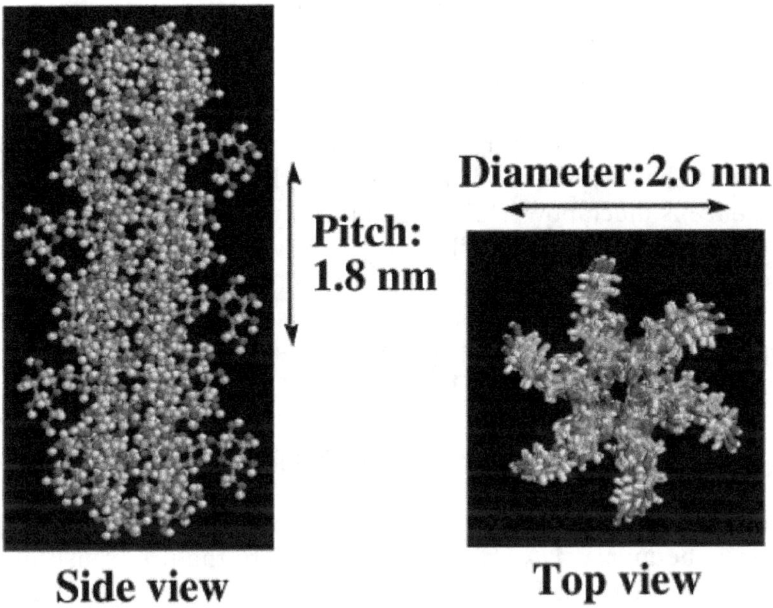

Figure 8.2 Typical model of SPG triple helix. Reproduced from Reference 37 with permission from American Chemical Society.

8.4 Properties

8.4.1 Concentrated Solution Behavior

The steady shear viscosity of the polymer solution exhibits a Newtonian plateau at reduced shear rate. Nevertheless, it starts to reduce at a specific critical shear rate value. Shear thinning reduction in viscosity with enhancement in shear rate is more obvious with developing polymer chain's stiffness [40]. Mostly, with enhancement in concentration, the critical shear rate, at which the viscosity starts to reduce, transfers to lower values. Enomoto *et al.* [41] examined the shear rate reliance of SPG aqueous solution from 0.000257 g/cm^3 to 0.00411 g/cm^3 concentration within shear rate range of 10^{-4}–10 s^{-1} by utilizing a rotational viscometer suitable for rigid molecular chains like schizophyllan. Viscosity was observed to be not dependent on the shear rate over the analyzed range, with the value of viscosity contributing to the zero-shear viscosity. The authors also reported the concentration reliance of the zero-shear viscosity of schizophyllan as $\eta_o \sim C^5$ or C^8. For the rigid rod system, the molecular weight dependency of the zero-shear viscosity of schizophyllan was noted to be $\eta_o \sim M^{6.8}$ or $M^{8.8}$. However, for the semi-flexible specimens, the exponent was almost 5, which revealed that the molecular weight reliance turned to be weaker with enhancing flexibility. Nevertheless, it was also noted that the molecular weight dependence is much powerful in rod like polymers like schizophyllan, when compared to the flexible polymers where $\eta_o \sim M^{3.4}$. In addition, it should be also noted that stiff chains in schizophyllan generate a liquid crystal beyond a specific concentration and the anisotropic phase's concentration reliance of the viscosity is dissimilar to that in isotropic phase [42,43,44]. Van *et al.* [45] noticed a cholesteric mesophase in 14.3% schizophyllan solution, while the 2.33% solution stayed isotropic at the ambient temperature. Further, the authors noted a quick variation in the optical rotatory dispersion while the temperature of isotropic solution was reduced to the isotropic-biphasic boundary, i.e., 5°C, which was due to the pre-transition to the cholesteric phase from the isotropic phase.

As the schizophyllan triple helices are denatured into single coils by enhancing the temperature [46,47] or by the incorporation of polar or strong basic solvents [48,49], the chains are not rebounded to the original triple helices when returned to the initial state. Zentz *et al.* [46] utilized low angle laser light scattering as well as size exclusion chromatography for examining the schizophyllan morphology during thermal operation in aerated solution (temperature greater than 100 °C). The thermal denaturation of the inherent triple helices to individual chains was established at a temperature beyond 135 °C and was completed at 160 °C in 10 minutes. In the temperature range 130-140 °C, both conformations were able to coexist. Both forms of the biopolymer experienced extreme thermal degradation in the presence of oxygen. The degradation rate was observed to be independent of chain length as well as conformation. The examination of chromatographic patterns revealed that the degradation took place by way of an "all-or-none" technique. Young and Jacobs [49] also revealed that the fluorescence magnitude of the aniline blue bound to the biopolymer SPG could be utilized to examine the comparative quantity of single helix transforming to triple helix during various stages of a renature-denature cycle [49]. The study contributed a technique for analyzing conformation variation which is simpler as well as easier to carry out when compared to other methods. It was noticed that the aniline blue did not stain native schizophyllan, however, stained the renatured NaOH treated schizophyllan. Several research studies have examined the chain conformation as well as morphology of denatured-renatured schizophyllan [8,50,51]. The denaturation-renaturation (DR) techniques employed for the biopolymer schizophyllan include (1) dissolution in NaOH and subsequent neutralization with HCl, (2) dissolving schizophyllan in dimethyl sulfoxide (DMSO), followed by dialyzing against water and (3) heating above 135 °C, followed by cooling. Stokke *et al.* [51] proved experimentally that DR schizophyllan can exist in a triple stranded circular pattern. In addition to this circular form, it was also noticed that the DR schizophyllan can exist in different forms such as hairpin, linear and granular forms as well as complex clusters [52,53]. The denaturation-renaturation process of schizophyllan having a molecular weight of $2.6\text{x}10^6$ (termed as SPG-1) has been analyzed by using various intrinsic viscosity tests and rheological techniques [51,54]. Using $[OH^-]$=0.2 mol/L, the SPG-1 was initially made basic, followed by neutralization to pH=7 for generating DRSPG-1. In the concentration range 0.65-0.75 wt%, a sol to weak gel change took place. In case the concentration was higher than 0.75 wt%, the aqueous systems of DRSPG-1 exhibited weak gel rheological behavior.

Yoshiba *et al.* [55] examined the cholesteric pitch (P) of a triple helix schizophyllan as a function of concentration and temperature. It was noticed that the value of cholesteric pitch changed with temperature as well as concentration. It also exhibited varying concentration reliance at higher and reduced temperatures, with a quick reduction in between. This behavior was because of the order-disorder change in SPG solutions in D_2O at a temperature of approx. 17 °C. Two schizophyllan specimens (fractionated) were generated for the cholesteric pitch analysis: specimen KR-1A (M_W = $23.7\text{x}10^4$) and specimen S21 (M_V = $20.8\text{x}10^4$). The P value of schizophyllan in

D_2O was analyzed by 633 nm and 488 nm light diffraction as well as polarized microscopy using a thermostatic cell holder. Figure 8.3 also provides the comparison of the concentration dependence of cholesteric pitch for H_2O as well as D_2O solutions of SPG at temperatures 10 °C and 30 °C. The H_2O solution displayed the order-disorder transition at almost 7 °C, indicating that the triple helical structure of SPG was in the state of disorder in H_2O at 10 °C as well as 30 °C.

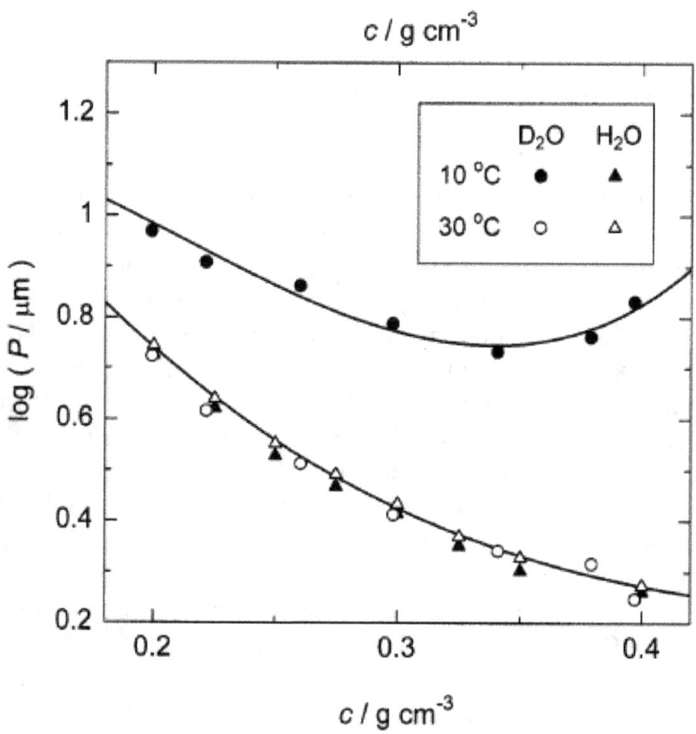

Figure 8.3 Comparison of the concentration dependence of cholesteric pitch for H_2O as well as D_2O solutions of SPG at temperatures 10 °C and 30 °C. Reproduced from Reference 55 with permission from American Chemical Society.

8.4.2 Gelation Behavior

As the aqueous solution is cooled below 6 °C, schizophyllan is able to generate only a weakened gel [56]. The incorporation of certain small molecules, such as sorbitol or borax, has been reported to cause the generation of comparatively stronger schizophyllan gels [57,58,59]. In the case of borax system, this is because of the complexation process occurring between the borate ions and hydroxyl groups present in the schizophyllan side chain, as noted for scleroglucan [60]. The influence of borax and schizophyllan concentrations, temperature as well as pH on the gelation characteristics of SPG-borax system were qualitatively analyzed by Grisell and Muller [59]. The authors observed that in the presence of borate ions, schizophyllan exhibited a significant development in the viscoelastic characteristics. Under suitable conditions, the physical gel framework could be associated with the ionic complexation of schizophyllan. The temperature and salt (nature, content) greatly influenced the rheological behavior of the mixtures. A conventional phase diagram of the system in the presence of sodium chloride (0.5 mol/L) at a temperature 25 °C is presented in Figure 8.4. It was noted that the pH controlled the range of borate ions present in the media and

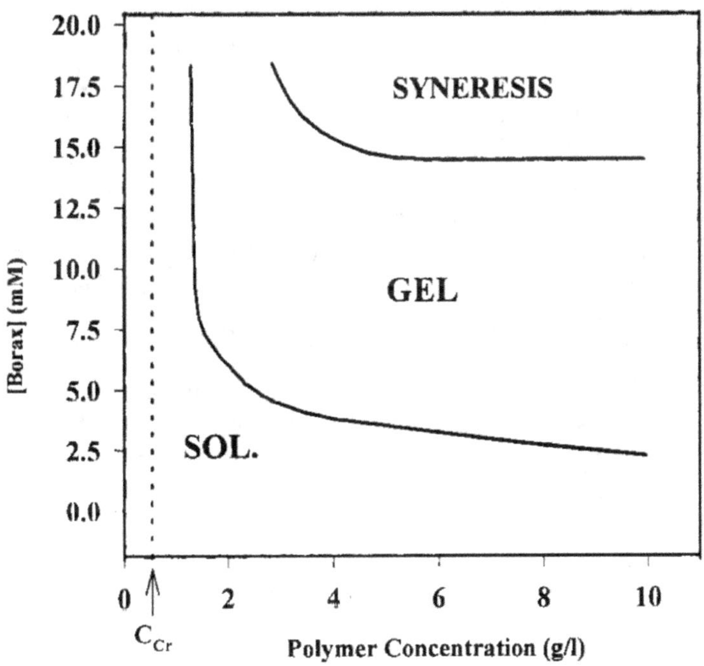

Figure 8.4 Phase diagram of the borax-schizophyllan system in the presence of sodium chloride (0.5 mol/L) at 25 °C. Reproduced from Reference 59 with permission from American Chemical Society.

influenced the phase separation. Gel behavior was consistently noticed succeeding a time setback, thereby, demonstrating the gelification process, representing a connectivity transition. At slower reaction rate, it was feasible to employ multi-frequency oscillation tests for examining the system and to explain the characteristics of the SPG-borax system in the course of the gelification process. Generally, higher the concentration of cross-linker existing in the medium, the quicker the sol-to-gel conversion took place. Also, as the concentration was reduced, the free borate species present in the solution formed greater inter-molecular links between the polysaccharide chains according to the complexation thermodynamics of the corresponding ions.

Fang *et al.* [54] also carried out dynamic viscoelastic analysis for examining the gelation behavior of schizophyllan with the cross-linker borax. The gelation mechanism of the system could be explained by the first order kinetic model. Different factors such as concentration of SPG, salinity, content of borax, temperature, strain and salt types were observed to influence the gelation. It was noted that the gelation process advanced more quickly at elevated content of borax, however, no gelation was observed under its critical content. In Figure 8.5, the plots of gelation rate ($1/\tau_c$) and induction time (t_0) versus the concentration of schizophyllan (C_p) are presented. It is evident that when C_p was increased, t_0 became smaller. Further, it was noted that $1/\tau_c$ enhanced quickly at reduced concentrations and evenly at elevated concentrations. A linear relationship existed between the concentration of schizophyllan and storage modulus. From the temperature dependencies of storage modulus and rate of gelation, the cross-linking enthalpy and apparent activation energy were observed to be -32.4 kJ/mol and -74.5 kJ/mol, respectively. As compared to the galactomannan-borax and poly(vinyl alcohol)-borax systems, the cross-linking enthalpy of the schizophyllan-borax system was observed to be comparatively higher and cross-links were found to be more stable. Further, it was noted that the rate of gelation enhanced with increase in the concentration of salt. The stimulating influence of the monovalent cations followed the order $Cs^+ > K^+ > Na^+ > Li^+$, because of the greater binding magnitude for smaller cations. Nevertheless, gelation did not take place on the incorporation of the divalent cations (Ba^{2+} or Ca^{2+}), due the generation of less-soluble barium borate or calcium borate. The extent of applied strain also affected the gel strength, however, it did not cause variation in gelation rate and induction time.

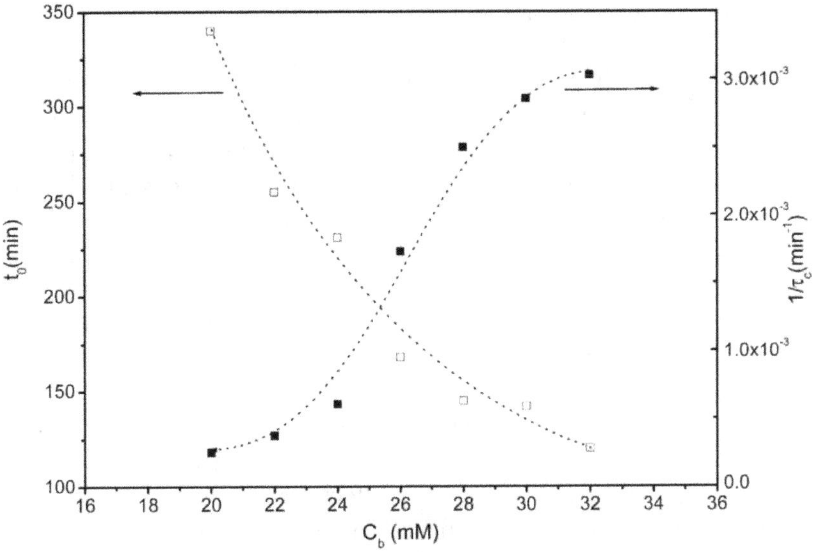

Figure 8.5 Gelation rate ($1/\tau_c$) and induction time (t_0) versus schizophyllan concentration (C_p). Reproduced from Reference 54 with permission from American Chemical Society.

The incorporation of D-sorbitol activates the thermo-reversible gelation process of SPG aqueous solution and the gel temperature is observed to depend only on the concentration of sorbitol [57]. Fuchs *et al.* [57] observed that sorbitol reduced water activity, thus, SPG triple helices combined to generate junction zones. In another study, Maeda *et al.* [58] studied the gel formation by SPG in aqueous solution of sorbitol. The structural variation in SPG brought about by sol-gel transition in aqueous solution of sorbitol was examined by small angle X-ray scattering (SAXS). The authors observed that the limited disentanglement of triple helical structure into distinct coils might be due to the sorbitol breaking the inter-molecular hydrogen bonding which stabilized the rigid triple stranded structure of SPG. The sorbitol melted the triple helices moderately, however, certainly not from the ends of the chain. The decline in water activity brought about the hydrogen bonding between the damaged portions of triple stranded helices and established the junction zone to generate a network structure. Thus, it was clear that the SPG network was not generated by the combination of triple stranded helices, however, the junction zone was comprised of hydrogen bonded as well as disentangled SPG chains.

8.4.3 Interfacial Properties

The interfacial characteristics are necessary for estimating the properties as well as morphology of the phase-separating systems. Chen *et al.* [61] carried out the analysis of interfacial tension (γ) between the co-existing cholesteric and isotropic phases of aqueous solution of SPG using the sessile drop technique at 25 °C. The aqueous solution of schizophyllan experienced a phase separation into the cholesteric and isotropic phases within a confined concentration range of the polymer. The cholesteric-isotropic interfacial area was monitored by introducing a compensator having a retardation of 530 nm aligned at 45° to the crossed polarizers' direction. Figure 8.6 presents the polarizing micrograph of a biphasic SPG aqueous solution having viscosity average M_w 13.1x10^4, which was equilibrated at a temperature of 4 °C for an extended time. The top section with the interference shade is considered to be the isotropic phase, while the lower part having a fingerprint pattern is regarded as the cholesteric phase. The dark and light lines present in the cholesteric phase lined up relatively parallel to the interface, revealing that the cholesteric axis existed normal to interface.

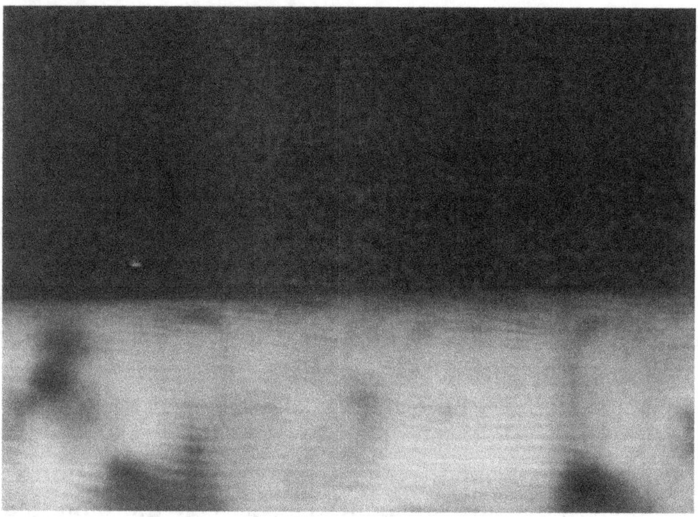

Figure 8.6 Polarizing micrograph of a biphasic schizophyllan aqueous solution having viscosity average M_w 13.1x10^4, close to cholesteric-isotropic interface at 4 °C and average SPG concentration 0.243 g/cm^3. Reproduced from Reference 61 with permission from American Chemical Society.

8.4.4 Dilute Solution Properties

The solution conformation of schizophyllan has been widely studied [6,62,63,48]. As mentioned earlier, the polymer dissolves in water like a rod like triple helix structure, however, it disperses in DMSO to a distinct random coil swollen by the expelled volume impact, generally seen in flexible linear polymers. Both the z-average molar mass ratio M_z(water)/M_z(DMSO) as well as weight average molar mass ratio M_w(water)/M_w(DMSO) were reported to be approximately 3, by using light scattering and sedimentation equilibrium [6]. Kasihiwagi *et al.* [63] carried out the viscosity and light scattering analysis of a dilute solution of schizophyllan in NaOH (0.01 M), which entirely dissolved the biopolymer without causing any destruction in the triple helix framework and eliminated the issue of removal of micro-gels. The pitch per glucose debris and the persistence length in the schizophyllan triple helical structure were noticed to be 0.30 nm and 180 nm, respectively. These values were congruent with the values attained in water. Yanaki *et al.* [62] attained the limiting sedimentation coefficients as well as intrinsic viscosity of the schizophyllan specimens by utilizing different viscometers, suitable for the sample's molecular weight. The results confirmed that the triple helical structure of the biopolymer was essentially rigid till a molecular weight of 5×10^5 g/mol, however, it attained flexibility at greater molecular weights. The triple helices of SPG dissociated almost entirely into single chains in water-DMSO system, when the content of DMSO reached above 87% [48,64]. Further, by the incorporation of NaOH to schizophyllan solution with molar concentration greater than 0.2 mol/L at room temperature or by heating the aqueous solution of schizophyllan higher than 135 °C, the conformational change from triple helices to distinct chains could also be generated [47].

The solutions of schizophyllan exhibit two unified conformational changes on heating, one at approximately 6 °C and the other at almost 135 °C [64,47,65]. The order-disorder variation of schizophyllan at temperature 6-7 °C in water was also observed to change to 17-18 °C in heavy water [66,67]. Further, it was noted that the transition at greater temperature is considered to be thermally irreversible and is characterized by the dissociation of schizophyllan triple helix into single chains [64,47]. The reduced temperature analog of the schizophyllan helices, termed as triple helix I, is regarded to be extra ordered when compared to the greater temperature species, i.e., triple helices II. The two triple helices vary in the aspect of synergy of the β-glucosyl side groups having molecules of water. Itou *et al.* [68] suggested a molecular model for the triple helix I structure, where the side groups bearing on the same helical strand were linked collectively into a helix framework. In this manner, three analogous side group helices were generated and performed the function of an external shell, thereby, enhancing the schizophyllan chain stiffness as well as diameter. As the temperature was increased, the ordered shell framework melted and the transformation occurred from triple helix I to triple helix II.

Depending on the heat capacity evaluation and dielectric relaxation analysis, Yoshiba *et al.* [65,69] noticed that the molecules of water in the triple helical I condition remained in various structural forms such as structured water, bound water, loosely structured water as well as free water in enhancing order of motility. The authors conducted the heat capacity analysis on aqueous solution of triple helix SPG by means of precision adiabatic calorimetry at a temperature range of 5-15 K over a broad range of concentrations [65]. The obtained heat capacity curves were classified into four groups contingent upon the weight fraction of SPG as regions I, II, III and IV. In area I, the triple helices having the covering of structured water, bound water and loosely structured water generating layers over the helical core were ingrained in the free water. However, in the case of region II, free water was absent and the loosely structured water remained constant with enhancement in weight fraction. In area III, the bound water continued to be unaffected, nevertheless, the structured water reduced with enrichment in weight fraction, by overlying with each other. Ultimately, bound water and schizophyllan existed in region IV, where the bound water reduced with increase in weight fraction. Figure 8.7 also presents the cross-sectional outlook of triple helical framework of schizophyllan in water. The recurrent unit of β-1,3-D-glucan of SPG consists of three glucose residues as well as one side chain glucose residue.

8.5 Application in Enhanced Oil Recovery

For majority of the oilfields in the world, usually half of the available oil is left behind in the formations after the primary and secondary recovery (water flooding) processes. Water is introduced into underground reservoir rocks for maintaining the reservoir pressure and to sweep oil into the production wells. This procedure is termed

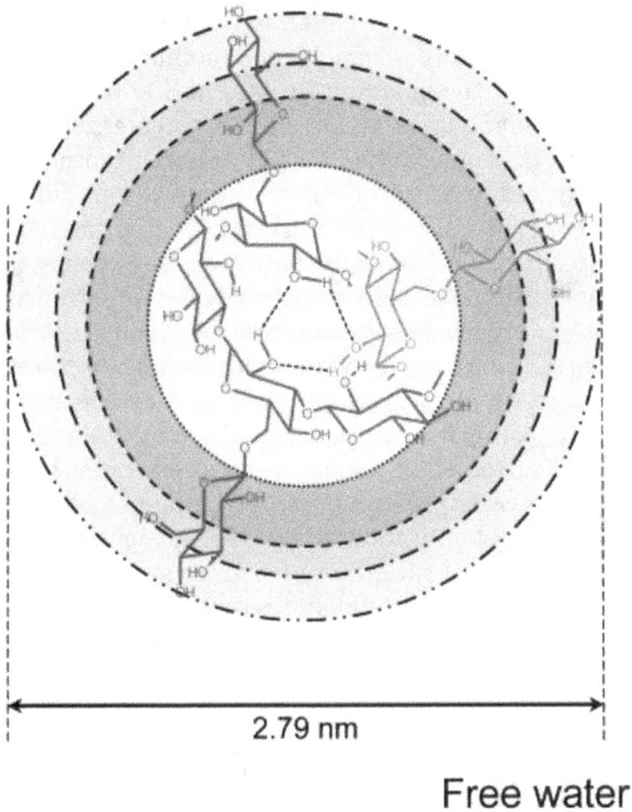

Figure 8.7 Cross-sectional outlook of triple helical framework of schizophyllan in water. Reproduced from Reference 65 with permission from American Chemical Society.

as water flooding and is one of the most prominent secondary recovery methods because of its comparatively low cost. In most of the cases, water has comparably greater mobility as compared to oil, hence, the water bypasses the oil and flows through the passage of minimum resistance [70]. However, this process recovers lesser than 50% of the original oil in place (OOIP). Studies have recommended different techniques for enhancing the oil recovery, which include chemical, thermal and gas injection methods [71]. Polymer flooding is the process of introduction of polymer solutions into the reservoirs and is considered as an effective chemical enhanced oil recovery (EOR) approach. Since 1990s, several large-scale polymer flooding works have been executed [72]. The two major mechanisms that increase oil production in the course of polymer flooding are enhanced reservoir conformance as the polymer solution passes via the pores and a more advantageous mobility ratio of water-oil due to the polymer. The polymers utilized in EOR operations can be categorized as biopolymers and synthetic polymers. Polymers like partially hydrolyzed polyacrylamide (HPAM) and xanthan gum have been demonstrated to be promising candidates for use in enhancing the oil recovery [70]. For the commercial EOR applications, HPAM is one of the most frequently injected polymers. Nevertheless, HPAM is considerably sensitive to salinity as well as temperature [72,73]. Thus, HPAM undergoes degradation when introduced into the high temperature reservoirs having water with significant salinity. Other issues related to the utilization of HPAM include separation of oil-water, toxicity concerns associated with the HPAM degradation products and the need of treatment of produced water. These limitations restrict the widespread industrial acceptance of this polymer category in the EOR application. In the recent years, many other biopolymers have gained attention of petroleum industries as EOR candidates which can tolerate higher temperature as well as higher salinity [74,75]. Recent research studies revealed that biopolymers like schizophyllan are immensely useful under such harsh conditions [76].

Microbial enhanced oil recovery (MEOR) is a useful technique for extending the lifetime of production wells in a declining oil reservoir [77]. This method is based on making use of bio-products as well as metabolites of microorganisms for recovering incremental or additional oil from a reservoir. The usage of biopolymers such as SPG in such EOR application results in beneficial features such as permeability advancement for improving the water flood sweep effectiveness and careful plugging of greater permeability zones, which is termed as the thief zones [78]. However, the possible utilization of schizophyllan in the petroleum industry is yet to be examined in significant detail [79]. In one such study, Gao *et al.* [2] analyzed the rheological behavior of schizophyllan under high temperature as well as high salinity conditions. Schizophyllan was combined with base brine and the viscosity was measured at a temperature of 55 °C using a shearing-type viscometer. It was noted that the viscosity of HPAM solution was lower than schizophyllan under identical conditions. Further, the viscosity of SPG reduced mildly with increase in temperature, whereas the viscosity of HPAM and xanthan gum degraded more critically. As an illustration, schizophyllan at 90 °C retained 90% of its original viscosity observed at 30 °C. However, the xanthan gum and HPAM at 90 °C lost 50% and 32% of their initial viscosity, respectively. The authors also observed that the viscosity of schizophyllan did not reduce remarkably subsequent to testing for two months. From the resistance factor test analysis, it was noted that the injection of schizophyllan elevated the pressure drop across the rock core. Nevertheless, a quick decline in the resistance factor was seen at greater injection velocity. However, the residual resistance factor continued to be at lesser value under both high as well as low fluid velocity. The recovery analysis proved that the water flooding caused oil recovery of 65%, following the introduction of 5 pore volumes of brine. In the case of recovery experiment with schizophyllan solution, stable state was attained succeeding 3 pore volumes of SPG solution incorporation and the recovery enhanced to 75% after 4 pore volumes of injection of solution.

Quadri *et al.* [80] also examined the influence of salinity and temperature on the rheological behavior of schizophyllan. The polymer was observed to exhibit good temperature stability as well as salt tolerance with no reduction in the viscosity till 220 g/L salinity and 135 °C. The salt tolerance and temperature stability of schizophyllan are due to its triple helical structure as well as non-ionic nature. In addition, schizophyllan solution exhibited no reduction in viscosity when kept at 120 °C under anaerobic conditions over a period of eight months. From the rheological analysis, it was observed that the biopolymer exhibited a powerful shear thinning nature with enhancing shear rate for all concentrations. Low shear Newtonian region was observed for low polymer concentrations (200, 500 and 1000 ppm), however, for solutions with higher polymer concentration solutions (2500 and 5600 ppm), only shear thinning nature was noticed. This is because of the structural peculiarity of the schizophyllan chains, which are lined up along the flow streamline as shear rate enhances, whereas in the Newtonian region at lesser shear rates, the structural peculiarity has no influence on the viscosity. The authors also observed that schizophyllan solution viscosity was not sensitive to greater salinities and had distinct shear-thinning nature for the all salinities at a concentration of 500 ppm. A set of core flooding tests were carried out at 167 g/L salinity and 120 °C reservoir temperature to examine the efficiency of the polymer for EOR process in the carbonate reservoirs. The results confirmed that resistance factors (RF) enhanced and residual resistance factors (RRF) decreased with increase in permeability. Also, lesser adsorption was observed in the high permeability cores.

Compared to water flooding, the assessment of polymer flooding generally has an extra layer of complexity due to the fact that the injected fluid has non-Newtonian behavior. This results in the alteration of the existing techniques or the development of new assessment methods for examining the effectiveness of the flooding process. Modified Hall plots and falloff tests are some of the examples of the techniques which have been utilized in several analyses. Ogezi *et al.* [79] selected a compartment in the Bockstedt field, an oilfield in Northern Germany, for carrying out a pilot test with schizophyllan. The modified Hall plot, utilized for examining injectivity of polymer, did not exhibit any remarkable decrease in the injectivity. It was also predicted that after periods of extended shut-ins (unplanned or planned), injection of polymer could be restarted without any detrimental impact on the injectivity of polymer. The pressure falloff analysis is an easy reservoir surveillance method for examining the water injection wells. Analysis of the properties of the reservoir during the water injection phase and consequently in-situ polymer viscosity during polymer injection phase proved that schizophyllan did not exhibit shear degradation in the reservoir.

Joshi *et al.* [81] also carried out the structural analysis of SPG and evaluated its applications in MEOR. The authors reported that significant elasticity along with the shear thinning nature owing to the rigid triple helix structure of SPG make it suitable for the field utilization where the viscosity of the injected SPG solution at turbulent flow might lower the greater back-pressure and will avoid the damage to the formation. The viscosity was observed to be 21535 cP at 0.1 s^{-1} shear rate and 43 cP at 100.0 s^{-1} shear rate, after incubation for 14 days. From the thermo-gravimetric analysis (TGA), it was noted that the polymer was stable in the temperature range 100-200 °C, however, it stared melting at a temperature >250 °C. From the core-flooding analysis, it was noted that 28% extra oil recovery took place by employing schizophyllan.

8.6 Other Developments

As mentioned earlier, in alkaline media or DMSO, SPG exhibits single chain (s-SPG) morphology [82]. By neutralizing alkaline solution with acid or by exchanging DMSO with water, the single chain SPG is able to recapture the original triple helical structure (renaturation procedure). In the course of the renaturation procedure, schizophyllan generates a one-dimensional (1-D) hydrophobic cavity inside its helical superstructure and accepts molecular assemblies, functional polymers and nanoparticles in an induced way to generate 1-D water soluble nanocomposites. Several studies have confirmed that the schizophyllan-based 1-D hosts can include different guest molecules for generating exclusive fibrous assemblies of these entities [83,84,85]. Numata *et al.* [83] reported the generation of water soluble polyaniline (PANI) nanofiber attained by employing the polymer wrapping method. Figure 8.8 illustrates the entrapping of polyaniline into schizophyllan hydrophobic cavity during renaturation process.

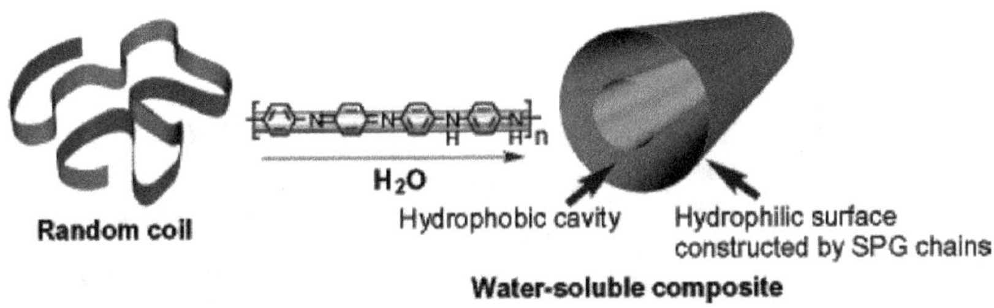

Figure 8.8 Entrapment of polyaniline into hydrophobic cavity of schizophyllan during renaturation process. Reproduced from Reference 83 with permission from American Chemical Society.

Sc3p is one of the main hydrophobins generated as well as excreted by the fungus *Schizophyllum commune*, when matured as a monokaryon in the culture. Martin *et al.* [86] reported a new method for isolating the hydrophobin, without the need of treatment with strong organic acids. The authors also confirmed that SPG stabilized the minute multimers of hydrophobin in the aqueous solution. Further, it was noted that SPG was required for the consecutive assembly of hydrophobin into films on the hydrophilic exterior. The authors proposed a model where the hydrophobin small multimers or/and unimers achieved stabilization in solution by combining with higher molecular weight SPG, avoiding the self-aggregation, as shown in Figure 8.9. More specifically, the stabilization was generated by means of hydrogen bonding or different synergies along the lengthened chains of SPG, e.g., the triple helix segment, as shown is Figure 8.9A. By preference, a flexible SPG conformation might also permit the SPG chains to perform as a typical steric stabilizer (Figure 8.9B). The shearing of SPG/hydrophobin complex in the presence of a liquid hydrophobe (e.g. oil) uncovered the hydrophobic residues for anchoring of the surface, thereby, leading to sterically stabilized droplets of oil.

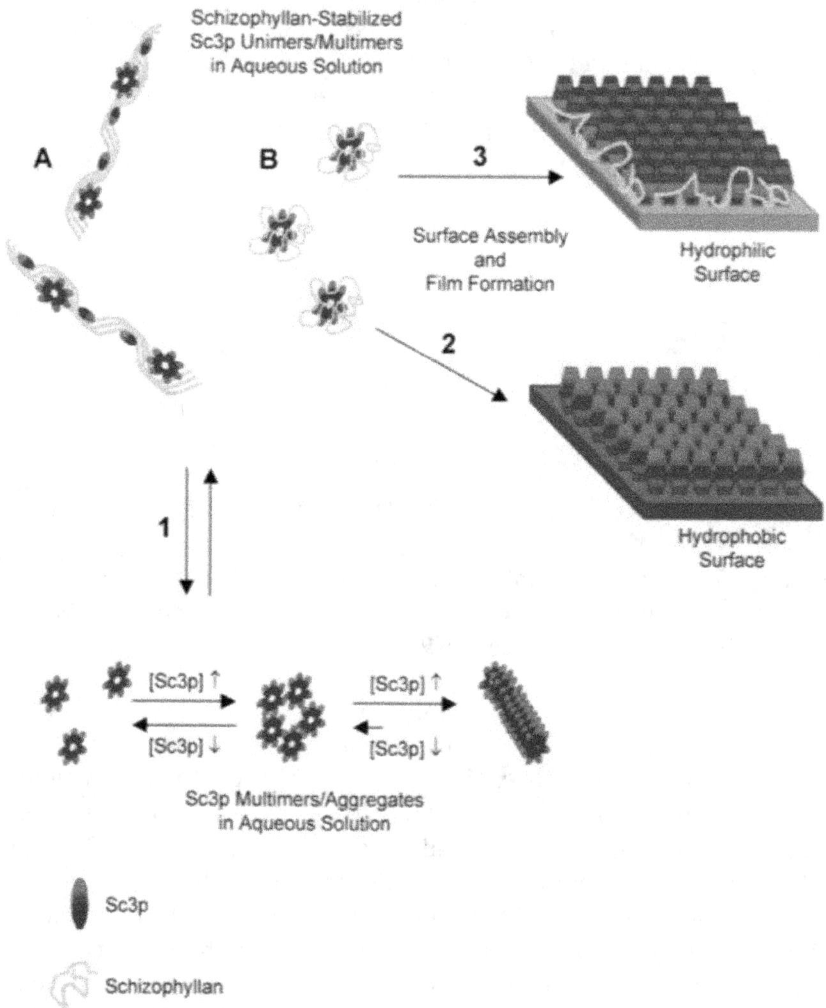

Figure 8.9 Model of hydrophobin/SPG synergy in the aqueous solution as well as film generation at hydrophilic or hydrophobic interfaces. Reproduced from Reference 86 with permission from American Chemical Society.

The metal nanoparticles (NPs) as well as their polymer composites contribute to a broad range of engineering applications. Due to the higher surface area of the nanoparticles, their uniform dispersion in the polymers remains one of the major challenges. Significant aggregation occurs when the NPs are employed alone, thereby, causing a decline in their surface activity [87]. One of the efficient methods for lowering the NP aggregation is to generate these nanoparticles by in-situ approaches in the presence of an appropriate polymer matrix. Abdel-Mohsen *et al.* [88] generated silver nanoparticles by employing schizophyllan both as stabilizing as well as reducing agent. In order to obtain uniform dispersion of silver NPs, the concentrations of SPG and silver nitrate, temperature, pH, reaction mixture and pH were optimized. The resultant silver NPs were spherical in morphology with an average size of about 6±2 nm. The SPG as well as SPG/silver-NP nanocomposite were observed to be non-toxic in nature over a broad range of concentrations with time.

8.7 Conclusions

Schizophyllan is a high-performance biopolymer with significant potential for use in a variety of advanced applications, especially enhanced oil recovery, owing to the combination of useful characteristics. Preliminary studies on schizophyllan have confirmed its superiority over other polymers as a candidate for high temperature and high salinity polymer flooding applications. Ongoing efforts in this direction for further understanding the polymer behavior and tuning of structure-property correlations would subsequently accelerate the application of schizophyllan in the oil wells for enhanced oil recovery.

References

1. Rau, U. (1999) Production of schizophyllan. In: *Carbohydrate Biotechnology Protocols*, Bucke, C. (ed.), 10th volume, Humana Press, USA, pp. 43-55.
2. Gao, C. (2016) Application of a novel biopolymer to enhance oil recovery. *Journal of Petroleum Exploration and Production Technology*, **6**(4), 749-753.
3. Zhang, Y., Kong, H., Fang, Y., Nishinari, K., and Phillips, G. O. (2013) Schizophyllan: a review on its structure, properties, bioactivities and recent developments. *Bioactive Carbohydrates and Dietary Fibre,* **1**(1), 53-71.
4. Fang, Y., Takahashi, R., and Nishinari, K. (2005) Protein/polysacch-

aride cogel formation based on gelatin and chemically modified schizophyllan. *Biomacromolecules,* **6**(6), 3202-3208.

5. Rau, U., Muller, R. J., Cordes, K., and Klein, J. (1990) Process and molecular data of branched 1,3-β-D-glucans in comparison with Xanthan. *Bioprocess and Biosystems Engineering,* **5**(2), 89-93.
6. Norisuye, T., Yanaki, T., and Fujita, H. (1980) Triple helix of a Schizophyllum commune polysaccharide in aqueous solution. *Journal of Polymer Science: Polymer Physics Edition,* **18**(3), 547-558.
7. Ferretti, R., Stoll, S., Zhang, J., and Buffle, J. (2003) Flocculation of hematite particles by a comparatively large rigid polysaccharide: schizophyllan. *Journal of colloid and interface science,* **266**(2), 328-338.
8. Kitamura, S., Hirano, T., Takeo, K., Fukada, H., Takahashi, K., Falch, B. H., and Stokke, B. T. (1996) Conformational transitions of schizophyllan in aqueous alkaline solution. *Biopolymers,* **39**(3), 407-416.
9. Bae, A. H., Lee, S. W., Ikeda, M., Sano, M., Shinkai, S., and Sakurai, K. (2004) Rod-like architecture and helicity of the poly (C)/schizophyllan complex observed by AFM and SEM. *Carbohydrate research,* **339**(2), 251-258.
10. Rau, U., Gura, E., Olszewski, E., and Wagner, F. (1992) Enhanced glucan formation of filamentous fungi by effective mixing, oxygen limitation and fed-batch processing. *Journal of Industrial Microbiology & Biotechnology,* **9**(1), 19-25.
11. Gawronski, M., Conrad, H., Springer, T., and Stahmann, K. P. (1996) Conformational changes of the polysaccharide cinerean in different solvents from scattering methods. *Macromolecules,* **29**(24), 7820-7825.
12. Pilz, F., Auling, G., Rau, U., and Wagner, F. (1991) A high-affinity Zn^{2+} uptake system controls growth and biosynthesis of an extracellular, branched β-1,3-β-1,6-glucan in Sclerotium rolfsii ATCC 15205. *Experimental Mycology,* **15**(3), 181-192.
13. Rau, U., Haarstrick, A., and Wagner, F. (1992). Suitability of schizophyllan solutions for polymer flooding of oil-fields at high-temperature and salinity. *Chemie Ingenieur Technik,* **64**(6), 576-577.
14. Schulz, D., Rau, U., and Wagner, F. (1992) Characteristics of films prepared from native and modified branched β-1,3-D-glucans. *Carbohydrate polymers,* **18**(4), 295-299.
15. Sutivisedsak, N., Leathers, T. D., Biresaw, G., Nunnally, M. S., and Bischoff, K. M. (2016) Simplified process for preparation of schizophyllan solutions for biomaterial applications. *Preparative Biochemistry and Biotechnology,* **46**(3), 313-319.
16. Kikumoto, S., Miyajima, T., Kimura, K., Okubo, S., and Komatsu, N.

(1971) Polysaccharide produced by Schizophyllum commune. II. Chemical structure of an extracellular polysaccharide. *Nippon Nōgeikagaku Kaishi*, **45**, 162-168.

17. Kikumoto, S., Miyajima, T., Yoshizumi, S., Fujimoto, S., and Kimura, K. (1970) Polysaccharide produced by schizophyllum commune. 1. Formation and some properties of an extracellular polysaccharide. *Nippon Nōgeikagaku Kaishi*, **44**, 337-342.
18. Steiner, W., Lafferty, R. M., Gomes, I., and Esterbauer, H. (1987) Studies on a wild strain of Schizophyllum commune: cellulase and xylanase production and formation of the extracellular polysaccharide schizophyllan. *Biotechnology and Bioengineering*, **30**(2), 169-178.
19. Leathers, T. D., Nunnally, M. S., and Price, N. P. (2006). Co-production of schizophyllan and arabinoxylan from corn fiber. *Biotechnology Letters*, **28**(9), 623-626.
20. Rosentrater, K. A., Ileleji, K., and Johnston, D. B. (2011). Manufacturing of fuel ethanol and distillers grains: Current and evolving processes. In: *Distillers Grains: Production, Properties, and Utilization*, Liu, K. and Rosentrater, K. A. (eds.), AOCS Publishing, USA, pp. 73-102.
21. Sutivisedsak, N., Leathers, T. D., Nunnally, M. S., Price, N. P., and Biresaw, G. (2013) Utilization of agricultural biomass in the production of the biopolymer schizophyllan. *Journal of Industrial Microbiology & Biotechnology*, **40**(1), 105-112.
22. Prokop, A., Rapp, P., and Wagner, F. (1992) Production of extracellular β-1,3-/β-1,6-glucan by mono-and dikaryons of Schizophyllum commune. *Experimental Mycology*, **16**(3), 197-206.
23. Kumari, M., Survase, S. A., and Singhal, R. S. (2008) Production of schizophyllan using Schizophyllum commune NRCM. *Bioresource Technology*, **99**(5), 1036-1043.
24. Sutivisedsak, N., Leathers, T. D., and Price, N. P. (2013) Production of schizophyllan from distiller's dried grains with solubles by diverse strains of Schizophyllum commune. *Springerplus*, **2**(1), 476.
25. Ohm, R. A., De Jong, J. F., Lugones, L. G., Aerts, A., Kothe, E., Stajich, J. E., De Vries, R. P., Record, E., Levasseur, A., Baker, S. E., and Bartholomew, K. A. (2010) Genome sequence of the model mushroom Schizophyllum commune. *Nature Biotechnology*, **28**(9), 957-963.
26. Shu, C. H., and Hsu, H. J. (2011). Production of schizophyllan glucan by Schizophyllum commune ATCC 38548 from detoxificated hydrolysate of rice hull. *Journal of the Taiwan Institute of Chemical Engineers*, **42**(3), 387-393.
27. Gao, L. Z., and Zhou, S. M. (2008) Optimization of extraction technology of schizophyllan produced by solid fermentation. *Science and Technology of Food Industry*, **2**, 064.
28. Desai, K. M., Survase, S. A., Saudagar, P. S., Lele, S. S., and Singhal, R.

S. (2008) Comparison of artificial neural network (ANN) and response surface methodology (RSM) in fermentation media optimization: case study of fermentative production of scleroglucan. *Biochemical Engineering Journal*, **41**(3), 266-273.

29. Zhang, M., Cui, S. W., Cheung, P. C. K., and Wang, Q. (2007) Antitumor polysaccharides from mushrooms: a review on their isolation process, structural characteristics and antitumor activity. *Trends in Food Science & Technology*, **18**(1), 4-19.
30. Chenghua, D., Xiangliang, Y., Xiaoman, G., Yan, W., Jingyan, Z., and Huibi, X. (2000) A β-d-glucan from the sclerotia of Pleurotus tuber-regium (Fr.) Sing. *Carbohydrate Research*, **328**(4), 629-633.
31. Chaubey, M., and Kapoor, V. P. (2001) Structure of a galactomannan from the seeds of Cassia angustifolia Vahl. *Carbohydrate Research*, **332**(4), 439-444.
32. Johnson, J., Kirkwood, S., Misaki, A., Nelson, T. E., Scaletti, J. V., and Smith, F. (1963) Structure of a new glucan. *Chemistry & Industry*, 820-822.
33. Tombs, M. P., and Harding, S. E. (1998) *An Introduction to Polysaccharide Biotechnology*, Taylor and Francis, UK.
34. Bluhm, T. L., Deslandes, Y., Marchessault, R. H., Perez, S., & Rinaudo, M. (1982) Solid-state and solution conformation of scleroglucan. *Carbohydrate Research*, **100**(1), 117-130.
35. Miyoshi, K., Uezu, K., Sakurai, K., and Shinkai, S. (2004) Proposal of a new hydrogen-bonding form to maintain curdlan triple helix. *Chemistry & Biodiversity*, **1**(6), 916-924.
36. Deslandes, Y., Marchessault, R. H., and Sarko, A. (1980) Triple-helical structure of (1→3)-β-D-glucan. *Macromolecules*, **13**(6), 1466-1471.
37. Numata, M., Asai, M., Kaneko, K., Bae, A. H., Hasegawa, T., Sakurai, K., and Shinkai, S. (2005) Inclusion of cut and as-grown single-walled carbon nanotubes in the helical superstructure of schizophyllan and curdlan (β-1,3-glucans). *Journal of the American Chemical Society*, **127**(16), 5875-5884.
38. Okumura, H., Kawaguchi, Y., and Harada, A. (2003) Preparation and characterization of the inclusion complexes of poly (dimethylsilane)s with cyclodextrins. *Macromolecules*, **36**(17), 6422-6429.
39. McIntire, T. M., and Brant, D. A. (1998) Observations of the (1→3)-β-d-glucan linear triple helix to macrocycle interconversion using noncontact atomic force microscopy. *Journal of the American Chemical Society*, **120**(28), 6909-6919.
40. Gidley, M. J., and Nishinari, K. (2009) Physico-chemistry of (1,3)-β-Glucans. In: *Chemistry, Biochemistry and Biology of (1,3)-Beta Glucans and Related Polysaccharides*, Bacic, A., Fincher, G., and Stone, B. (eds.), Academic Press, USA, pp. 47-118.

41. Enomoto, H., Einaga, Y., and Teramoto, A. (1985) Viscosity of concentrated solutions of rodlike polymers. *Macromolecules*, **18**(12), 2695-2702.
42. Lee, H. C., and Brant, D. A. (2002) Rheology of concentrated isotropic and anisotropic xanthan solutions. 1. A rod like low molecular weight sample. *Macromolecules*, **35**(6), 2212-2222.
43. Lee, H. C., and Brant, D. A. (2002) Rheology of concentrated isotropic and anisotropic xanthan solutions. 2. A semiflexible wormlike intermediate molecular weight sample. *Macromolecules*, **35**(6), 2223-2234.
44. Lee, H. C., and Brant, D. A. (2002) Rheology of concentrated isotropic and anisotropic xanthan solutions: 3. Temperature dependence. *Biomacromolecules*, **3**(4), 742-753.
45. Van, K., and Teramoto, A. (1982) Isotropic-liquid crystal phase equilibrium in aqueous solutions of a triple-helical polysaccharide schizophyllan. *Polymer Journal*, **14**(12), 999-1005.
46. Zentz, F., Verchere, J. F., and Muller, G. (1992) Thermal denaturation and degradation of schizophyllan. *Carbohydrate Polymers*, **17**(4), 289-297.
47. Yanaki, T., Tabata, K., and Kojima, T. (1985) Melting behaviour of a triple helical polysaccharide schizophyllan in aqueous solution. *Carbohydrate Polymers*, **5**(4), 275-283.
48. Sato, T., Sakurai, K., Norisuye, T., and Fujita, H. (1983) Collapse of randomly coiled Schizophyllan in mixtures of water and dimethylsulfoxide. *Polymer Journal*, **15**(1), 87-96.
49. Young, S. H., and Jacobs, R. R. (1998) Sodium hydroxide-induced conformational change in schizophyllan detected by the fluorescence dye, aniline blue. *Carbohydrate Research*, **310**(1), 91-99.
50. Vuppu, A. K., Garcia, A. A., and Vernia, C. (1997) Tapping mode atomic force microscopy of scleroglucan networks. *Biopolymers*, **42**(1), 89-100.
51. Stokke, B. T., Elgsaeter, A., Brant, D. A., and Kitamura, S. (1991) Supercoiling in circular triple-helical polysaccharides. *Macromolecules*, **24**(23), 6349-6351.
52. Stokke, B. T., Elgsaeter, A., Brant, D. A., Kuge, T., and Kitamura, S. (1993) Macromolecular cyclization of (1→6)-branched-(1→3)-β-D-glucans observed after denaturation-renaturation of the triple-helical structure. *Biopolymers*, **33**(1), 193-198.
53. Falch, B. H., Elgsaeter, A., and Stokke, B. T. (1999) Exploring the (1→3)-β-D-glucan conformational phase diagrams to optimize the linear to macrocycle conversion of the triple-helical polysaccharide scleroglucan. *Biopolymers*, **50**(5), 496-512.
54. Fang, Y., Takahashi, R., and Nishinari, K. (2004) A gel network constituted by rigid schizophyllan chains and nonpermanent crosslinks. *Biomacromolecules*, **5**(1), 126-136.

55. Yoshiba, K., Teramoto, A., Nakamura, N., and Sato, T. (2003) Chiral interactions in polymer liquid crystals reflecting polymer conformations: triple-helical polysaccharide schizophyllan and poly (γ-benzyl L-glutamate). *Macromolecules,* **36**(6), 2108-2113.
56. Bot, A., Smorenburg, H. E., Vreeker, R., Paques, M., and Clark, A. H. (2001) Melting behaviour of schizophyllan extracellular polysaccharide gels in the temperature range between 5 and 20 C. *Carbohydrate Polymers,* **45**(4), 363-372.
57. Fuchs, T., Richtering, W., Burchard, W., Kajiwara, K., and Kitamura, S. (1998) Gel point in physical gels: rheology and light scattering from thermoreversibly gelling schizophyllan. *Polymer Gels and Networks,* **5**(6), 541-559.
58. Maeda, H., Yuguchi, Y., Kitamura, S., Urakawa, H., Kajiwara, K., Richtering, W., Fuchs, T., and Burchard, W. (1999) Structural aspect of gelation in schizophyllan/sorbitol aqueous solution. *Polymer Journal,* **31**(6), 530-534.
59. Grisel, M., and Muller, G. (1998) Rheological properties of the schizophyllan-borax system. *Macromolecules,* **31**(13), 4277-4281.
60. Fang, Y., and Nishinari, K. (2004) Gelation behaviors of schizophyllan-sorbitol aqueous solutions. *Biopolymers,* **73**(1), 44-60.
61. Chen, W. L., Sato, T., and Teramoto, A. (1999) Interfacial tension between coexisting isotropic and cholesteric phases for aqueous solutions of schizophyllan. *Macromolecules,* **32**(5), 1549-1553.
62. Yanaki, T., Norisuye, T., and Fujita, H. (1980) Triple helix of Schizophyllum commune polysaccharide in dilute solution. 3. Hydrodynamic properties in water. *Macromolecules,* **13**(6), 1462-1466.
63. Kashiwagi, Y., Norisuye, T., and Fujita, H. (1981) Triple helix of Schizophyllum commune polysaccharide in dilute solution. 4. Light scattering and viscosity in dilute aqueous sodium hydroxide. *Macromolecules*, **14**(5), 1220-1225.
64. Kitamura, S., and Kuge, T. (1989) A differential scanning calorimetric study of the conformational transitions of schizophyllan in mixtures of water and dimethylsulfoxide. *Biopolymers,* **28**(2), 639-654.
65. Yoshiba, K., Teramoto, A., Nakamura, N., Kikuchi, K., Miyazaki, Y., and Sorai, M. (2003) Static water structure detected by heat capacity measurements on aqueous solutions of a triple-helical polysaccharide schizophyllan. *Biomacromolecules,* **4**(5), 1348-1356.
66. Itou, T., Teramoto, A., Matsuo, T., and Suga, H. (1986) Ordered structure in aqueous polysaccharide. 5. Cooperative order-disorder transition in aqueous schizophyllan. *Macromolecules,* **19**(4), 1234-1240.
67. Hayashi, Y., Shinyashiki, N., Yagihara, S., Yoshiba, K., Teramoto, A., Nakamura, N., Miyazaki Y., Sorai, M., and Wang, Q. (2002). Ordering in aqueous polysaccharide solutions. I. Dielectric relaxation in

aqueous solutions of a triple-helical polysaccharide schizophyllan. *Biopolymers*, **63**(1), 21-31.

68. Itou, T., Teramoto, A., Matsuo, T., and Suga, H. (1987) Isotope effect on the order-disorder transition in aqueous schizophyllan. *Carbohydrate Research*, **160**, 243-257.
69. Yoshiba, K., Ishino, T., Teramoto, A., Nakamura, N., Miyazaki, Y., Sorai, M., Wang Q., Hayashi Y., Shinyashiki, N., and Yagihara, S. (2002) Ordering in aqueous polysaccharide solutions. II. Optical rotation and heat capacity of aqueous solutions of a triple-helical polysaccharide schizophyllan. *Biopolymers*, **63**(6), 370-381.
70. Sheng, J. (2010) *Modern Chemical Enhanced Oil Recovery: Theory and Practice*, 1st edition, Gulf Professional Publishing, UK.
71. Gao, C. H. (2011) Scientific research and field applications of polymer flooding in heavy oil recovery. *Journal of Petroleum Exploration and Production Technology*, **1**(2-4), 65-70.
72. Gao, C. (2014) Empirical correlations for viscosity of partially hydrolyzed Polyacrylamide. *Journal of Petroleum Exploration and Production Technology*, **4**(2), 209-213.
73. Gao, C. (2013) Viscosity of partially hydrolyzed polyacrylamide under shearing and heat. *Journal of Petroleum Exploration and Production Technology*, **3**(3), 203-206.
74. Xu, L., Xu, G., Yu, L., Gong, H., Dong, M., and Li, Y. (2014) The displacement efficiency and rheology of welan gum for enhanced heavy oil recovery. *Polymers for Advanced Technologies*, **25**(10), 1122-1129.
75. Gao, C. (2015) Potential of welan gum to enhance oil recovery. *Journal of Petroleum Exploration and Production Technology*, **5**(2), 197-200.
76. Leonhardt, B., Ernst, B., Reimann, S., Steigerwald, A., and Lehr, F. (2014) Field Testing the Polysaccharide Schizophyllan: Results of the First Year. *SPE Improved Oil Recovery Symposium*, USA. Online: https://www.onepetro.org/conference-paper/SPE-169032-MS (assessed 12th April 2017).
77. Al-Sulaimani, H., Joshi, S., Al-Wahaibi, Y., Al-Bahry, S., Elshafie, A., and Al-Bemani, A. (2011) Microbial biotechnology for enhancing oil recovery: Current developments and future prospects. *Biotechnol. Bioinf. Bioeng.*, **1**(2), 147-158.
78. Shabani Afrapoli, M., Crescente, C. M., Li, S., Alipour, S., and Torsater, O. (2012) Simulation Study of Displacement Mechanisms in Microbial Improved Oil Recovery Experiments. *SPE EOR Conference at Oil and Gas West Asia*, Oman. Online: https://www.onepetro.org/conference-paper/SPE-153323-MS (assessed 16th April 2017).
79. Ogezi, O., Strobel, J., Egbuniwe, D., and Leonhardt, B. (2014) Operational Aspects of a Biopolymer Flood in a Mature Oilfield. *SPE Im-*

proved Oil Recovery Symposium, USA. Online: https://www.onepetro.org/conference-paper/SPE-169158-MS (assessed 12th April 2017).

80. Quadri, S. M. R., Jiran, L., Shoaib, M., Hashmet, M. R., AlSumaiti, A. M., and Alhassan, S. M. (2015) Application of Biopolymer to Improve Oil Recovery in High Temperature High Salinity Carbonate Reservoirs. *Abu Dhabi International Petroleum Exhibition and Conference,* UAE. Online: https://www.onepetro.org/conference-paper/SPE-177915-MS (assessed 9th November 2017).
81. Joshi, S. J., Al-Wahaibi, Y. M., Al-Bahry, S., Elshafie, A., Al-Bemani, A. S., Al-Hashmi, A., Samuel P., Sassi, M., Al-Farsi, H., and Al-Mandhari, M. S. (2016) Production and Application of Schizophyllan in Microbial Enhanced Heavy Oil Recovery. *SPE EOR Conference at Oil and Gas West Asia,* Oman. Online: https://www.onepetro.org/conference-paper/SPE-179775-MS (assessed 21st March 2017).
82. Tabata, K., Ito, W., Kojima, T., Kawabata, S., and Misaki, A. (1981) Ultrasonic degradation of schizophyllan, an antitumor polysaccharide produced by Schizophyllum commune Fries. *Carbohydrate Research,* **89**(1), 121-135.
83. Numata, M., Hasegawa, T., Fujisawa, T., Sakurai, K., and Shinkai, S. (2004) β-1,3-glucan (schizophyllan) can act as a one-dimensional host for creation of novel poly(aniline) nanofiber structures. *Organic Letters,* **6**(24), 4447-4450.
84. Sakurai, K., Uezu, K., Numata, M., Hasegawa, T., Li, C., Kaneko, K., and Shinkai, S. (2005) β-1,3-Glucan polysaccharides as novel one-dimensional hosts for DNA/RNA, conjugated polymers and nanoparticles. *Chemical Communications,* 4383-4398.
85. Tamesue, S., Numata, M., Kaneko, K., James, T. D., and Shinkai, S. (2008) Hierarchical carbon nanotube assemblies created by sugar-boric or boronic acid interactions. *Chemical Communications,* 4478-4480.
86. Martin, G. G., Cannon, G. C., and McCormick, C. L. (2000) Sc3p hydrophobin organization in aqueous media and assembly onto surfaces as mediated by the associated polysaccharide schizophyllan. *Biomacromolecules,* **1**(1), 49-60.
87. Zhu, H. Y., Orthman, J. A., Li, J. Y., Zhao, J. C., Churchman, G. J., and Vansant, E. F. (2002) Novel composites of TiO_2 (anatase) and silicate nanoparticles. *Chemistry of Materials,* **14**(12), 5037-5044.
88. Abdel-Mohsen, A. M., Abdel-Rahman, R. M., Fouda, M. M., Vojtova, L., Uhrova, L., Hassan, A. F., Al-Deyab S.S., El-Shamy I. E., and Jancar, J. (2014) Preparation, characterization and cytotoxicity of schizophyllan/silver nanoparticle composite. *Carbohydrate Polymers,* **102**, 238-245.

9

PEBAX-MOF Membranes

9.1 Introduction

Polymeric membranes are micro-porous films that work as semi-permeable barriers for the separation of different gases. Polymer membranes based gas separation technology has various advantages, including low energy consumption, mechanical simplicity, easy scale up and smaller footprint [1]. An additional advantage of the polymeric membrane technology is the possibility to control the density, size, size distribution, shape and vertical alignment of membrane pores which is not possible in other types of membranes technologies [2]. However, polymer membranes suffer from a trade-off between permeability and selectivity, i.e., the polymers with high gas permeability generally have low gas selectivity and vice versa [1,3]. The simultaneous enhancement of the two phenomena is generally very challenging. In order to overcome this issue, several inorganic fillers have been incorporated in the organic polymer matrices to generate mixed matrix membranes (MMMs), which possess merits of easy processability, superior permeability and high selectivity, owing to the reinforcement effect induced by the dispersed phase. Recently, metal organic frameworks (MOFs) have been considered as outstanding candidates to be used as fillers for MMMs. Several MOFs, such as MIL-53, MIL-101, ZIF-7, ZIF-8, ZIF-90, UiO-66, UiO-67, CuBTC and CuTPA have been identified for this purpose. Among various MOFs, HKUST-1 is a preferred material for the development of MMMs due to its three-dimensional morphology with interconnected porous network, which is beneficial for attaining an effective degree of gas separation performance.

Among the various polymer matrices used to develop membranes, PEBAX-4533 thermoplastic elastomer has been widely employed. It is a block copolymer composed of covalently bonded hard segments of polyamide (PA), which provide mechanical support to the polymer, and soft segments of polyether (PE) with high free volume, which offer transport channels for effective gas permeation

Sara Alkhoori, M. R. Vengatesan, G. Karanikolos and Vikas Mittal, Khalifa University of Science and Technology, Abu Dhabi, UAE

[4,5]. Different PA:PE ratios result in different PEBAX grades. PEBAX polymers have wide range of applications as they combine properties of both thermoplastics and elastomers.

Previous works have reported MMMs developed using different PEBAX grades containing altered PE and PA block lengths, along with various additives and nanofillers. Sheth *et al.* [1] studied four different series of PEBAX containing nylon 12 and poly(tetramethylene oxide). PEBAX with higher PA content was observed to exhibit enhanced Young's modulus. However, a decrease in elongation at break and ultimate tensile strength were accompanied due to the rigid PA chains. Thermal analysis indicated an enhancement in thermal stability of PEBAX with higher PA content, and a shift of glass transition temperature to higher values was observed. Moreover, PEBAX with higher PA content showed two melting and crystallization peaks. A study by Murali *et al.* [6] investigated the effect of multi-walled carbon nanotubes (MWCNTs) (from 0 to 5 wt%) and crosslinking on the permeation behavior of PEBAX-1675 (60% PE and 40% PA). Incorporation of MWCNTs increased the inter-segmental spacing of the polymer. Gas permeation analysis revealed that the filler improved the permeation of the polymer, specifically toward CO_2. The effect of different graphene fillers on the poly(vinyl alcohol) (PVA) membranes was also studied by Ovid'ko [3]. The author observed that graphene reinforcement was superior than carbon nanotubes due high modulus and 2D geometry. Mittal *et al.* [7] also carried out a similar study by using both carbon nanotubes and graphene as fillers for polymeric membranes. The authors underlined three key factors of filler dispersion, interaction between the filler and polymer matrix and alignment of the filler in the matrix. It was observed that using graphene nanoplatelets (GNPs) provided better mechanical properties than CNTs due to 2D morphology. Also, the high surface to volume ratio in GNPs allowed better stress transfer in the polymer matrix during loading.

Mixed matrix membranes of PEBAX-1657 loaded with 4A zeolite for gaseous separations were also developed by Murali *et al.* [8], and the main objective was to overcome the trade-off relationship limitations in polymeric membranes by utilizing the properties of organic polymer (good permeability) and inorganic zeolite (high selectivity due to small pore size and thermal stability). Other studies have also mentioned zeolites to be excellent fillers for separation purposes due to their molecular sieve ability and high thermal, mechanical and chemical stability [4]. A good dispersion of zeolite filler was observed for lower filler fractions. Also, corrugations were observed in the cross-sectional images which corresponded to the formation of void morphology due to the placement of the filler particles in between the polymer layers. Diffraction analysis exhibited an improvement in the amorphous nature (permeability) of the polymer, along with improved crystallinity of zeolite (which was indicated by sharper peaks), with increasing zeolite loading. Polymer membranes filled with 10 wt% zeolite showed good permeability and selectivity. Moreover, permeability of CO_2 was observed to improve by increasing the pressure.

Different MOF based MMMs have also been developed in recent years [9,10]. The incorporation of HKUST-1 in the polymeric membranes has been observed to result in improved properties. To further explore the usefulness of MOFs for developing high performance membranes, MMMs involving PEBAX and HKUST-1 (in varying weight fractions) were developed and analyzed for their mechanical and structural characteristics in the current study.

9.2 Experimental

9.2.1 Materials

PEBAX-4533 was provided by ARKEMA in granular form. HKUST-1 MOF was synthesized at the Petroleum Institute laboratories by following the literature reported protocols. The synthesized HKUST-1 was a fine blue powder with a purity of 99%. The solvents used in the study were acquired from Sigma Aldrich.

9.2.2 Optimization of PEBAX-4533 Dissolution and Membrane Preparation

The formation of fine membranes is directly related to the effective dissolution of PEBAX. The dissolution of PEBAX-4533 was studied using different solvents such as NMP/ethanol, dimethyl sulfoxide (DMSO) and dimethylformamide (DMF). Around 10 mL of the solvent was used to dissolve 1 g of PEBAX-4533. In order to achieve a homogeneous solution, it was observed that the polymer granules should be stirred in the solvent at 80 °C for 4 h. PEBAX was observed to dissolve uniformly in most of the solvents. After several experiments, the PEBAX dissolution protocols were optimized with DMF as per the following steps:

- Stir the polymer solution at 80 °C for 4 h.
- Pre-heat the casting dish before casting the solution. This step is to prevent the formation of lumps within the membrane due to cooling down of the polymer solution during casting process.
- Solvent evaporation in a pre-heated oven at 85 °C for 24 h.

9.2.3 Preparation of PEBAX-4533/HKUST-1 based MMMs

MMM formulations were prepared by stirring a specific amount of HKUST-1 in 1 g of PEBAX-4533 in DMF at 80 °C for 4 h. Similar casting procedure, as used for pure PEBAX membranes, was employed for generating MMMs. MMMs with different filler fractions of 0, 0.25, 0.5, 1 and 3 wt% were generated.

9.2.4 Characterization

Tensile analysis was carried out using Instron 5567 Universal Testing Machine following ASTM D-638 standard. The dumbbell shaped samples of dimensions 17 * 3 * 0.1 mm (length, width and thickness respectively) were analyzed at a strain rate of 10 mm/min. Five samples for each membrane were tested, and the mean was taken.

Calorimetric properties of the samples were analyzed using differential scanning calorimeter (DSC) from TA Instruments. 3-5 mg of the samples were used for the analysis. The heating cycle ranged from the ambient temperature up to 250 °C to remove previous processing history, followed by cooling to -50 °C, using a rate of 10°C/min under dry nitrogen atmosphere. The calorimetric data was taken from the second heating and cooling cycles.

Thermogravimetric analysis (TGA) was used to study the thermal stability of the polymer and MMMs using a TA discovery TGA employing TRIOS software. 3-5 mg of the samples were heated to 700 °C using a heating rate of 10 °C/min under inert gas conditions, and weight loss versus temperature curves were established.

Wide angle X-ray diffraction (WAXD) was utilized for the evaluation of the crystallinity of the polymeric membranes by applying Bragg's Law and the microstructure morphology (including the crystallites size) by the application of Scherrer's equation. The analysis was carried out on Panalytical X-Pert Pro Diffractometer using monochromatic CuKα radiation in reflection mode with filament current of 40 mA and accelerated voltage of 40 kV, by scanning between 5°

and 60° 2θ at room temperature. Zero background holder was used for lowering the noise.

Compositional analysis was performed at room temperature using attenuated total reflection Fourier transform infrared spectroscopy (ATR-FTR) between 4000–500 cm^{-1} on BRUKER Tensor II Platinum ATR with mid-infrared range detector.

Contact angle measurements were carried out using Rame-hartINC goniometer employing 5 μL deionized water (DIW) and 5 μL diiodomethane (DIM), both with known surface tension and contributions of polar and dispersive interactions (Table 9.1).

Table 9.1 Surface tension and corresponding contributions of polar and dispersive interactions

	Surface tension ($\frac{mJ}{m^2}$)	**Dispersive portion ($\frac{mJ}{m^2}$)**	**Polar portion ($\frac{mJ}{m^2}$)**
DIM	50.8	50.8	0
DIW	72.8	21.8	51

9.3 Results and Discussion

Figures 9.1-9.3 illustrate the tensile results of the MMM samples. Pure

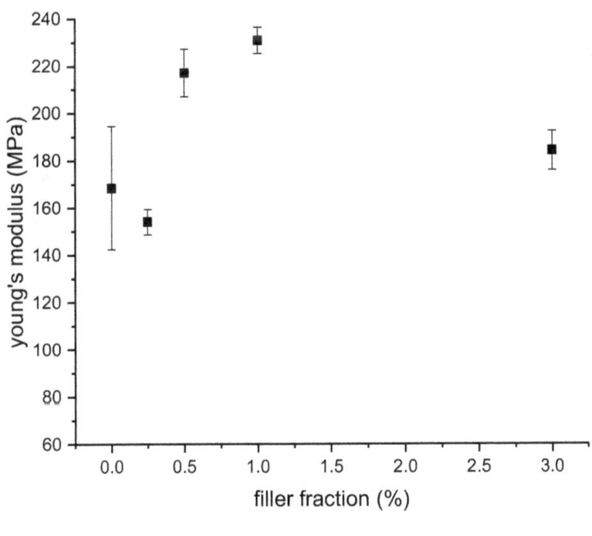

Figure 9.1 Tensile modulus of PEBAX and MMMs.

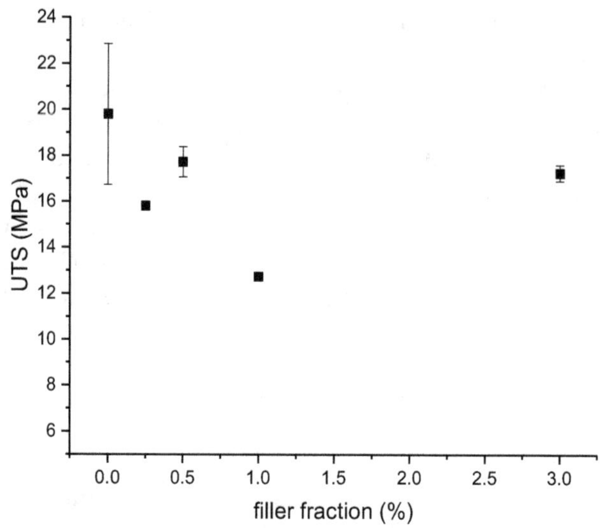

Figure 9.2 Ultimate tensile strength (UTS) of PEBAX and MMMs.

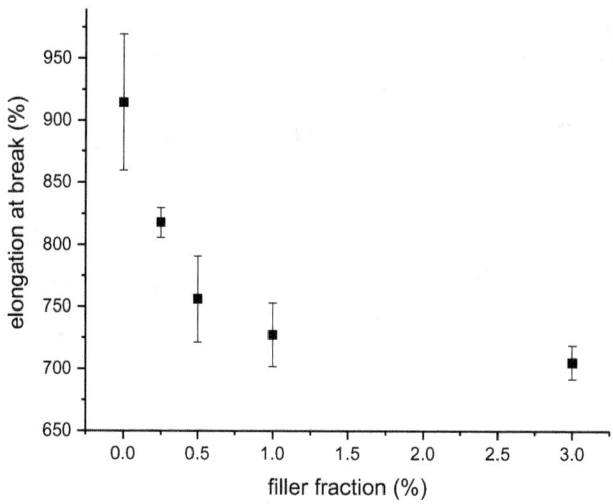

Figure 9.3 Elongation at break of PEBAX and MMMs.

PEBAX film was observed to have high elongation at break, moderate Young's modulus and low ultimate tensile strength (UTS), indicating a tough and non-rigid behavior similar to the soft PE segments. The incorporation of MOF nanofiller in PEBAX resulted in a decrease in both elongation at break and UTS, though the decrease in UTS was not significant. It indicated that the MOF filler acted as stress concentrator in the matrix resulting in a decrease in elongation at break, a common issue faced by glassy polymers due to their rigid structure [4,10-12]. The Young's modulus of the nanocomposite films was observed to enhance with filler fractions higher than 1%. This suggested a good degree of interaction and strass transfer between the filler and matrix. The high surface area of the filler allowed better stress transfer from the polymer matrix during loading. In contrast, it is interesting to note that the membrane with 0.25% loading resulted in a decrease in Young's modulus as compared to the pure polymer. This is due to the lower content of the MOF acting as a solid state plasticizer. Additionally, at a higher filler content of 3%, the modulus of the MMM exhibited a decrease due to a small degree of agglomeration of the filler in the polymer matrix, though the modulus was still higher than the pure polymer.

The melting peaks exhibited broad temperature range (Figure 9.4), around 50 °C, which is expected due to the wide range of crystallite size distribution [13]. PEBAX-4533 pellets exhibited one melting peak at 147 °C owing to prevalent thermal history (Table 9.2), whereas two melting peaks were observed for the PEBAX membrane. Also, in PEBAX samples with higher PA content, only one melting peak is generally observed in DSC, while the second melting peak at lower temperatures is related to the PE content of the copolymer in

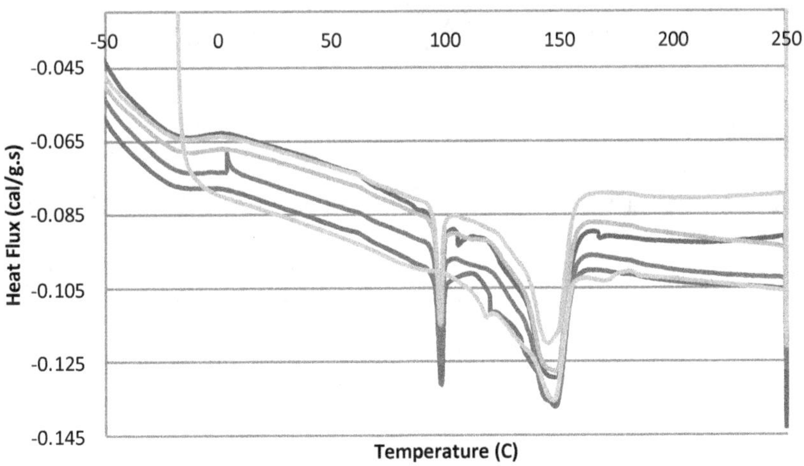

Figure 9.4 DSC second heating thermograms. The curves from bottom to top indicate PEBEX pellets, PEBAX film in DMF and nanocomposite films with 0.25, 0.5, 1 and 3% filler.

Table 9.2 DSC melting temperature and enthalpy of melting

Filler fraction (wt%)	PEBAX pellets	0	0.25%	0.5%	1%	3%
Lower bound T_m (°C)	None	98	98	98	98	98
Lower bound $\Delta H_{melting}$ $(\frac{J}{g})$	None	1.8	2.5	2.4	2.4	2.5
Upper bound T_m (°C)	147	148	149	145	148	145
Upper bound $\Delta H_{melting}(\frac{J}{g})$	20	21	21	22	21	22

higher PE content grades [1]. At higher filler fractions, somewhat higher enthalpy values were observed probably due to the filler particles affecting the crystallinity of the resultant membranes.

Similar to the heating thermal scans, the cooling scans for the pure polymer pallets exhibited one crystallization dip at 75 °C, while all other samples had two crystallization dips at higher and lower temperature ranges due to the crystallization of PA segments at higher temperatures and PE segments at lower temperatures (Table 9.3).

Table 9.3 DSC crystallization temperature and enthalpy of crystallization

Filler fraction (wt%)	PEBAX pellets	0	0.25	0.5	2	3
Lower bound T_c (°C)	None	174	174	173	174	174
Lower bound $\Delta H_{crystallization}$ $(\frac{J}{g})$	None	0.1	0.1	0.2	0.2	0.2
Upper bound T_c (°C)	75	93	110	111	117	111
Upper bound $\Delta H_{crystallization}(\frac{J}{g})$	22.2	21.0	23.1	23.2	23.4	23.0

The addition of MOF exhibited higher T_c at upper temperature range indicating fast crystallization in the MMMs. The enthalpy of crystallization was also observed to increase with increasing the filler fractions. The crystalline morphology of MOF particles acts as a nucleation center for the crystallization of the polymer chains, allowing them to arrange themselves into more perfect crystalline lamellae swiftly at higher temperature.

Also, the differences in the melting and crystallization behaviors of the PEBAX pellets and PEBAX film indicated that the orientation of crystals was more uniform in the film.

From the second heating thermograms, glass transition temperature (T_g) values of the MMMs were also recorded, as shown in Table 9.4. The T_g values of the MMMs increased on enhancing the filler content, which indicated a good degree of compatibility between the filler and polymer matrix.

Table 9.4 Glass transition temperature values of PEBAX and MMMs

Filler fraction (wt%)	T_g(°C)
0	-33.5
0.25	-31.8
0.5	-33.5
1	-32.9
3	-31.5

High thermal stability of HKUST-1 for high temperature applications has been confirmed by other studies. HKUST-1 degradation temperature varies based on the synthesis method. Some HKUST-1 degradation temperatures reported in literature are 220 °C [14], 240 °C [14,15], 300 °C [9,16], 310 °C [17], 350 °C [18] and 376 °C [19], thus, verifying the high thermal stability of the filler. The presence of impurities within the pore structure of MOF is indicated by the char yield or total weight loss from the 'true' weight of the solid. The TGA thermograms of the pure polymer and the MMMs exhibited a single degradation step. The materials exhibited the degradation in the temperature range 260-360 °C, with the thermal stability of the membranes slightly enhanced with the addition of filler. Table 9.5 summarizes the degradation temperatures of the membranes with different filler fractions. Table 9.6 also shows the char yield, taken at 700 °C, and the limited oxygen index (LOI) values calculated for the PEBAX and membrane samples.

Table 9.5 TGA degradation temperatures of membranes

	0%	0.25%	0.5%	1%	3%
5% weight loss	351	354	353	354	353
10% weight loss	363	367	366	368	366
50% weigh loss	402	404	404	406	406

Table 9.6 Char yield and the LOI of PEBAX and MMMs

Filler fraction (wt%)	0	0.25	0.5	1	3
Char yield %	1.48	1.32	1.86	2.09	2.00
LOI	18.1	18.0	18.2	18.3	18.3

XRD of HKUST-1 is presented in Figure 9.5, and it was in good agreement with the reported literature. HKUST-1 is highly crystalline material possessing face-centered cubic crystalline structure, and it shows several sharp narrow peaks with high intensities at 2θ angles

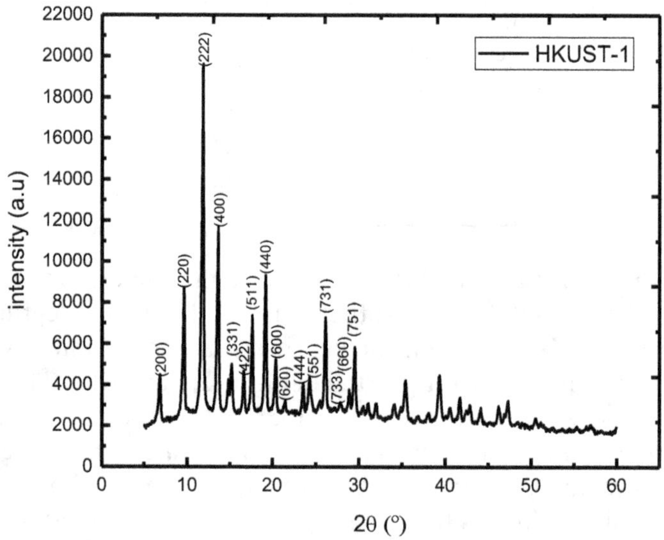

Figure 9.5 WAXRD pattern of HKUST-1.

around 7°, 12°, 14°, 18°, 19° and 26°, which evidence the microporous nature of the MOF. Moreover, the characteristic peak at around 9.6° 2θ corresponded to the *d*-spacing of 9.5 Å for HKUST-1's main pores [6,8,15,20]. Diffractograms of pure PEBAX-4533 and MMMs showed semi-crystalline nature, as shown in Figure 9.6. Pure PEBAX exhibited a narrow peak around 2θ = 22.7°. The MMMs exhibited similar XRD patterns as pure PEBAX, with a slight shifting of the peak to higher angles, indicating that the addition of MOF had no major effect on the crystalline structure of PEBAX. The absence of any characteristic peak of the filler particles in the composite membranes also indicated a uniform dispersion of the filler in the polymer matrix, irrespective of the filler concentration.

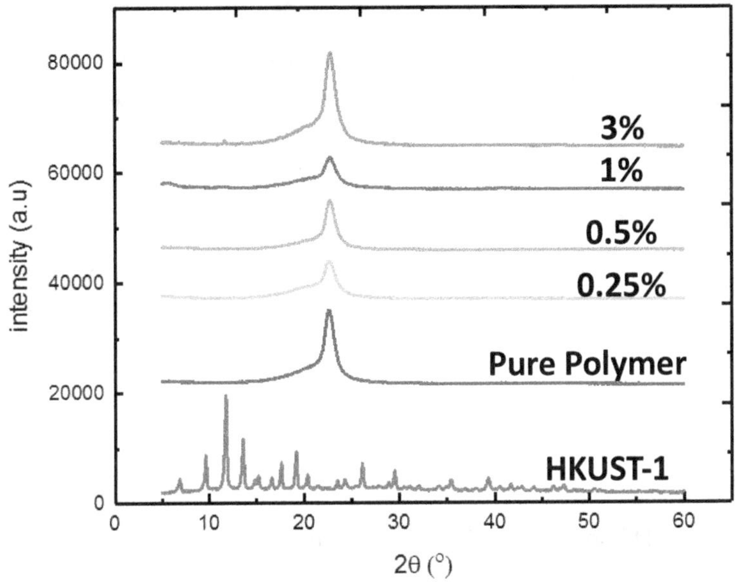

Figure 9.6 WAXD patterns of PEBAX-4533 and MMMs.

The relative crystallinity of different MMMs was also estimated from the ratio of the peak intensities of MMMs and pure polymer, as shown in Table 9.7. From the table, it is clear that the addition of HKUST-1 gradually increased the crystallinity of PEBAX on enhancing the filler fraction, thus, coinciding with DSC findings.

Table 9.7 Relative crystallinities of pure PEBAX-4533 and MMMs

Filler fraction (%)	Relative crystallinity (%)
0	-
0.25	1.25
0.5	1.57
1	1.79
3	2.33

PEBAX comprises of amide and ether segments linked by ester groups, with –OH ends. Among the polymer chains, the amide blocks are linked with strong hydrogen bonds, while the covalent bonds in the polyether repeating units are rather stretchable and bendable. The FTIR spectrum of PEBAX in Figure 9.7 exhibited two characterization peak at 1734 and 1637 cm^{-1} corresponding to C=O stretching vibration of free amide and the hydrogen-bonded amide molecules

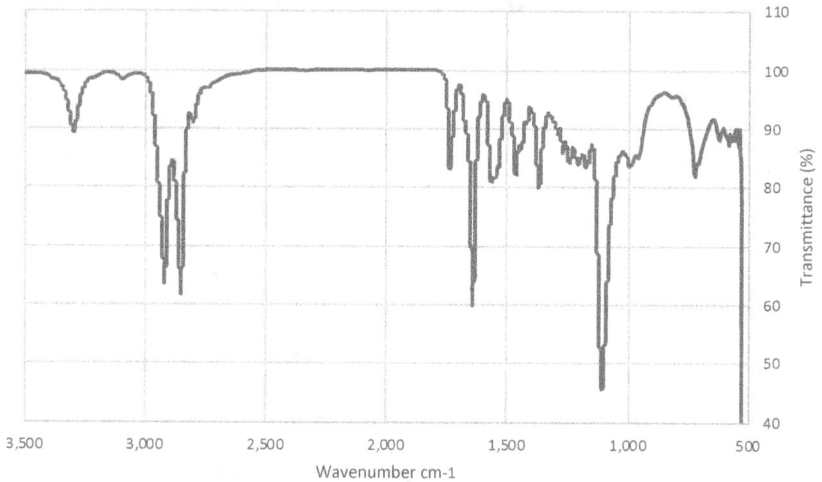

Figure 9.7 FTIR spectrum of PEBAX-4533 film.

respectively. Another peak at 3295 cm^{-1} indicated the presence of N-H group of amide units. For the ether units, a characteristic absorbance band appeared at 1107 cm^{-1} corresponding to the C-O vibration. Moreover, C-H aliphatic groups had a peak at 2918 cm^{-1}. A small bulge appeared around 2796 cm^{-1} which might correspond to the hydrogen bonded -OH groups. The ester units had characteristic peaks at 1734 cm^{-1} for C=O stretching. Figure 9.8 demonstrates the FTIR spectrum

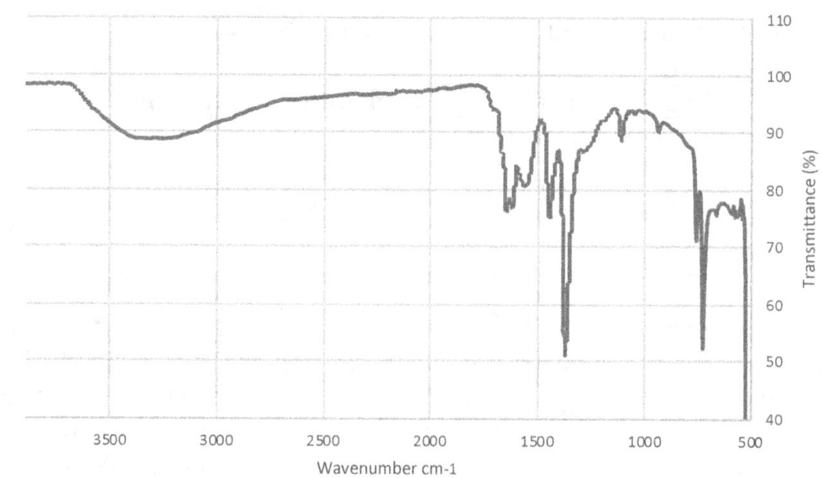

Figure 9.8 FTIR spectrum of HKUST-1 powder.

of HKUST-1. Small peaks around 1668-1643 cm^{-1} indicated C=O groups of deprotonated benzene tricarboxylic acid. Carboxylate groups have two characteristic peaks owing to two types of stretching: symmetrical with sharp peak at 1370 cm^{-1} for C=O and anti-symmetrical with wide peak at 1556 cm^{-1} for O-H. This is due to the aromatic stretching of the para-distributed phenyl groups. The stretching of C-C groups in the ring was observed at 1370 cm^{-1}. Moreover, the presence of a wide band between 1600 and 1500 cm^{-1} was due to the carboxylate groups forming bidentate inter-cluster bridges that can cause aggregation of the filler particles [15,20]. For the organic ligands, absorbance bands around 1700 cm^{-1} are related to C=O and C-OH stretching vibration in the free BTC ligands, which were absence in the spectrum. This indicated that the protonated free ligands did not exist in the sample. Several peaks between 1300-500 cm^{-1} can be attributed to out-of-plane vibrations of the BTC groups such as the stretching of C-O at 111 cm^{-1} and C-COO at 729 cm^{-1}, 760 cm^{-1} and 940 cm^{-1}.

Figure 9.9 compares the FTIR spectra of pure PEBAX, HKUST-1 and 3% MMM to analyze the effect of filler on the polymer matrix. It is noticeable that the MMM spectrum overlapped with the spectrum of pure PEBAX. The characteristic peak of HKUST-1 was observed at 1367 cm^{-1}. In general, no new peak was detected in the MMM spectrum, indicating no chemical interactions between the filler and polymer chains. However, the finger-print region in the 500-1000 cm^{-1}

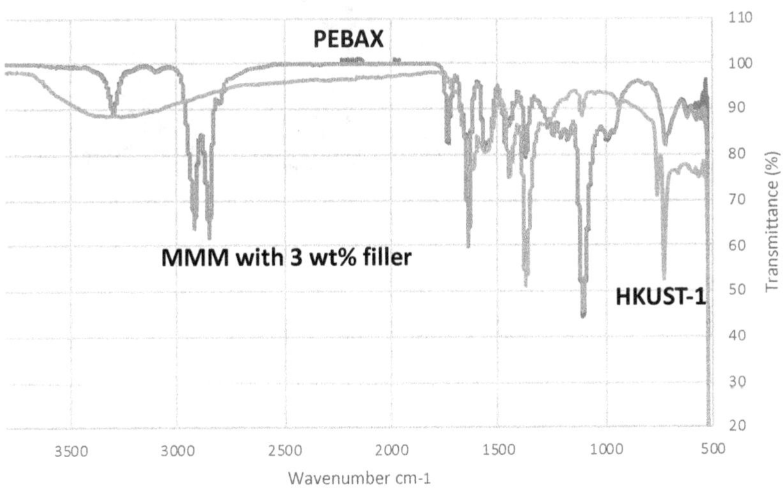

Figure 9.9 FTIR spectra of PEBAX, 3 wt% MMM and HKUST-1.

range indicated the emergence of small peaks, increase in the intensity of the peaks and slight peak shifting, thus, indicating a small degree of physical interactions among the components.

Tables 9.8 and 9.9 summarize the results of the contact angle and surface fee energy analysis (SFE) in DIW and DIM. The results showed

Table 9.8 Images of the contact angle of the droplets on the surface of PEBAX and MMMs with DIM and DIW

MMM filler fraction (%)	DIM	DIW
0		
0.25		
0.5		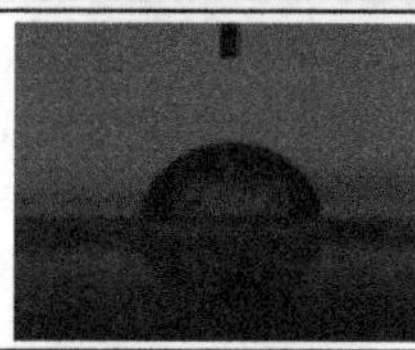
3	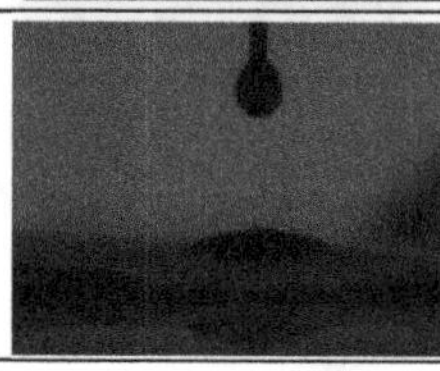	

Table 9.9 Contact angle and corresponding surface free energy of PEBAX and MMMs

	Contact angle (°)		Surface Free Energy ($\frac{mJ}{m^2}$)		
	DIW	DIM	r	r (d)	r (p)
Pure PEBAX	61.8	32.5	53.5	43.1	10.3
0.25% MMM	64.5	29.3	53.1	44.5	8.6
0.5% MMM	63.6	21.2	55.6	47.4	8.2
3% MMM	64.1	22.7	55.0	47.0	8.1

hydrophilic nature of PEBAX-4533 with relatively low SFE. The MMMs exhibited higher contact angles with water and lower contact angle with DIM, indicating an increase in hydrophobicity due to the non-polar nature of HKUST-1. The MMMs with 0.5 and 3 wt% filler resulted in higher SFE values, while the membrane with 0.25 wt% filler fraction showed a lower SFE. This indicated lower degree of interfacial interactions between the phases in the case of lower filler loadings [21,22]. The increase in SFE of the MMMs is attributed to the higher content of dispersive portions of the filler in the polymer matrix. Oppositely, the polar portions of the SFE were observed to decrease significantly. However, it should be noted that the contact angle analysis should not be taken as a definite measurement due to the limitation associated with polymer morphology. The membrane preparation protocols such as the concentration of the polymer in the solution, type of solvent and thickness of the film can influence the contact angle values. It should be noted that the highly porous films can result in low contact angle values even though the polymer may be highly hydrophobic.

9.4 Conclusions

Defect-free MMMs were successfully developed using PEBAX-4533 with different weight fractions of HKUST-1 through solvent-evaporation method. Mechanical analysis revealed that the addition of filler decreased both elongation at break and the ultimate tensile strength, however, improvements in Young's modulus were observed with filler addition. The addition of highly crystalline HKUST-1 enhanced the crystallinity of PEBAX and shifted the glass transition temperature to higher values. Furthermore, TGA analysis showed that the MMMs exhibited enhanced thermal stability. FTIR spectra of the MMMs further confirmed the absence of any chemical interactions between the filler and polymer phases. Contact angle measurements showed that the MMMs with high filler content exhibited higher surface free energy.

References

1. Sheth, J. P., Xu, J., and Wilkes, G. L. (2003) Sold state structure property behavior of semicrystalline poly(ether-block-amide) PEBAX (R) thermoplastic elastomers. *Polymer*, **44**(3), 743-756.
2. Solanki S. J., and Desai, R. N. (2013) Polymeric membrane technol-

ogy. *International Journal of Engineering Science and Innovative Technology*, **2**(2), 400-403.
3. Ovid'ko, I. A. (2013) Enhanced mechanical properties of polymer matrix nanocomposites reinforced by graphene inclusions: A review. *Review on Advanced Materials Science*, **34**, 19-25.
4. Venna, S. R., and Carreon, M. A. (2015) Metal organic framework membranes for carbon dioxide separation. *Chemical Engineering Science*, **124**, 3-19.
5. Li, H., Haas-Santo, K., Schygulla, U., and Dittmeyer, R. (2015) Inorganic microporous membranes for H_2 and CO_2 separation - Review of experimental and modeling progress. *Chemical Engineering Science*, **127**, 401-417.
6. Murali, R. S., Sridhar, S., Sankarshana, T., and Ravikumar, Y. V. L. (2010) Gas permeation behavior of PEBAX-1657 nanocomposite membrane incorporated with multiwalled carbon nanotubes. *Industrial & Engineering Chemistry Research*, **49**(194), 6530-6538.
7. Mittal, G., Dhand, V., Rhee, K. Y., Park, S.-J., and Lee, W. R. (2015) A review on carbon nanotubes and graphene as fillers in reinforced polymer nanocomposites. *Journal of Industrial and Engineering Chemistry*, **21**, 11-25.
8. Murali, R. S., Ismail, A. F., Rahman, M. A., and Sridhar, S. (2014) Mixed matrix membranes of Pebax-1657loaded with 4A zeolite for gaseous separations. *Separation and Purification Technology*, **129**, 1-8.
9. Aroon, M. A., Ismail, A. F., Matsuura, T., and Montazer-Rahmati, M. M. (2010) Performance studies of mixed matrix membranes for gas separation: A review. *Separation and Purification Technology*, **75**, 229-242.
10. Basu, S., Cano-Odena, A., and Vankelecom, I. F. J. (2011) MOF-containing mixed-matrix membranes for CO_2/CH_4 and CO_2/N_2 binary gas mixture separations. *Separation and Purification Technology*, **81**(1), 31-40.
11. Lim, S. Y., Choi, J., Kim, H.-Y., Kim, Y., Kim, S.-J., Kang, Y. S., and Won, J. (2014) New CO_2 separation membranes containing gas-selective Cu-MOFs. *Journal of Membrane Science*, **467**, 67-72.
12. Zornoza, B., Tellez, C., Coronas, J., Gascon, J., and Kapteijn, F. (2013) Metal organic framework based mixed matrix membranes: An increasingly important field of research with a large application potential. *Microporous and Mesoporous Materials*, **166**, 67-78.
13. *Thermal Analysis of Polymers - Selected Applications*, Mettler Toledo. Online: https://www.mt.com/dam/Analytical/ThermalAnalysi/TA-PDF/Part%20of%20Polymers-Selected%20Applications.pdf [accessed 16th June 2019].
14. Wee, L. H., Lohe, M. R., Janssens, N., Kaskel, S., and Martens, J. A. (2012) Fine tuning of the metal–organic framework $Cu_3(BTC)_2$

HKUST-1 crystal size in the 100 nm to 5 micron range. *Journal of Materials Chemistry*, **22**, 13742-13746.

15. Sorribas, S., Kudashiva, A., Scholes, C. A., Almedro, E., Zornoza, B., de la Iglesia, O., Tellez, C., and Coronas, J. (2015) Pervaporation and membrane reactor performance of polyimide based mixed matrix membranes containing MOF HKUST-1. *Chemical Engineering Science*, **124**, 37-44.
16. Chui, S. S.-Y., Lo, S. M.-F., Charmant, J. P. H., Orpen, A. G., and Williams, I. D. (1999) A chemically functionalizable nanoporous material $[Cu_3(TMA)_2(H_2O)_3]_n$. *Science*, **283**, 1148-1150.
17. Shahid, S., and Nijmeijer, K. (2014) Performance and plasticization behavior of polymer-MOF membranes for gas separation at elevated pressures. *Journal of Membrane Science*, **470**, 166-177.
18. Sun, B., Kayal, S., and Chakraborty, A. (2014) Study of HKUST (copper benzene-1,3,5-tricarboxylate, Cu-BTC MOF)-1 metal organic frameworks for CH_4 adsorption: An experimental Investigation with GCMC (grand canonical Monte-Carlo) simulation. *Energy*, **76**, 419-427.
19. Chowdhury, P., Bikkina, C., Meister, D., Dreisbach, F., and Gumma, S. (2009) Comparison of adsorption isotherms on Cu-BTC metal organic frameworks synthesized from different routes. *Microporous and Mesoporous Materials*, **117**, 406-413.
20. Li, L., Liu, X. L., Geng, H. Y., Hu, B., Song, G. W., and Xu, Z. S. (2013) A MOF/graphene oxide hybrid (MOF:HKUST-1) material for adsorption of methylene blue for aqueous solution. *Journal of Materials Chemistry A*, **1**, 10292-10299.
21. Selvi, M., Vengatesan, M. R., Devaraju, S., Kumar, M., and Alagar, M. (2014) In situ sol–gel synthesis of silica reinforced polybenzoxazine hybrid materials with low surface free energy. *RSC Advances*, **4**, 8446-8452.
22. Jainczuk, B., and Bialopiotrowicz, T. (1990) The total surface free energy and the contact angle in the case of low energetic solids. *Journal of Colloid and Interface Science*, **140**, 362-372.

10

PEBAX-MOF Membranes for CO_2 Separation

10.1 Introduction

Currently, a significant research attention has been devoted to the development of polymeric membranes as an alternative to the conventional gas separation technologies. In this respect, efforts have been carried out to enhance the chemical and/or physical properties of various polymers. Polymeric membrane separation technology has meritorious properties, including low energy consumption, mechanical simplicity, easy scale up and smaller footprint [1,2]. However, polymeric membranes suffer from a trade-off between permeability and selectivity, i.e., polymers with high gas permeability generally have low gas selectivity and vice versa [2-4]. The limited choice of the membrane materials may also limit their application, especially for challenging conditions. In order to overcome this issue, several inorganic fillers have been incorporated in the organic polymers to generate mixed matrix membranes (MMMs), which synergistically combine the easy processability of polymers with superior permeability and selectivity of inorganic materials.

Recently, metal organic frameworks (MOFs) have been considered as outstanding candidates to be used as fillers for MMMs owing to their superior properties. Specifically, the tunability of the structure of such materials is beneficial to achieve the required performance. Based on this, several MOFs, such as MIL-53, MIL-101, ZIF-7, ZIF-8, ZIF-90, UiO-66, UiO-67, CuBTC and CuTPA, have been specifically identified for use in gas separation membranes. Among these materials, HKUST-1 is a facile cost-effective MOF filler material due to its three-dimensional morphology with interconnected porous network structure, which is immensely beneficial for effective gas separation.

Permeation of gases through the polymeric membranes is based on solution-diffusion mechanism [5-8]. Permeation of a gas through a membrane (P_e) depends on its solubility in the membrane (S) and

Sara Alkhoori, Muthukumaraswamy R. Vengatesan, Georgios Karanikolos and Vikas Mittal, Khalifa University of Science and Technology, Abu Dhabi, UAE

its diffusion though the boundary layer of the membrane (D). The gas molecules travel from the feed side across the membrane thickness l to the permeate side based on the chemical potential gradient across the membrane (from high pressure feed gas side $P_{i,f}$ to lower pressure permeate side $P_{i,p}$). The rate of permeation or flux is given by the equation:

$$J = \frac{DS\left(P_{i,f} - P_{i,p}\right)}{l}$$

The overall selectivity coefficient α_{12} is given by the following equation:

$$\alpha_{12} = \frac{S_1 D_1}{S_2 D_2} = \frac{P_1}{P_2}$$

where 1 and 2 are the faster and slower components respectively. In the case of several component mixture, selectivity is given in terms of the slowest permeating gas for the sake of simplicity.

One of the arising applications of polymeric membranes is the separation of CO_2 from natural gas. A study on the effect of greenhouse gases mentioned that CO_2 level has increased by 30% since the pre-industrial time and if it continues at the same rate, it can lead to significant global warming. One way to solve this problem is by sequestering the CO_2 emissions [9]. Additionally, the separation of CO_2 from natural gas is very important for controlling corrosion during transport and distribution as well as catalyst poisoning. Previously, the most commonly used technology for CO_2 capture from natural gas has been amine treatment, however, its operational difficulties and challenges of keeping the solvent clean make polymeric membranes an attractive alternative. However, many challenges have to be overcome for the membrane technology to be economically viable. In gas processing, for instance, the main priorities are (1) to control hydrocarbon recovery and (2) to meet sales gas specifications. The minimum CO_2 content specification in the sales gas is < 2%. Since membranes cannot accomplish such high level of separation, multiple stages might be required. This would increase the complexity, energy consumption and cost. Comparing with amine unit in terms of hydrocarbon losses, Sridhar *et al.* [1] mentioned that for a single-stage membrane unit, the losses range from 2% to 10%, while the use of two-stage membrane unit decreases the loss to 2%-5%. Losses in the amine unit are less than 1% and are due to the entrainment of the hydrocarbons with flashed streams. A good membrane should have high selectivity and high permeability, however, there is a trade-off relationship between the two, which introduces another challenge. This can be seen by plotting log (α_{ij}) vs. log (P_i), and the performance of polymeric membranes usually falls within this trade-off of Robeson upper-bound. For the industry, materials that have separation properties above this limit are favored. In addition, it is important for the membranes to have high mass transfer area, fouling resistance and mechanical stability, along with reproducibility and low manufacturing costs. For this, a good understanding of the properties and gas transport behavior of the polymers is important [1].

Among the various polymer matrices used to develop membranes, thermoplastic elastomer PEBAX-4533 is widely used in commercial applications. It is a block copolymer composed of covalently bonded hard segments of polyamide (PA), which provide the mechanical support to the polymer, and soft segments of polyether (PE) with high free volume, which offer transport channels for gas permeation [5,6]. The general chemical structure of PEBAX is

$$\mathrm{HO - [CO - (PA)_x - CO - O - (PE_y) - O]_n - H}$$

Different PA:PE ratios result in different PEBAX grades. PEBAX polymers have wide range of applications as they combine properties of both thermoplastics and elastomers. One of the key factors affecting the performance of the membranes is the molecular structure, i.e. the arrangement of the chemical groups in the polymer chain. In the PEBAX polymer, PE units result in the amorphous phase which improves the permeability and the PA units lead to a crystalline phase which would enhance the stiffness, tensile strength and glass transition temperature (T_g). Thus, the ratio between the PE and PA blocks determines the performance of the MMM [10]. The covalent bonding between the PA and PE blocks prevents intermolecular mixing and a 'micro-phase separation' may arise. This is an important phenomenon of the block copolymers [3,4]. When micro-phase separation happens, the nanometer sized structures form uniquely arranged morphologies resulting in a micro-lattice structure. These arrangements can create micelles, spheres, cylinders, lamellae or surface patterning structure based on the relative length between the blocks.

Previous studies have reported different types of MMMs using different PEBAX grades containing altered PE and PA block lengths, along with a variety of nanofillers. A study by Murali *et al.* [3] investigated the effect of multi-walled carbon nanotubes (MWCNTs) (from 0 to 5 wt%) and crosslinking on the permeation behavior of PEBAX-1675 (60% PE and 40% PA). Incorporation of MWCNTs increased the intersegmental spacing of the polymer. Gas permeation analysis revealed that the filler improved the permeation of the polymer, specifically toward CO_2. Mixed matrix membranes of PEBAX-1657 loaded with 4A zeolite for gaseous separations were also developed by Murali *et al.* [6], and the main objective was to overcome the trade-off relationship limitations in polymeric membranes by utilizing the properties of organic polymer (good permeability) and inorganic zeolite (high selectivity due to small pore size and thermal stability). Other studies have also mentioned zeolites to be excellent fillers for separation purposes due to their molecular sieve ability and high thermal, mechanical and chemical stability [5]. A good dispersion of zeolite filler was observed for lower filler fractions. Also, corrugations were observed in the cross-sectional images which corresponded to the formation of void morphology due to the placement of the filler particles in between the polymer layers. Diffraction analysis exhibited an improvement in the amorphous nature (permeability) of the polymer, along with improved crystallinity of zeolite which was indicated by sharper peaks, with increasing zeolite loading. Polymer membranes filled with 10 wt% zeolite showed good permeability and selectivity. Moreover, permeability of CO_2 was observed to improve by increasing the pressure. Several literature studies have also carried out the permeability analysis of MMMs with HKUST-1 and most of them agreed that MOF enhanced the CO_2 separation performance of the polymeric membranes [11-14].

To further explore the usefulness of MOFs for gas separation membranes, MMMs involving PEBAX and HKUST-1 (in varying weight fractions) were developed and analyzed for their environmental and water stability as well as CO_2 permeation characteristics.

10.2 Experimental

10.2.1 Materials

PEBAX-4533 was kindly provided by ARKEMA in granular form. HKUST-1 MOF was synthesized at the Petroleum Institute laboratories by following the literature reported protocols. The synthesized HKUST-1 was a fine blue powder with a purity of 99%. The solvents used for the study were acquired from Sigma Aldrich.

10.2.2 Optimization of PEBAX-4533 Dissolution and Membrane Preparation

The dissolution of PEBAX-4533 was studied using different solvents such as N-methylpyrrolidone (NMP)/ethanol, dimethyl sulfoxide (DMSO) and dimethylformamide (DMF), along with optimum temperature and time. Around 10 mL of the solvent was used to dissolve 1 g of PEBAX-4533. In order to achieve a homogeneous solution, it was observed that the polymer granules should be stirred in the solvent at 80 °C for 4 h. It was observed that PEBAX dissolved uniformly in the most of the solvents. Overall, the PEBAX dissolution and membrane casting protocol was optimized with DMF as follows:

- Stir the polymer solution at 80 °C for 4 h.
- Pre-heat the casting dish before casting the solution. This step is to prevent the formation of lumps within the membrane film due to cooling down of the polymer solution while casting.
- Solvent evaporation in a pre-heated oven at 85 °C for 24 h.

10.2.3 Preparation of PEBAX-4533/HKUST-1 based MMMs

MMM formulations were prepared by stirring a specific amount of HKUST-1 in 1 g of PEBAX-4533 in DMF at 80 °C for 4 h. Similar casting procedure, as used for pure PEBAX membranes, was employed for generating MMMs. MMMs with different filler fractions of 0, 0.25, 0.5, 1 and 3 wt% were generated.

10.2.4 Characterization

The developed MMMs were tested for their water and environmental stability. For the environmental stability test, the samples were exposed directly to the outdoor environment without shade, while, for the water test, the samples were immersed in water in closed individual bottles at room temperature. In both tests, the samples were cut into 1 x 1 cm^2 size and were left for two weeks. The samples were characterized with thermogravimetric analysis (TGA), Fourier transform infra-red spectroscopy (FTIR) and scanning electron microscopy (SEM) before and after subjecting to environment and water conditions to study the (1) changes in the hydrogen bonding in the MMMs and (2) effect on the interfacial interactions between the MOF and polymer. TGA of the samples was carried out on a Discovery TGA from TA Instruments using nitrogen as carrier gas. 3-7 mg of the sample was heated from room temperature to 700 °C at a heating rate of 10 °C/min, and the weight loss as a function of temperature and time was recorded. For FTIR analysis, the samples were scanned between 4000-500 cm^{-1} with an accuracy of 0.27 cm^{-1} on a BRUKER Tensor II, Platinum ATR with mid-infrared range (MIR) detector at room temperature. For SEM, the membrane surface was analyzed in Hitachi field emission in-lens S-900 high resolution scanning electron microscope at accelerating voltages of 10-20 kV.

Permeability values were calculated using the steady state increase in the pressure with respect to time at a constant flow rate, as given by the equation:

$$P_i = \frac{J_i * l}{A * (P_{if} - P_{ip})}$$

where P_i is the permeability coefficient of the component i in $\frac{cm^3(STP)cm}{s\ cm^2\ cmHg}$, J_i is the volumetric flow rate of permeated component in $\frac{cm^3}{s}$, l is the thickness of the membrane selective layer in cm, A is the effective membrane area per gas permeation in cm^2, P_{if} is the partial pressure of the component i at feed side and P_{ip} is the partial pressure of the component i at permeate side in $cmHg$.

Gases permeate through the membrane by solution-diffusion mechanism. At low pressures, CO_2 is mainly adsorbed on the metal centers of the MOF first followed by filling the small cages around the metal and organic linkers. In contrast, at higher pressures, adsorption is mainly in the vicinity of the rings and windows linking the pores [11-13].

Pure gas permeability analysis was carried out at room temperature with high-pressure gas permeation manifold built in the laboratory with nut and ferrule fittings for leak-free transportation of the gas streams. Figures 10.1 and 10.2 represent the in-house developed permeation manifold. The flow was controlled through needle valves with $\frac{1}{4}$ inch stainless steel 4BKT end connections. Before running the test, all lines along with the separation cell were flushed with pure CO_2. Pure CO_2 was used in all experiments at feed pressures of P_{if} =

3, 5 and 10 bar, while the permeate side pressure P_{ip} was kept at vacuum to deliver a driving force for the gas flow. MMM films were inserted in-between the stainless steel separation cell with an effective surface area A = 0.048 cm^2. Subsequently, the feed gas was allowed to follow through the separation cell and into the permeate side. After reaching a steady state, the data from the permeate side were collected through the Data Acquisition-Agilent 34972 and were transferred to the Benchlink software for producing a plot of pressure with respect to time.

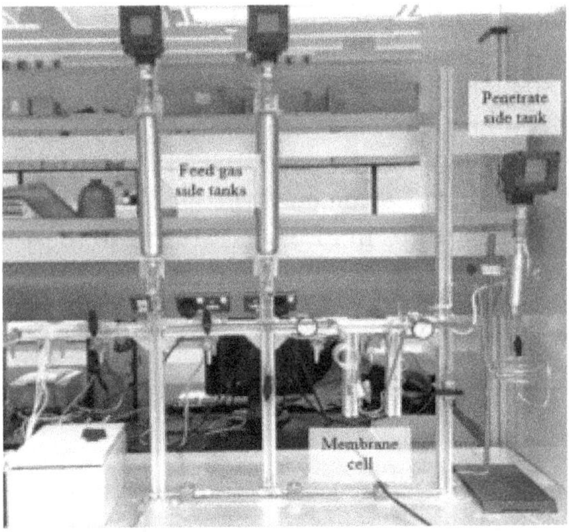

Figure 10.1 Gas permeation rig as well as feed gas and penetrate tanks.

Figure 10.2 a) Gas permeation rig, b) CO_2 feed and c) pump used for the permeation analysis.

10.3 Results and Discussion

From Figure 10.3 and Table 10.1, it can be seen that for the environmental stability test, the initial degradation of pure PEBAX as well as 0.25 wt% and 0.5 wt% composite samples occurred at significantly lower temperatures as compared to the original samples. This can be due to the harsh environment of high humidity and sunlight. However, for the membranes with filler fraction of 1 wt% and 3 wt%, the initial degradation was shifted to higher temperatures. The membranes were slightly hydrophilic in nature and underwent intermolecular interaction with the atmospheric moisture. For both environmental and water stability, the pure PEBAX membrane and 3 wt% MOF filled composite showed higher char yield due to the formation of H-bonded complex network. However, the lower filler fraction loaded composites exhibited uneven char yields. The higher filler loading also accelerated stronger hydrogen bonding between the MOF and the amide groups in PEBAX, resulting in improved thermal stability and char yield. The H-bonding in the nanocomposite was further confirmed by the FTIR analysis.

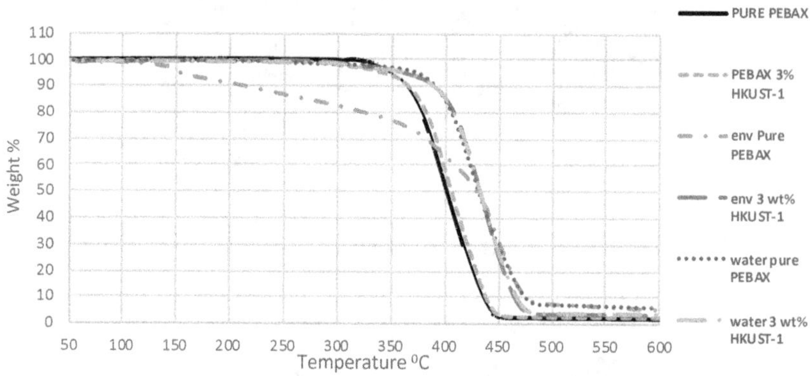

Figure 10.3 TGA plots of the MMMs before and after the environmental and water stability tests.

Hydrogen bonding plays an important role for achieving the stability in PEBAX based membranes. It has been reported that the addition of fillers can affect the hydrogen bonding state in the polymer as a result of the interfacial interaction between the MOF and polymer matrix, in addition to the effect of MOF on the polymer chain alignment and mobility [15,16]. More stable hydrogen bonding is observed to result in effective thermal stability of the MMMs. It has also been reported that the hydrogen bonding improves the structural order of crystalline polymers. PEBAX has highly polar carbonyl and amide groups that attribute to the hydrogen bonding in the polymer. Hydrogen bonds are the dominant interfaces holding the neighboring polymer chains together resulting in highly crystalline structure.

Table 10.1 TGA results of MMMs before and after environmental and water stability tests

Stability test	HKUST-1 wt%	5% degradation (°C)	10% degradation (°C)	Char yield (wt%)
Original samples	0	351	363	1.2
	0.25	353	366	1.6
	0.5	354	368	2.2
	1	354	367	2.5
	3	343	366	2.3
Environmental test	0	153	214	5.4
	0.25	19	24	-
	0.5	178	255	2.2
	1	373	393	1.1
	3	362	390	3.3
Water test	0	375	391	6.1
	0.25	375	393	-
	0.5	382	396	1.6-
	1	373	389	-
	3	365	389	3.7

FTIR spectra of MMMs before the stability tests showed that the free amide vibrations around 1734 cm^{-1} remained unchanged, which indicated that the addition of MOF did not disturb the H-bonding of the amide groups. FTIR spectroscopy was subsequently carried out for the stability test samples to study the changes in hydrogen bonding. There are three bands for amide stretching in PEBAX: (i) hydrogen bonded carbonyl groups which are ordered crystalline domain, (ii) hydrogen bonded disordered amorphous conformation and (iii) non-hydrogen bonded free carbonyl groups. Therefore, the behavior of both amorphous and crystalline phases of PEBAX should be considered separately. Basically, if the amide characteristic bands are shifted to higher wavenumber or/and lower intensity, it indicates the disassociation of hydrogen bonding. This can occur only if the samples are exposed to higher temperatures such as in the case of environmental test. Moreover, the presence of peaks around 1680 cm^{-1}

and 1650 cm^{-1} is an indication of free and disordered hydrogen bonds of the carbonyl groups in the polymer. Figure 10.4 shows the FTIR spectroscopy of the water stability test samples plotted against the pure PEBAX membrane before the stability test.

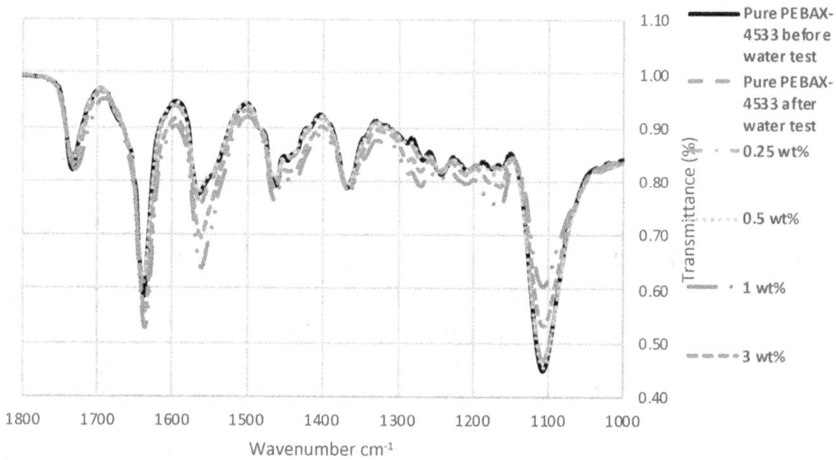

Figure 10.4 FTIR spectroscopy analysis of the water stability test samples and the pure PEBAX membrane before the stability test.

PEBAX characteristic peak around 1640 cm^{-1}, corresponding to the hydrogen-bonded amide groups, was slightly shifted to lower wavenumbers and increased in intensity for membranes with filler fractions of 1 and 3 wt%, indicating higher extent of hydrogen bonding. The other samples exhibited slightly lower intensity and also marginally shifted to lower wavenumbers.

For the free amide vibrations of $-C = O$, originally observed at 1734 cm^{-1}, the peak was observed to be broader with higher intensity for 1 wt% filler fraction along with a shift to lower wavenumber, thus, indicating the reconstruction of hydrogen bonding that hindered the chain mobility in the amorphous conformations, thereby, enhancing the thermal stability. However, the peaks for the rest of the samples were shifted slightly to lower wavenumbers and had decreased intensities.

Additionally, the FTIR analysis of two-week environmental stability test samples (pure PEBAX and composites with 0.5 and 1 wt% filler fraction) was carried out, as shown in Figure 10.5. The amide characteristic peaks around 1640 cm^{-1} for both composite membranes were slightly broadened and decreased in absolute intensity with no visible shift. These changes in the peaks demonstrated that the overall thermal stability and char yield of the membranes reduced after the environmental test.

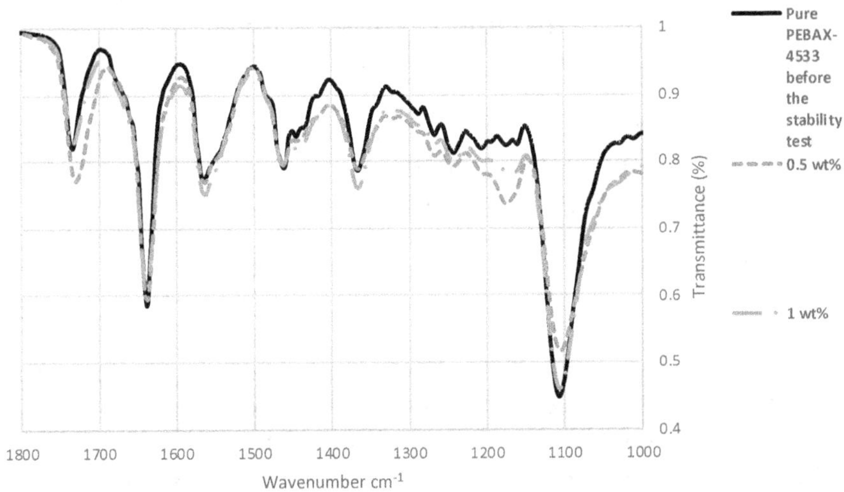

Figure 10.5 FTIR spectroscopy of the environmental stability test samples and the pure PEBAX membrane before the stability test.

For the free amide vibrations at 1734 cm^{-1}, peaks of both composite membranes were slightly broadened and shifted to lower wavenumbers. The peak corresponding to the 0.5 wt% membrane increased in absolute intensity suggesting the development of hydrogen bonding, while the intensity of the peak corresponding to 1 wt% membrane was slightly decreased.

Considering the hydrogen bonding in the crystalline domains of the polymer, it has been mentioned that the higher temperatures cause the crystalline lattice to expand, thus, increasing the distance between the chains. Therefore, the crystalline structure falls apart causing the ordered hydrogen bonds to turn into disordered bonds, thereby, decreasing the thermal stability. Hydrogen bonds in the crystalline domains were enhanced by effective interfacial interaction between the MOF and polymer matrix which was also revealed by SEM analysis, discussed later in this section. It was also shown by FTIR analysis that the interaction between the MOF and polymer was only physical, thus, the MOF acted like a physical crosslinking between the polymer chains, thereby, hindering the chain mobility and consequentially enhancing the thermal stability.

On the other hand, in the case of disordered amorphous domains in the polymers, increasing the temperature causes thermal expansion that weakens the strength of disordered hydrogen bonds. At highly elevated temperatures, these bonds may start to disassociate. It should be noted that the disordered hydrogen bonding in the amorphous phase is a dynamic equilibrium at which

- at lower temperatures, the rate of dissociation is much slower that the rate of construction
- at higher temperatures, the rate of reconstruction is faster since it is an endothermic process, thus, resulting in shifting the equilibrium to the dissociation. Accordingly, free hydrogen bonds are formed.

Also, other factors affect the direction shift of the equilibrium such as the presence of large number of highly polar groups (N – H and C = O) which shift the equilibrium towards reconstruction. Moreover, effective interaction between the filler and polymer matrix provides more surface area for the MOF to form hydrogen bonds with the amide groups in PEBAX.

The microstructure of MOF and MMMs was studied using SEM. From the SEM image of pure MOF (Figure 10.6), it can be seen that HKUST-1 exhibited a distinctive crystalline structure as octahedral shaped particles, and some of the particles are elongated towards a hexagonal symmetry [17]. The HKUST-1 crystals were also observed to aggregate with irregular morphology and size.

Figure 10.6 SEM micrograph of HKUST-1.

SEM images of the surface morphology of the pure PEBAX exhibited homogeneous and clear surface with no defects, cracks or plastic deformation (Figure 10.7a). On the other hand, the surface morphology of the MMMs (Figures 10.7b,c) exhibited the presence of sub-micron particles in the polymer matrix. The images showed that HKUST-1 particles were uniformly dispersed throughout the membrane without significant aggregations or defects. At higher magnifications, the polymer chains were observed to be wrapped around the filler crystals, suggesting a good interfacial interaction and adhesion between the phases.

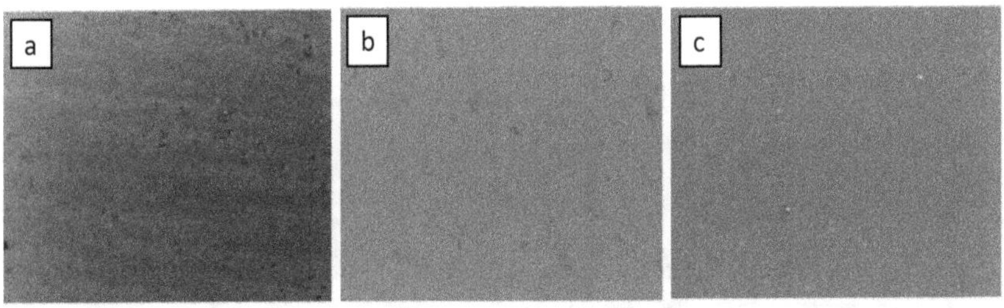

Figure 10.7 SEM surface images of a) pure PEBAX, b) 0.25 wt% MOF/PEBAX and c) 3.0 wt% MOF/PEBAX. The image width corresponds to 500 μm.

In the cross-sectional image analysis, PEBAX revealed a smooth horizontal layer morphology (Figure 10.8a). The morphology of the MMMs at lower filler fractions (Figure 10.8b) showed homogeneously distributed MOF within the polymer matrix. Even at higher loadings of MOF (Figure 10.8c), no significant cluster formation and aggregation were observed. In addition, no filler pull out was observed from the composite surface.

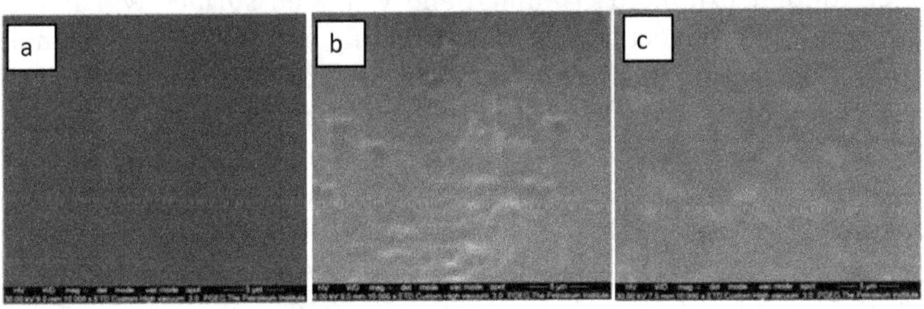

Figure 10.8 SEM cross-sectional images of a) pure PEBAX, b) 0.25 wt% MOF/PEBAX and c) 3.0 wt% MOF/PEBAX.

SEM analysis was also carried out for the water and environmental stability test samples in order to analyze the effect of nanoscale spatially constraining effect of MOF on the polymer chains. In the SEM images of the environmental stability test samples (0.5 wt% and 1 wt%) in Figure 10.9, MOF particles were observed to be randomly dispersed within the polymer matrix producing micro-void structure. As mentioned earlier, the effective interaction of the MOF particles with the polymer created physical crosslinking owing to the hydrogen bonding between the MOF and polymer chains, thus, hindering the chain mobility. The interaction was more pronounced in the 1 wt% membrane, which was also in good agreement with the TGA and FTIR results.

Figure 10.9 SEM micrographs of environmental stability test samples: a) 0.5 wt% and b) 1 wt% membranes.

Figure 10.10 represents the SEM images of the water stability test samples. The surface of the pure PEBAX membrane exhibited a rough morphology. The nanocomposite membranes with both low and high filler content exhibited crumbled and wrinkled surface which was due to the intermolecular hydrogen bonding between the water molecules, PEBAX and MOF. SEM micrographs exhibited an effective interfacial interaction between the MOF and polymer matrix for the 3 wt% membrane, which was in good agreement with the high thermal stability observed in TGA and induced hydrogen bond reconstruction observed in FTIR. Membrane with 0.25 wt% filler fraction also exhibited enhanced properties, however, at a lower magnitude as compared to other samples. This could be due to the lower filler loading, thus, lesser number of polar groups, which was not optimal to increase the reconstruction rate of the hydrogen bonds.

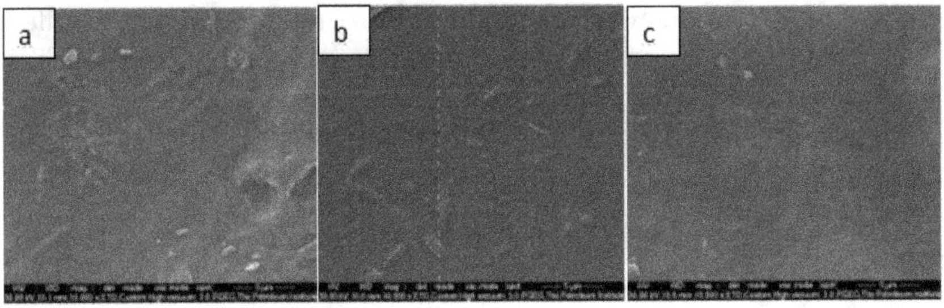

Figure 10.10 SEM micrographs of water stability test samples: a) pure PEBAX membrane, b) 0.25 wt% membrane and c) 3 wt% membrane.

PEBAX is a rubbery polymer, and the gas separation process is mainly through solution-diffusion mechanism since the volume fraction of the polymer matrix is much larger than that of HKUST-1 [5,12,13]. Permeability comprises of solubility, related to the compatibility of the filler and polymer matrix, and diffusivity through the membrane thickness [5]. Assuming good compatibility, diffusivity has a higher impact on the permeability as compared to solubility [5,11]. Incorporation of MOF enhances the CO_2 sorption and diffusion, along with delaying the plasticization owing to HKUST-1 characteristics. Diffusivity is highly dependent on the penetrate kinetic diameter, with CO_2 having a relatively small kinetic diameter of 3.3 Å, thus, leading to high solubility in PEBAX [13].

It should be noted that the permeation occurs only through the amorphous phase of the polymeric membrane. From the XRD analysis, it was confirmed that the pure PEBAX had amorphous and as well as crystalline phases. The incorporation of MOF did not exhibit any change in the crystalline phase in the composite, however, the flexible organic ligands of the filler may have interacted physically with the blocks of the copolymer, thus, resulting in an increase in the *d*-spacing and amorphous region of PEBAX. This indicated that the filler owing to its uniform distribution in PEBAX enhanced the permeability.

Moreover, CO_2 has high affinity towards MOF due to high quadrupole moment ($-1.4 * 10^{-35}$ C m) of CO_2, consequently resulting in the induction of molecular interactions between CO_2 and unsaturated Cu sites in the MOF, leading to the formation of H-bonds [11,14,18]. MOF increases the solubility of the condensable and polar gases in the polymer matrix by increasing the number of $-OH$ polar groups and altering the morphology at the interface between the permeate gas and polymer. This was further evidenced by the SEM and FTIR analysis of water and environmental stability studies, which revealed that a good interaction between the polymer matrix and HKUST-1 was due to the enhanced H-bonding.

Figure 10.11 and Table 10.2 show the permeation test results for the membranes at different CO_2 feed gas pressures of 3, 5 and 10 bar at 22 °C. First, in order to study the effect of MOF loading and the feed gas pressure on CO_2 permeability, the true permeability coefficients of the MMMs were plotted against the varying feed gas pressures. The pure PEBAX membrane exhibited the least permeability values as compared to the MOF filled matrix membranes. The permeability values of the MMMs were corrected with respect to MOF content. The membranes with 0.25 and 0.5 wt% filler fractions exhibited higher permeability than the pure PEBAX. This was attributed to the hindrance to the PEBAX chain mobility and possibility of the creation

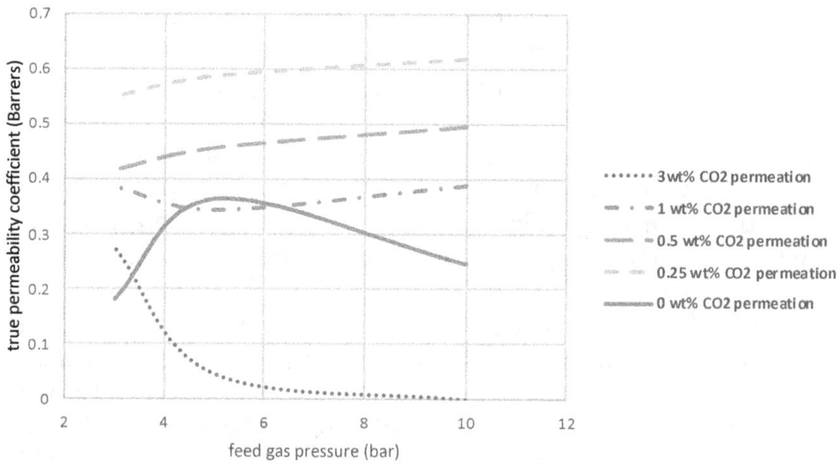

Figure 10.11 Effect of HKUST-1 loading and CO_2 feed gas pressure on permeation.

Table 10.2 True permeability coefficients (Barrer) of MMMs at different CO_2 feed gas pressures

pressure(bar)	0 wt%	0.25 wt%	0.5 wt%	1 wt%	3 wt%
10	0.246	0.619	0.494	0.39	0.168
5	0.364	0.587	0.457	0.344	0.0468
3	0.181	0.549	0.418	0.388	0.276

of voids at the interface. The membrane with 3.0 wt% HUKST-1 possessed lower permeability due to the filler aggregation at the microscale. Thus, the higher filler loadings blocked the pores available for permeation. HKUST-1 filled MMMs, as compared with other types of fillers, have higher possibility of decreasing the permeability due to the dominant molecular sieving behavior of HKUST-1. Furthermore, the penetration of the polymer chains may narrow the pore openings, thus, resulting in further reduction in diffusivity. Due to the flexible nature of the MOF organic ligands, these introduce the breathing effect, thus, improving the permeability, but, this can affect the separation performance negatively by limiting the molecular sieving phenomena [5,6,13,14]. Results from literature studies have shown that the breathing effect allows the HKUST-1 framework to contract and expand, consequently reducing or increasing the pore dimensions [5,11]. The enhancement in permeability indicates the absence of non-selective voids and good interaction between the filler and polymer matrix. It also suggests that composite membranes incorporated with 0.25 and 0.5 wt% HKUST-1 may be good candidates for CO_2 separation.

Secondly, in order to study the effect of CO_2 feed gas pressure, the permeability of each MMMs has been varied with feed gas pressure. In the case of 0.25 and 0.5 wt% MOF filled membranes, the permeability increased gradually with increasing the feed gas pressure and showed the highest values over the whole pressure range. Similar behavior has been reported in literature with PEBAX-1675 [13]. 1 wt% HKUST-1 filled membrane showed a 'dual-mode' sorption behavior which is also similar to the reported results [8,11]. A drop in permeability occurred at 5 bar pressure, which was attributed to the plasticization pressure level. The membranes with 0.25, 0.5 and 1.0 wt% MOF exhibited effective permeability at higher feed gas pressure of 10 bar of CO_2. At low feed gas pressure, CO_2 adsorption was preferentially around Cu sites followed by the vicinity around the metal center and organic ligands which resulted in a reduction in the solubility. On the other hand, at higher pressures, CO_2 diffused faster and filled the vicinity of the areas connecting the pores. It can be seen that the membrane with 3 wt% of HKUST-1 showed a significant decrease in permeability values throughout the whole pressure range probably owing to the aggregation of the filler particles, as mentioned earlier. It should also be noted that the trend at 10 bar pressure could be due to the non-ideal conditions.

As no experimental studies on PEBAX-4533/HKUST-1 MMMs

have been reported in the literature, thus, the comparisons were made with the nanocomposite membranes generated using other PEBAX grades. Table 10.3 summarizes permeability coefficients reported in literature for PEBAX-1675 based MMMs with different fillers at similar experimental conditions of pressure and temperature.

Table 10.3 Permeability values of PEBAX-1675 nanocomposites

Filler	wt%	P (*bar*)	T (°C)	CO_2 permeability (Barrer)	Ref.
MWNT	0	10	30	55.9	3
	2			329	
	5			262	
	0	30		183.8	
Nano-Silica	0	10	30	1.1	14
	0.05			1.9	
	0.1			3.2	
	0.2			5.4	
	0.3			8.9	
H-Mordenite	0.05			1.6	
	0.1			2.1	
	0.2			3	
	0.3			5.4	
ZIF-8	2, 4,	2		464	19
	8, 16	8		758	
ZIF-8	0	4	25	55	20
	11			76	
	18			178	
	21			136	
	33			84	
4A zeolite	0	5	30	55	21
	5			71	
	10			97	
	20			113	
	30			155	
	10	25		130	
A4 zeolite	0	5	30	55.8	12
	5			71.4	
	10			97	
	20			113.7	
	30			155.8	

Pure PEBAX-1675 membranes have permeability coefficients around 55 Barrer at pressures ranging from 2 to 10 bar [3,12,14,19-

21], while the values obtained for pure PEBAX-4533 membranes in this study were 0.181, 0.36 and 0.246 Barrer at feed gas pressures of 3, 5 and 10 bar respectively. It is, thus, evident that PEBAX-1675 separation performance is superior to PEBAX-4533.

With incorporation of fillers (e.g., MWCNTs, nano-silica, H-mordenite and zeolites), different permeability values were obtained for PEBAX-1675. Overall, the separation performance was observed to generally enhance with the incorporation of fillers. In zeolite filled MMMs, the permeability coefficients increased with increasing both filler loading and the feed gas pressure [12,19-21]. Significant increase up to 464 Barrer was attained with zeolite imidazolate framework-8 (ZIF-8) at 2 Bar, followed by 758 Barrer at 8 bar [19,20]. Higher feed gas pressure enhanced the driving force for gas flux though the membrane which led to a higher extent of CO_2 adsorption. Additionally, both studies have found that after certain filler loading, a decrease in permeability values occurred. It was attributed to the effect of higher filler loadings that hindered the polymer chain mobility. This observation is in agreement with the permeation behavior of 3 wt% PEBAX-4533/HKUST-1 membrane, which showed a significant decrease in permeability. On the other hand, the permeability coefficient of pure PEBAX-1675 at 10 bar was observed to be as low as 1.1 Barrer [14]. Even with the addition of fillers (nano-silica and H-mordenite), the enhancement was insignificant.

Comparing the permeability coefficients of CuBTC MOF filled MMMs, it is observed that the membranes with CuBTC generally exhibit better separation performance than other fillers such as zeolites, CuBTC/zeolite/MIL-53 hybrid, etc. [7,22,23]. Comparing the polymers, matrimid CuBTC MMMs have been reported to exhibit higher permeability coefficients than PEBAX-4533/CuBTC MMMs. On the other hand, a study by Basu *et al.* [24] reported the permeability coefficient of 0.2 Barrer at 5 bar feed gas pressure for matrimid/CuBTC MMMs, which was lower than 0.36 Barrer observed in this study for PEBAX-4533/CuBTC MMMs at the same pressure.

10.4 Conclusions

In this study, defect-free MMMs were successfully developed using PEBAX-4533 with different weight fractions of HKUST-1 through solvent-evaporation method. The MMMs exhibited effective permeability for CO_2 gas at different pressure ranges. The permeability of N_2 as well as CH_4 in addition to CO_2/N_2 and CO_2/CH_4 mixed-gas systems is required to be investigated further to gain insights about the diffusivity and selectivity of the prepared MMMs.

References

1. Sridhar, S., Smitha, B., and Aminabhavi, T. M. (2007) Separation of carbon dioxide from natural gas mixtures through polymeric membranes - A review. *Separation and Purification Reviews*, **36**(2), 113-174.
2. Sheth, J. P., Xu, J., and Wilkes, G. L. (2003) Sold state structure property behavior of semicrystalline poly(ether-block-amide) PEBAX (R) thermoplastic elastomers. *Polymer*, **44**(3), 743-756.
3. Murali, R. S., Sridhar, S., Sankarshana, T., and Ravikumar, Y. V. L. (2010) Gas permeation behavior of PEBAX-1657 nanocomposite membrane incorporated with multiwalled carbon nanotubes. *Industrial & Engineering Chemistry Research*, **49**(194), 6530-6538.
4. Ovid'ko, I. A. (2013) Enhanced mechanical properties of polymer matrix nanocomposites reinforced by graphene inclusions: A review. *Review on Advanced Materials Science*, **34**, 19-25.
5. Venna, S. R., and Carreon, M. A. (2015) Metal organic framework membranes for carbon dioxide separation. *Chemical Engineering Science*, **124**, 3-19.
6. Li, H., Haas-Santo, K., Schygulla, U., and Dittmeyer, R. (2015) Inorganic microporous membranes for H_2 and CO_2 separation - Review of experimental and modeling progress. *Chemical Engineering Science*, **127**, 401-417.
7. Kanehashi, S., Chen, G. Q., Scholes, C. A., Ozcelik, B., Hua, C., Ciddor, L., Shouthon, P. D., D'Alessandro, D. M., and Kentish, S. E. (2015) Enhancing gas permeability in mixed matric membranes through tuning the nanoparticle properties. *Journal of Membrane Science*, **482**, 49-55.
8. Lim, S. Y., Choi, J., Kim, H.-Y., Kim, Y., Kim, S.-J., Kang, Y. S., and Won, J. (2014) New CO_2 separation membranes containing gas-selective Cu-MOFs. *Journal of Membrane Science*, **467**, 67-72.
9. Leal, O., Bolivar, C., Ovalles, C., Urbina, A., Revette, J., and Jose, J. (1996) *Carbon Dioxide Removal from Natural Gas using Amine Surface Bonded Adsorbents*. Online: https://web.anl.gov/PCS/acsfuel/preprint%20archive/Files/Merge/Vol-41_4-0003.pdf [accessed 15th February 2019].
10. *Synthetic Membranes*, Wikipedia (2017). Online: http://en.wikipedia.org/wiki/Synthetic_membrane [accessed 18th April 2017].
11. Shahid, S., and Nijmeijer, K. (2014) Performance and plasticization behavior of polymer-MOF membranes for gas separation at eleva-

ted pressures. *Journal of Membrane Science*, **470**, 166-177.

12. Murali, R. S., Ismail, A. F., Rahman, M. A., and Sridhar, S. (2014) Mixed matrix membranes of Pebax-1657loaded with 4A zeolite for gaseous separations. *Separation and Purification Technology*, **129**, 1-8.
13. Gue, X., Huang, H., Ban, Y., Yang, Q., Xiao, Y., Li, Y., Yang, W., and Zhong, C. (2015) Mixed matric membranes incorporated with amine-functionalized titanium-based metal-organic framework for CO_2/CH_4 separation. *Journal of Membrane Science*, **478**, 130-139.
14. Murali, R. S., Kumar, K. P., Ismail, A. F., and Sridhar, S. (2014) Nano-silica and H-mordenite incorporated poly(ether-block-amide)-1657 membranes for gaseous separations. *Microporous and Mesoporous Materials*, **197**, 291-298.
15. Lia, L., and Yang, G. (2009) Variable-temperature FTIR studies on thermal stability of hydrogen bonding in nylon6/mesoporous silica nanocomposite. *Polymer International*, **58**(5), 503-510.
16. Zhang, L., Ruesch, M., Zhang, X., Bai, Z., and Liu, L. (2015) Tuning thermal conductivity of crystalline polymer nanofibers by inter-chain hydrogen bonding. *RSC Advances*, **5**, 87981-87986.
17. Sun, B., Kayal, S., and Chakraborty, A. (2014) Study of HKUST (copper benzene-1,3,5-tricarboxylate, Cu-BTC MOF)-1 metal organic frameworks for CH_4 adsorption: An experimental Investigation with GCMC (grand canonical Monte-Carlo) simulation. *Energy*, **76**, 419-427.
18. Sorribas, S., Kudashiva, A., Scholes, C. A., Almedro, E., Zornoza, B., de la Iglesia, O., Tellez, C., and Coronas, J. (2015) Pervaporation and membrane reactor performance of polyimide based mixed matrix membranes containing MOF HKUST-1. *Chemical Engineering Science*, **124**, 37-44.
19. Jomekian, A., Behbahani, R. M., Mohammadi, T., and Kargari, A. (2016) CO_2/CH_4 separation by high performance co-casted ZIF-8/Pebax 1657/PES mixed matrix membrane. *Journal of Natural Gas Science and Engineering*, **31**, 562-574.
20. Xu, L., Xiang, L., Wang, C., Yu, J., Zhang, L., and Pan, Y. (2017) Enhanced permeation performance of polyether-polyamide block copolymer membranes through incorporating ZIF-8 nanocrystals. *Chinese Journal of Chemical Engineering*, **25**(7), 882-891.
21. Mittal, G., Dhand, V., Rhee, K. Y., Park, S.-J., and Lee, W. R. (2015) A review on carbon nanotubes and graphene as fillers in reinforced polymer nanocomposites. *Journal of Industrial and Engineering Chemistry*, **21**, 11-25.
22. Zornoza, B., Tellez, C., Coronas, J., Gascon, J., and Kapteijn, F. (2013) Metal organic framework based mixed matrix membranes: An increasingly important field of research with a large application potential. *Microporous and Mesoporous Materials*, **166**, 67-78.

23. Li, W., Zhang, Y., Li, Q., and Zhang, G. (2015) Metal-organic framework composite membranes; synthesis and separation applications. *Chemical Engineering Science*, **135**, 232-257.
24. Basu, S., Cano-Odena, A., and Vankelecom, I. F. J. (2011) MOF-containing mixed-matrix membranes for CO_2/CH_4 and CO_2/N_2 binary gas mixture separations. *Separation and Purification Technology*, **81**(1), 31-40.

Index

A

B

C

D

Index

H

I

L

M

N

O

P

R

S

T

U

V

W

X

Y

Z

www.ingramcontent.com/pod-product-compliance
Lightning Source LLC
LaVergne TN
LVHW011526200926
848484LV00001B/169